POST TRAUMATIC STRESS DISORDERS IN A GLOBAL CONTEXT

Edited by **Emilio Ovuga**

Post Traumatic Stress Disorders in a Global Context
Edited by Emilio Ovuga

Published by InTech

Janeza Trdine 9, 51000 Rijeka, Croatia

Publishing Process Manager Igor Babic

Technical Editor Teodora Smiljanic

Cover Designer InTech Design Team

Additional hard copies can be obtained from orders@intechopen.com

Post Traumatic Stress Disorders in a Global Context, Edited by Emilio Ovuga
p. cm.
ISBN 978-953-307-825-0

For EU product safety concerns:
IN TECH d.o.o., Prolaz Marije Krucifikse Kozulić 3, 51000 Rijeka, Croatia,
info@intechopen.com or visit our website at intechopen.com.

Meet the editor

Dr Emilio Ovuga, PhD is Professor of Mental Health and Dean of the Faculty of Medicine at Gulu University in northern Uganda, and Professor Extraordinarius at the University of South Africa (UNISA). Dr Ovuga completed his undergraduate medical education at Makerere University in 1976, and postgraduate training in psychiatry at the same university in 1981. Dr Ovuga holds a joint PhD of Karolinska Institute, Sweden and Makerere University, Uganda in suicidology and psychiatric epidemiology. Dr Ovuga worked at Makerere University, serving that institution in various capacities including being Head of the Department of Psychiatry and member of several committees of Senate from 1989 to 2006. Outside his home country, Uganda, Dr Ovuga worked as a psychiatrist in Kenya from 1981 to 1984 and the Transkei homeland of South Africa from 1984 to 1989. Apart from active administrative duties, Dr Ovuga is also currently working on the development of a vibrant research ethics review system in Uganda where he is Chair of the Forum of Institutional Review Committees in Uganda (FICU).

Contents

Preface

One of my most nurse clients once told me in therapy; "When you have an abscess you know the source of the pain, you can get at it and remove it at source. When you have a wound on the body you know the source of the pain and you can get at the source and remove it. This pain of mental illness is different. You know the pain is there and it is real but you do not know where the source is and cannot get at it." Yes the pain of emotional illness is real and it is there deep inside our clients. Psychiatrists and other mental health professionals grapple with the problem of emotional pain on a daily basis. Unfortunately our understanding of how mental health problems arise is not yet adequate. Our only diagnostic gold standard in psychiatry remains the psychiatric interview and this can unfortunately be very subjective. There are multiple theories of causation and there is no single proven magic therapeutic bullet that can help our clients. In this volume titled: *Post-traumatic Stress Disorder in a Global Context*, authors from around the world share their valuable insights, knowledge, experience, and research in the hope that mental health practitioners gain a better understanding of this special mental disorder, post-traumatic stress disorder (PTSD) that arises after individuals are exposed to severe, terrifying, and horrifying experiences that threaten their safety, life, or integrity.

The pain of what has evolved to be known as post-traumatic stress disorder is peculiar. A common factor in the origins of PTSD appears to be an injury to the inner integrity of its victims, which psychoanalysts refer to as narcissistic injury. Author Corrine Anna Bissouma from Ivory Coast refers to this narcissistic injury as a "wound in the mind" to make it easier for her community of former child soldiers to understand what she means in her communications with them and their caregivers. What makes PTSD peculiar is the fact that it is often caused by people who are close to us; those we rely on everyday in our households; the people that we place our inner most trust in; the people that we believe cannot betray us; the people who are significant to us and are the reason we possibly exist. As humans even though most people may not be close to us in terms of their not being our family members, we still expect that they should not violate our integrity and cause emotional pain in us for whatever reason.

In the course of time psychiatrists from as early as Sigmund Freud have identified two major origins of PTSD; i.e. those arising secondary to the experience of natural disasters, and the ones that result from traumatic experiences in our interactions with each other. Freud particularly traced the origins of neurosis to the sexual violations of young children by their own adult kin. The commonplace experience of traumatic experiences, whether natural or manmade, makes one wonder if traumatic stress is in fact not the initial common pathway through which all major psychiatric disorders including post-traumatic stress disorder arise. After all according to Sigmund Freud, neurosis arising from childhood experience of sexual abuse, and sexual and physical childhood abuse, continues to be cited as one of the leading causes of mental health problems such as depression, suicide behavior, and post-traumatic stress disorder.

The authors of *post-traumatic Stress Disorder in a Global Context* present a wide array of information that practitioners will find useful in understanding PTSD in practice. Carefully chosen, each chapter blends in with the others without unnecessary repetition and redundant overlaps. The book is divided into four sections. Section I provides an overview of PTSD as is currently understood. Dr Prasad Amarendra provides an overview of clinical features and current management approaches while Banister and associates present the history, prevalence, etiology, treatment, and comorbid disorders in PTSD related to combat situations. Dr Richardson and associates describe the military related psychiatric management of PTSD. Section II describes etiologic theories and risk factors for PTSD. Dr Kevin provides an interesting animal model of how avoidant behavior in PTSD is actively acquired, and highlights strain and sex differences in the acquisition process. Dr Yukata brings research-based data indicating the need for clinicians to screen accident victims for PTSD, as the emotional reactions of accident victims to their experience and their heart rate predict the potential development of post-traumatic stress disorder after motor vehicle accidents. Dr Ask and Dr Christiansen describe sex differences in the manifestation of post-traumatic stress disorder among males and females while Dr Chou attributes the development of post-traumatic stress disorder to the complex interaction of individual's biologic, psychological, environmental, and social factors, and low resource availability that predispose the vulnerable individual with sub-threshold psychiatric disorder to the full manifestation of post-traumatic stress disorder. In section III Dr Bissouma describes the invaluable role of social and community support in the rehabilitation of former child soldiers in Ivory Coast. Dr Ovuga and Dr Larroque describe the precarious situation of children in northern Uganda who live in abject poverty and social adversity with significant levels of aggression in the daily lives of the children four years after active war in the region ceased. Dr Ovuga and Dr Larroque further provide possible evidence of post-traumatic stress disorder across the lifespan related to exposure to trauma not only in military but also civilian situations in northern Uganda and Southern Africa. Using clinical vignettes, Dr Ovuga and Dr Larroque describe the difficulties in the recognition of post-traumatic stress disorder in clinical practice. In Section IV research in special situations suggests that post-traumatic stress disorder can indeed present in malignant disease (Dr Tacon), after stroke (Dr. Norman) and after exposure to earthquake (Dr Wang). Awareness of any malignancy as a significant cause of traumatic stress for the sufferer, family and significant other members of the individual's immediate community is essential; recognition and management of PTSD at the time of cancer diagnosis as a vital component of palliative care is recommended. The development of PTSD in stroke as a significant cause of sudden and unexpected disability and potential lifelong dependence on other individuals is understandable. Finally Dr Wang has presented a carefully chosen set of research results that clearly highlight the long-term effects of exposure to earthquake on individuals and offspring of victims many years after exposure.

Emilio Ovuga, MD PhD
Professor of Mental Health and Dean,
Gulu University Faculty of Medicine, Northern Uganda, Gulu,
Uganda

Part 1

Overview of Clinical Aspects

1

Post Traumatic Stress Disorder – An Overview

Amarendra Narayan Prasad
Ministry of Defence (Indian Army), India

1. Introduction

Posttraumatic stress disorder (PTSD) is a severe anxiety disorder that can develop after exposure to any event that result in psychological trauma. This event may involve the threat of death to oneself or to someone else, or to one's own or someone else's physical, sexual, or psychological integrity, overwhelming the individual's ability to cope. As an effect of psychological trauma, PTSD is less frequent but more enduring than the more commonly seen acute stress response. Post-traumatic Stress Disorder (PTSD) is a persistent and sometimes crippling condition and develops in a significant proportion of individuals exposed to trauma, and untreated, can continue for years. Its symptoms can affect every life domain - physiological, psychological, occupational, and social.

Posttraumatic stress disorder (PTSD) was first introduced into the Diagnostic and Statistical Manual of Mental Disorders (DSM) in 1980, making it one of the more recently accepted psychiatric disorders. PTSD is one of the few DSM diagnoses to have a recognizable etiologic agent, in that it must develop in direct response to a severe (sudden, terrifying, or shocking) life event (American Psychiatric Association 2000). Since the introduction of PTSD into DSM-III (American Psychiatric Association 1980), the disorder has been documented in children exposed to traumas such as domestic violence, natural disasters, medical trauma (such as hospitalization or medical procedures performed on children), war, terrorism, and community violence.

According to the American Psychological Association, posttraumatic stress disorder (PTSD) is defined as "an anxiety disorder that can develop after exposure to a terrifying event or ordeal in which grave physical harm occurred or was threatened. Traumatic events that may trigger PTSD include violent personal assaults, natural or human-caused disasters, such as terrorist attacks, motor vehicle accidents, rape, physical and sexual abuse, and other crimes, or military combat [1]."

PTSD is a problem in which the human brain continues to react with nervousness after the horrific trauma even though the original trauma is over. Brain can react by staying in "overdrive" and being hyperalert in preparation for the next possible trauma. Sometimes the brain continues to "remember" the trauma by having "flashbacks" about the event or nightmares even though the trauma was in the past.

2. Historical background

Reports of battle-associated stress reactions appeared as early as the 6th century BC. One of the first descriptions of PTSD was made by the Greek historian Herodotus. In 490 BC

Herodotus described, during the Battle of Marathon, an Athenian soldier who suffered no injury from war but became permanently blind after witnessing the death of a fellow soldier. In the early 19th century military medical doctors started diagnosing soldiers with "exhaustion" after the stress of battle. This "exhaustion" was characterized by mental shutdown due to individual or group trauma. Soldiers during the 19th century were not supposed to be scared or show any fear in the midst of battle. The only treatment for this "exhaustion" was to bring the afflicted back for a bit for a short term therapy and then send them back into battle. During the intense and frequently repeated stress, the soldiers became fatigued as a part of their body's natural shock reaction. According to Stéphane Audoin-Rouzeau and Annette Becker, "One-tenth of mobilized American men were hospitalized for mental disturbances between 1942 and 1945, and after thirty-five days of uninterrupted combat, 98% of them manifested psychiatric disturbances in varying degrees."

Previous diagnoses now considered historical equivalents of PTSD include railway spine, stress syndrome, shell shock, battle fatigue, or traumatic war neurosis. Although PTSD-like symptoms have also been recognized in combat veterans of many military conflicts, the modern understanding of PTSD dates from the 1970s, largely as a result of the problems that were still being experienced by US military veterans of the war in Vietnam. In its initial DSM-III (formulation 1980), a traumatic event was conceptualized as a catastrophic stressor that was outside the range of usual human experience. The framers of the original PTSD diagnosis had in mind events such as war, torture, rape, the Nazi Holocaust, the atomic bombings of Hiroshima and Nagasaki, natural disasters (such as earthquakes, hurricanes, and volcano eruptions) and human-made disasters (such as factory explosions, airplane crashes, and automobile accidents). They considered traumatic events as clearly different from the very painful stressors that constitute the normal vicissitudes of life such as divorce, failure, rejection, serious illness and financial reverses. (By this logic adverse psychological responses to such "ordinary stressors" would, in DSM-III terms, be characterized as Adjustment Disorders rather than PTSD.) This dichotomization between traumatic and other stressors was based on the assumption that although most individuals have the ability to cope with ordinary stress, their adaptive capacities are likely to be overwhelmed when confronted by a traumatic stressor.

The DSM-III diagnostic criteria for PTSD were revised in DSM-III-R (1987) and DSM-IV (1994) [2]. A very similar syndrome is classified in ICD-10 [3]. Since 1980 there has been a great deal of attention devoted to the development of instruments for assessing PTSD. Although an optimal evaluation of a patient for PTSD consists of a face-to-face interview by a mental health professional trained in diagnosing psychiatric disorders, several instruments are available to facilitate the diagnosis and assessment of posttraumatic stress disorder (PTSD). These include screening tools, diagnostic instruments, and trauma and symptom severity scales. For example, there are brief screening tools, such as the 4-item Primary Care PTSD Screen, developed by the Department of Veterans Affairs National Center for Posttraumatic Stress Disorder; self-report screening instruments, such as the Posttraumatic Diagnostic Scale; and structured or semi-structured interviews, such as the Clinician-Administered PTSD Scale (CAPS), the Structured Clinical Interview for DSM-IV (SCID), the Diagnostic Interview Schedule for DSM-IV (DIS-IV), and the Composite International Diagnostic Interview (CIDI), Acute Stress Disorder Interview (ASDI),

Posttraumatic Stress Disorder Checklist(PCL), Acute Stress Disorder Scale (ASDS), Acute Stress Checklist for Children (ASC-Kids), Child PTSD Symptom Scale (CPSS) and Reactions to Research Participation Questionnaires for Children and Parents (RRPQ-C and RRPQ-P)[4,5,6]. All these might be used prior to or as a complement to the clinical interview. Such measures are used most frequently in research settings, some might be used clinically to provide additional sources of documentation, and others might be given to veterans at a health facility prior to their first interview with health professional. Screening tools can be useful in initiating a conversation about exposure to traumatic events or possible PTSD symptoms. However, as noted by Briere (2004) "no psychological test can replace the focused attention, visible empathy, and extensive clinical experience of a well-trained and seasoned trauma clinician [7]." Working in Vietnam war-zone, veterans have developed both psychometric and psycho physiologic assessment techniques that have proven to be both reliable and valid. Other investigators have modified such assessment instruments and used them with natural disaster victims, rape/incest survivors, and other traumatized cohorts.

PTSD has been criticized from the perspective of cross-cultural psychology and medical anthropology, because it has usually been diagnosed by clinicians from Western industrialized nations working with patients from a similar background. . Despite these criticisms, PTSD is a real time mental disorder with devastating clinical, physical, social and economic consequences for sufferers. Though clinicians from developing countries continue to diagnose PTSD using diagnostic systems developed in industrialized countries, the major clinical features appear to be uniform across cultures. Major gaps remain in our understanding of the effects of ethnicity and culture on the clinical phenomenology of post-traumatic syndromes. We have only just begun to apply vigorous ethno cultural research strategies to delineate possible differences between Western and non-Western societies regarding the psychological impact of traumatic exposure and the clinical manifestations of such exposure.

3. Epidemiology and prevalence

The United Nations' World Health Organization publishes estimates of PTSD impact for each of its member states; the latest data available are for 2004. The age-standardised-disability adjusted life-year (DALY) rates for PTSD, per 100,000 inhabitants, in 10 most ranking countries is as table 1.

The National Comorbidity Survey has estimated that the lifetime prevalence of PTSD among adult Americans is 7.8%, with women (10.4%) twice as likely as men (5%) to have PTSD at some point in their lives. In the United States, 60% of men and 50% of women experience a traumatic event during their lifetimes. The rate is highest for soldiers. The United States Department of Veterans Affairs estimates that 830,000 Vietnam War veterans suffered symptoms of PTSD. The National Vietnam Veterans' Readjustment Study (NVVRS) found 15.2% of male and 8.5% of female Vietnam Vets to suffer from PTSD. Life-Time prevalence of PTSD was 30.9% for males and 26.9% for females. For soldiers who fought in the Iraq war in 2008, the prevalence of PTSD was 13.8%. The National Survey of Adolescents, which included a household probability sample of 4,023 adolescents between the ages of 12 and 17, found that using accepted diagnostic criteria for PTSD, the six-month prevalence was estimated to be 3.7% for boys and 6.3% for girls [8].

Country	PTSD DALY rate, overall	PTSD DALY rate, females	PTSD DALY rate, males
Thailand	59	86	30
Indonesia	58	86	30
Philippines	58	86	30
USA	58	86	30
Bangladesh	57	85	29
Egypt	56	83	30
India	56	85	29
Iran	56	83	30
Pakistan	56	85	29
Japan	55	80	31

Table 1.

4. Classification

Posttraumatic stress disorder is classified as an anxiety disorder, characterized by aversive anxiety-related experiences, behaviors, and physiological responses that develop after exposure to a psychologically traumatic event (sometimes months after). Its features persist for longer than 30 days, which distinguishes it from the briefer acute stress disorder. These persisting posttraumatic stress symptoms cause significant disruptions in one or more important areas of life function. It has three sub-forms: acute, chronic, and delayed-onset; based on onset of symptoms after the traumatic event. Acute is of < 1 month, chronic between 1-3 months and delayed is after 3 months.

Complex Post Traumatic Stress Disorder (C-PTSD) is a condition that results from chronic or long-term exposure to emotional trauma over which a victim has little or no control and from which there is little or no hope of escape, such as in cases of domestic emotional, physical or sexual abuse; childhood emotional, physical or sexual abuse; entrapment or kidnapping; slavery or enforced labor and long term imprisonment and torture. When people have been trapped in a situation over which they had little or no control at the beginning, middle or end, they can carry an intense sense of dread even after that situation is removed. This is because they know how bad things can possibly be. And they know that it could possibly happen again. And they know that if it ever does happen again, it might be worse than before. C-PTSD results more from chronic repetitive stress from which there is little chance of escape. PTSD can result from single events, or short term exposure to extreme stress or trauma.

5. Etiology and risk factors

PTSD is believed to be caused by either physical trauma or psychological trauma, or more frequently a combination of both. Possible sources of trauma include experiencing or witnessing childhood or adult physical, emotional or sexual abuse. Other recognized causes of PTSD include experiencing or witnessing an event perceived as life-threatening such as accidents, terminal illnesses, or employment in occupations exposed to war (such as soldiers) or disaster (such as emergency service workers). Traumatic events that may cause PTSD symptoms to develop include violent assault, kidnapping, sexual assault, torture, being a hostage, prisoner of war or concentration camp victim, experiencing a disaster, violent automobile accidents or getting a diagnosis of a life-threatening illness. Children or adults may develop PTSD symptoms by experiencing bullying or mob violence. Preliminary research suggests that child abuse may interact with mutations in a stress-related gene to increase the risk of PTSD in adults[9]. Multiple studies show that parental PTSD and other posttraumatic disturbances in parental psychological functioning can, despite a traumatized parent's best efforts, interfere with their response to their child as well as their child's response to trauma[10,11]. Parents with violence-related PTSD may, for example, inadvertently expose their children to developmentally inappropriate violent media due to their need to manage their own emotional dysregulation[12,13].

Military experience as risk factors for the development of PTSD include coming from an unstable family, being punished severely during childhood, childhood anti-social behavior and depression as pre-military factors, war-zone exposure, peri-traumatic dissociation, depression as military factors and recent stressful life events and depression as post-military factors[14]. Certain protective factors against PTSD in war-conditions include high school degree or college education, older age at entry to war, higher socioeconomic status, and positive paternal relationship as pre-military protective factors and social support at homecoming and current social support as post-military factors[15]. Research also indicates the protective effects of social support in averting and recovery from PTSD[16]. There may also be an attitudinal component; for example, a soldier who believes that they will not sustain injuries may be more likely to develop symptoms of PTSD than one who anticipates the possibility, should either be wounded[15]. Likewise, the later incidence of suicide among those injured in home fires above those injured in fires in the workplace suggests this possibility.

Posttraumatic stress responses have been documented in children who have suffered traumatic loss of their parents, siblings, and peers[17,18,19,20]. Results from a study indicated that knowing someone who was injured or killed, female gender, and bomb-related television viewing or other media exposure were associated with the most severe psychological reactions. Bereaved youths who suffered severe loss (e.g. a parent, sibling, close relative, or friend) as a result of the bombing were more likely to report posttraumatic stress symptoms than did children who did not experience this degree of loss.

Although most people (50-90%) encounter trauma over a lifetime, only about 8% develop full PTSD[21]. Vulnerability to PTSD presumably stems from an interaction of biological diathesis, early childhood developmental experiences, and trauma severity. Predictor models have consistently found that childhood trauma, chronic adversity, and familial stressors increase risk for PTSD as well as risk for biological markers of risk for PTSD after a traumatic event in adulthood[22]. This effect of childhood trauma, which is not well understood, may be a marker for both traumatic experiences and attachment problems. Proximity to, duration of, and severity of the trauma also make an impact; and interpersonal traumas cause more problems than impersonal ones[21]. People vary in susceptibility to PTSD. Genetic factors may play a significant role in susceptibility. Women develop PTSD at about twice the rate as men, even for the same crimes[21]. Individuals with a prior trauma history or multiple traumas are at increased risk[21]. A premorbid psychiatric history also increases the likelihood of developing the disorder[22]. It may be that people who have fewer supports and limited inter-personal coping skills are more likely to develop PTSD[21]. Studies of concentration camp survivors and prisoners of war suggest that even given sufficient trauma intensity and duration most of those who are exposed develop PTSD.

A positive relationship has been found between trauma intensity and the likelihood of PTSD[22]. People who have been injured or perceived the event as life threatening are more likely to develop PTSD than those with less severe trauma. Human caused traumatic events such as assaults and murder have a more powerful impact than accidents and natural disasters. Among crime victims, individuals who have suffered more brutal trauma have higher frequencies of PTSD – torture (54%), rape (49%); badly beaten (32%), and other sexual assault (24%)[21]. Dissociation during the trauma, peritraumatic dissociation, is associated with risk for PTSD[21].

There is evidence that susceptibility to PTSD is hereditary. For twin pairs exposed to combat in Vietnam, having a monozygotic (identical) twin with PTSD was associated with an increased risk of the co-twin having PTSD compared to twins that were dizygotic (non-identical twins)[23]. Recently, it has been found that several single-nucleotide polymorphisms (SNPs) in FK506 binding protein 5 (FKBP5) interact with childhood trauma to predict severity of adult PTSD[24]. These findings suggest that individuals with these SNPs who are abused as children are more susceptible to PTSD as adults. Another recent study found a single SNP in a putative estrogen response element on ADCYAP1R1 (encodes pituitary adenylate cyclase-activating polypeptide type I receptor or PAC1) to predict PTSD diagnosis and symptoms in females[25].

6. Neurobiology

Neurobiological research indicates that PTSD may be associated with stable neurobiological alterations in both the central and autonomic nervous systems[26]. Psycho physiological alterations associated with PTSD include hyper arousal of the sympathetic nervous system, increased sensitivity and augmentation of the acoustic-startle eye blink reflex, a reducer pattern of auditory evoked cortical potentials, and sleep abnormalities. Neuropharmacologic and neuroendocrine abnormalities have been detected in the noradrenergic, hypothalamic-pituitary-adrenocortical, and endogenous opioid systems[27]. There is increasing evidence that PTSD is associated with biological alterations or abnormalities. Individuals with PTSD have an atypical stress response. Instead of producing increases in cortisol, a stress related hormone, the usual hypothalamic-pituitary axis mechanisms are disrupted and result in lower than expected levels of the hormone[28]. PTSD symptoms may result when a traumatic event causes an overactive adrenaline response, which creates deep neurological patterns in the brain. These patterns can persist long after the event that triggered the fear, making an individual hyper-responsive to future fearful situations. Brain catecholamine levels are low, and corticotropin-releasing factor (CRF) concentrations are high. Together, these findings suggest abnormality in the hypothalamic-pituitary-adrenal (HPA) axis. Trauma victims who develop post-traumatic stress disorder often have higher levels of other stimulating hormones (catecholamines) under normal conditions in which the threat of trauma is not present as well as lower levels of cortisol. This combination of higher than normal arousal levels and lower than normal levels of the "calming" hormones of the changes creates the conditions for PTSD. The amygdala is the brain region that alerts the body to danger and activates hormonal systems.

After a month in this heightened state with stress hormones elevated and cortisol levels lowered, further physical changes, such as heightened hearing develop. This cascade of physical changes, one triggering another, suggests that early intervention may be the key to heading off the effects of post-traumatic stress disorder.

Given the strong cortisol suppression to dexamethasone in PTSD, HPA axis abnormalities are likely predicated on strong negative feedback inhibition of cortisol, itself likely due to an increased sensitivity of glucocorticoid receptors. Some researchers have associated the response to stress in PTSD with long-term exposure to high levels of norepinephrine and low levels of cortisol, a pattern associated with improved learning in animals. Translating this reaction to human conditions gives a pathophysiological explanation for PTSD by a maladaptive learning pathway to fear response through a hypersensitive, hyperreactive and hyperresponsive HPA axis. Low cortisol levels may predispose individuals to PTSD: Swedish soldiers serving in Bosnia and Herzegovina with low pre-service salivary cortisol levels had a higher risk of reacting with PTSD symptoms, following war trauma, than soldiers with normal pre-service levels[29]. Because cortisol is normally important in restoring homeostasis after the stress response, it is thought that trauma survivors with low cortisol experience a poorly contained—that is, longer and more distressing—response, setting the stage for PTSD.

However, there is considerable controversy within the medical community regarding the neurobiology of PTSD. A review of existing studies on this subject showed no clear relationship between cortisol levels and PTSD. Only a slight majority have found a decrease in cortisol levels while others have found no effect or even an increase. Decreased brain volume or volume of specific brain structures have been documented in some adults and children with PTSD [30,31]. The biologic correlates have not yet been fully explored, nor are the implications for intervention established.

Three areas of the brain whose function may be altered in PTSD have been identified: the prefrontal cortex, amygdala and hippocampus. Much of this research has utilised PTSD victims from the Vietnam War. For example, a prospective study using the Vietnam Head Injury Study showed that damage to the prefrontal cortex may actually be protective against later development of PTSD [32]. In a study by Gurvits et al, combat veterans of the Vietnam War with PTSD showed a 20% reduction in the volume of their hippocampus compared with veterans who suffered no such symptoms [33,34]. This finding could not be replicated in chronic PTSD patients traumatized at an air show plane crash in 1988 (Ramstein, Germany) [35]. In human studies, the amygdala has been shown to be strongly involved in the formation of emotional memories, especially fear-related memories. Neuroimaging studies in humans have revealed both morphological and functional aspects of PTSD. The amygdalocentric model of PTSD proposes that it is associated with hyperarousal of the amygdala and insufficient top-down control by the medial prefrontal cortex and the hippocampus particularly during extinction. This is consistent with an interpretation of PTSD as a syndrome of deficient extinction ability. Further animal and clinical research into the amygdala and fear conditioning may suggest additional treatments for the condition.

7. Clinical features

Describing children's responses to trauma, Terr(1991) presents four specific symptoms characteristic of childhood PTSD: repeatedly perceiving memories of the event through visualization, engaging in behavioral re-enactments and repetitive play related to the event, fears related to the trauma event, and pessimistic attitudes reflecting a sense of hopelessness about the future and life in general. The behavioral presentation of a child or adolescent experiencing PTSD or symptoms of PTSD may also include problems with verbalization and extremes of disconnections (no close relationships) or false connections (perceiving close relationships where none exist). Additionally, the diagnosis of PTSD cannot be made on the basis of the child's affective presentation alone (e.g. crying, sadness, or expressions of terror).

The symptoms of PTSD include:

- sleep problems including nightmares and waking early
- flashbacks and replays which you are unable to switch off
- impaired memory, forgetfulness
- inability to concentrate
- hyper vigilance (feels like but is not paranoia)
- exaggerated startle response
- irritability, sudden intense anger and occasional violent outbursts
- panic attacks
- hypersensitivity - almost every remark is perceived as critical
- obsessiveness - the experience takes over your life
- joint and muscle pains with no obvious cause
- feelings of nervousness and anxiety
- depression (reactive, not endogenous)
- excessive shame, embarrassment and guilt
- unnaturally high levels of fear
- low self-esteem, low self-confidence
- anhedonia, emotional numbness (inability to feel love or joy)
- detachment
- avoidance of anything that reminds you of the experience
- intense physiological reactivity and undue psychological distress at any reminder of the experience

Warning symptoms of PTSD: -

- Guilt about actions or shame over some failure
- Excessive drinking or drug use
- Uncontrolled or frequent crying and other extreme reactions to events that normally would be handled more calmly
- Sleep problems (too little, too much)
- Depression, anxiety, or anger
- Depending too much on others
- Verbal or physical family violence
- Stress-related physical illness (head and backache, intestinal problems, low energy)
- Inability to escape from horror scenes remembered from the war
- Difficulty concentrating
- Suicidal thoughts or plans

Diagnostic criteria: The diagnostic criteria for PTSD, stipulated in the Diagnostic and Statistical Manual of Mental Disorders IV (Text Revision) (DSM-IV-TR), may be summarized as [36,37]:

A: Exposure to a traumatic event

This must have involved both (a) loss of "physical integrity", or risk of serious injury or death, to self or others, and (b) an intense negative emotional response.

B: Persistent re-experiencing

One or more of these must be present in the victim: flashback memories, recurring distressing dreams, subjective re-experiencing of the traumatic event(s), or intense negative psychological or physiological response to any objective or subjective reminder of the traumatic event(s).

C: Persistent avoidance and emotional numbing

This involves a sufficient level of:

- avoidance of stimuli associated with the trauma, such as certain thoughts or feelings, or talking about the event(s);
- avoidance of behaviors, places, or people that might lead to distressing memories;
- inability to recall major parts of the trauma(s), or decreased involvement in significant life activities;
- decreased capacity (down to complete inability) to feel certain feelings;
- an expectation that one's future will be somehow constrained in ways not normal to other people.

D: Persistent symptoms of increased arousal not present before

These are all physiological response issues, such as difficulty falling or staying asleep, or problems with anger, concentration, or hypervigilance.

E: Duration of symptoms for more than 1 month

If all other criteria are present, but 30 days have not elapsed, the individual is diagnosed with 'acute stress disorder'.

F: Significant impairment

The symptoms reported must lead to "clinically significant distress or impairment" of major domains of life activity, such as social relations, occupational activities, or other "important areas of functioning".

In preparation for the May 2013 release of the DSM-5, the fifth version of the American Psychiatric Association's diagnostic manual, draft diagnostic criteria was released for public comment, followed by a two-year period of field testing. Proposed changes in DSM-5, to the criteria include:

- Criterion A (prior exposure to traumatic events) is more specifically stated, and evaluation of an individual's emotional response at the time (current criterion A2) is dropped.
- Several items in Criterion B (intrusion symptoms) are rewritten to add or augment certain distinctions now considered important.
- Special consideration is given to developmentally appropriate criteria for use with children and adolescents. This is especially evident in the restated Criterion B - intrusion symptoms. Development of age-specific criteria for diagnosis of PTSD is ongoing at this time.
- Criterion C (avoidance and numbing) has been split into "C" and "D":

- Criterion C (new version) now focuses solely on avoidance of behaviors or physical or temporal reminders of the traumatic experience(s). What were formerly two symptoms are now three, due to slight changes in descriptions.
- New Criterion D focuses on negative alterations in cognition and mood associated with the traumatic event(s), and contains two new symptoms, one expanded symptom, and four largely unchanged symptoms specified in the previous criteria.

- Criterion E (formerly "D"), which focuses on increased arousal and reactivity, contains one modestly revised, one entirely new, and four unchanged symptoms.
- Criterion F (formerly "E") still requires duration of symptoms to have been at least one month.
- Criterion G (formerly "F") stipulates symptom impact ("disturbance") in the same way as before.
- The "acute" vs. "delayed" distinction is dropped; the "delayed" specifier is considered appropriate if clinical symptom onset is no sooner than 6 months after the traumatic event(s).

PTSD is a clinical diagnosis; there are no laboratory tests or brain-imaging studies currently used in clinical practice to diagnose PTSD. Brain imaging studies are under way to learn more about the brain in the PTSD condition, but these are not used in everyday medical practice. A physical exam and some blood tests may be necessary to rule out medical conditions that may mimic PTSD, such as hyperthyroidism which can create an anxiety state.

8. Principles of management

There are various semi-structured diagnostic interviews schedules used in research, however, to date, there is no single instrument accepted as a "gold standard" for making the diagnosis of PTSD or monitoring symptoms.

8.1 Psychosocial treatment strategies

Four strategies have been distinguished by both empirical evaluation and the development of treatment manuals. Currently, only the cognitive-behavioral approaches have been investigated sufficiently to make empirically based recommendations. According to the State of Washington's Task Force on Promotion and Dissemination of Psychological procedures (1995), the four strategies that meet criteria for either "probably efficacious" or "well-established" are briefly described as follows [38,39,40]:

1. Prolonged Exposure (PE)

Prolonged Exposure is a standard technique that has been used with various anxiety disorders and has now been adapted for PTSD in rape victims (Foa & Rothbaum, 1998). PE involves repeated imaginal re-living of the traumatic experience. Then it is followed up with subsequent real life exposure to situations that are unpleasant reminders of the cause of the fear. The theory posits that repeated pairing of the emotional memories, with a non-dangerous environment will lead to reconditioning of the emotionally aversive associations to trauma memories [41]. Gradually being reminded or remembering the trauma will lose the intense negative quality. Breathing retraining to assist with relaxation is an initial component of the approach. The treatment ordinarily is carried out over ninety minute sessions that may occur twice a week. High-risk concerns such as psychosis, homicidal or suicidal tendencies should be addressed.

2. Cognitive Processing Therapy (CPT)

Cognitive Processing Therapy is an approach that focuses primarily on trauma-related attributions and cognition that are maladaptive. There is exposure to the trauma, but it occurs in a modulated fashion and is accomplished through having victims write descriptions of the trauma that are repeatedly reviewed and read. The description is analyzed to identify blocks and dysfunctional cognitions and cognitive therapy techniques are used to challenge and replace these distortions with more appropriate, accurate and adaptive views. Themes of safety, trust, power, esteem and intimacy are specifically addressed. Coping skills are taught to assist victims in predicting and managing stress responses. CPT has been proven effective with female rape victims. Resick and Schnicke (1995) provide the theory underlying the approach and a detailed description of the various techniques. The treatment occurs over 12 sessions.

3. Stress Inoculation Training (SIT)

SIT is a CBT approach that has a primary focus on teaching the identification and management of anxiety reactions to stressful situations. Michenbaum (1985) first developed this intervention for use with a wide variety of populations suffering from anxious response including trauma. SIT involved explaining the physical, cognitive and behavioral components of fear and anxiety reactions. Then victims are taught various coping strategies to address dysfunctional thoughts and unpleasant feelings that come up with exposure to certain trauma reminders. These include relaxation, shifting attention and self-coaching dialogues. The goal is that victims learn to manage trauma related anxiety with confidence and efficacy. SIT has been found effective with various stress-related conditions and for female rape victims. Typically this approach consists of 8-14 sessions.

4. Eye Movement Desensitization and Reprocessing (EMDR)

Shapiro (1995) developed the Eye Movement Desensitization and Reprocessing (EMDR) approach. Like SIT, this approach has been advocated as a treatment for a variety of psychological problems involving intense emotions and intrusive thoughts. It is generally considered a form of imaginal exposure accompanied by cognitive re-framing, which are standard elements of CBT. Victims are encouraged to imagine a stressful scene and replace dysfunctional cognitions with more adaptive ones while engaging in lateral eye movements. Therapists move fingers back and forth to facilitate this process. The unique aspect of the treatment is the eye movement component. The currently available research has established EMDR is as effective as CBT treatments [42]. However, the eye movements have not been found to be necessary and they do not explain symptom reduction. Initially, it was claimed that EMDR could cure PTSD in one or two sessions. The developer of the method now takes the position that up to 12 sessions may be necessary in some cases to achieve full effects.

8.2 Pharmacotherapy of adult PTSD

Though seldom the sole, or even primary treatment for PTSD, pharmacotherapy can alleviate suffering, help restore immediate functioning, and be a supportive adjunct to psychotherapy [43,44]. The scientific literature on PTSD pharmacology is relatively sparse. Most studies have been trials of different medications, only a few randomized trials have been conducted and they have had equivocal results. Treatment guidelines are largely developed on the basis of clinical experience and expert opinion. Antidepressants are the backbone of PTSD treatment; they are particularly useful for their anxiolytic qualities and ability to reduce arousal. The newer selective seratonin reuptake inhibitors and related medications are generally safer, better tolerated, and possibly more effective than older formulations. SSRIs (selective serotonin reuptake inhibitors) are considered to be a first-line drug treatment. SSRIs for which there are data to support use include: citalopram, escitalopram, fluoxetine, fluvoxamine, paroxetine and sertraline [45]. Atypical antidepressants like Nefazodone can be effective with sleep disturbance symptoms, and with secondary depression, anxiety, and sexual dysfunction symptoms. Trazodone can also reduce or eliminate problems with disturbed sleep, and with anger and anxiety. Heterocyclic/Tricyclic anti-depressants like Amitriptyline has shown benefit for positive distress symptoms, and for avoidance, and Imipramine has shown benefit for intrusive symptoms. Monoamine-oxidase inhibitors (MAOs) like Phenelzine has been observed to be effective with hyperarousal and depression, and is especially effective with nightmares.

A full psychopharmacologic approach can include the use of anticonvulsants and mood stabilizers, major tranquilizers and anti-psychotic medications of which newer drugs are well tolerated, and adrenaline blocking drugs [46]. Use of these combinations is usually best left to psychiatrists who are expert in the treatment of PTSD.

Beta blockers (Propranolol) has demonstrated possibilities in reducing hyperarousal symptoms, including sleep disturbances [47]. Also, post-stress high dose corticosterone administration was recently found to reduce 'PTSD-like' behaviors in a rat model of PTSD[48]. In this study, corticosterone impaired memory performance, suggesting that it may reduce risk for PTSD by interfering with consolidation of traumatic memories. Clinical trials evaluating methylenedioxymethamphetamine (MDMA, "Ecstasy") in conjunction with psychotherapy are being conducted in Switzerland and Israel.

Symptom prevention:- Some medications have shown benefit in preventing PTSD or reducing its incidence, when given in close proximity to a traumatic event. These medications include: Alpha-adrenergic antagonists (e.g. clonidine), Beta blockers (e.g. Propranolol),Glucocorticoids and Opiates [49].

8.3 General treatment components in children

When clinicians offer assistance to traumatized children and their families, they should begin with: (1) Establishing rapport with the child and caregiver(s) and (2) Providing a rationale for treatment. The clinician should keep the following points in mind when providing a rationale for treatment. The child and caregiver(s) should separately or together receive information regarding the purpose and process of treatment. Caregivers should be informed about the common effects of traumatic experiences on children; that children can have a variety of different reactions. Most children do not have lasting psychological effects (although with some experiences long term effects are more likely, e.g., abuse by the parent, long-term abuse). Treatment will most often be relatively short term and will involve talking about what happened, learning to express feelings appropriately, and gaining an accurate perception of the event. The treatment rationale and concrete goals of therapy should be presented to the child in a clear and simple manner. In the case of certain crimes, such as sexual abuse or physical abuse, where there may be misinformation about children's roles in what happened or offender patterns, it is important to provide corrective information. Educating caregivers and their children about healthy sexuality and personal safety skills is also important during the initial phase of treatment with victims of sexual abuse.

Empirical evidence from controlled treatment-outcome studies provides strongest support for the use of trauma-focused cognitive-behavioral treatment (CBT) to resolve PTSD symptoms in children[42]. Therefore, CBT may be considered as the first line approach, either alone or in conjunction with other forms of therapy. CBT usually involves the following components: direct discussion of the trauma, emotional and cognitive coping skills, corrective cognitive distortions, and contingency reinforcement programs for children displaying behavioral problems. The current consensus is that it is not necessary that children be diagnosed with PTSD to receive this treatment, only that they have identifiable posttraumatic stress symptoms that interfere with functioning. CBT approaches are based on the interrelationships between thoughts, feelings, and behaviors [50,51]. In many cases thoughts can lead to emotional states which in turn produce behavioral responses. For example, traumatized children may have over generalized or inaccurate beliefs derived from the traumatic stress experience that triggers anxiety responses. Anxiety is expressed as intensely uncomfortable or may be expressed in appropriate behaviors. In addition, avoidance coping may temporarily reduce anxiety but lead to maladaptive behavior patterns.

8.3.1 Teaching stress management techniques

Stress management techniques such as progressive muscle relaxation, thought-stopping, positive imagery, and controlled breathing are often taught to accompany direct trauma-focused discussion in treatment. It is usually recommended that these skills be taught to children prior to detailed discussions of the trauma. With practice, relaxation strategies can help the child gain confidence to approach the direct discussion of the trauma without overwhelming fear, as well as handle other stressful situations outside of the therapeutic context (i.e. flashbacks at school). Because stress management is a useful skill and is easy to master, this component of treatment can facilitate a more positive association to therapy to counterbalance some of the more difficult aspects.

1. Relaxation techniques

Systematic relaxation consists of a series of muscle tensing and relaxation exercises. progressive relaxation and guided tension releasing exercises are recommended for children above 10 years. Therapists may want to adapt exercises to the child's most problematic muscle groups or focus on head, torso and leg exercises separately. Image-induced relaxation is a strategy that may be more effective for younger children. They are taught to distinguish between tense and relaxed states. For example, a child is asked to stand like a "tin soldier" and conversely collapse like a "wet noodle" into a chair. Children are taught when confronted by distressing memories or cues to practice relaxed responding. Children are taught self-instruction such as "relax, hang loose, lighten up, or calm down" at these times and are encouraged to practice at home. Controlled/deep breathing consists of gradually breathing in and out on a count of four to restore normal breathing states and promote relaxation. This technique can be used in vivo for all types of stress inducing situations.

2. Cognitive coping techniques

Thought replacement consists of teaching children to interrupt upsetting or disturbing thoughts (e.g., imagines a stop sign and sub-vocalizes the word STOP), and focus on a positive experience or memory (e.g., getting hugged by a parent, going to Disneyland). Positive coping self-statements challenge the disturbing thoughts with self affirming or reassuring thoughts (e.g., I am strong, I can handle this situation, I am not really in danger now).

8.3.2 Direct exploration/discussion of the traumatic experience

There is a strong clinical consensus that addressing the traumatic experience, regardless of the specific methodology, is the core ingredient of effective treatment for PTSD in children. Exposure to the traumatic memories and feared reminders under safe circumstances serves to decondition these associations and reduce the use of avoidance coping. Safety does not just mean that the child has developed trust in the therapy environment. Most important is that the child is in a safe and supportive living situation. It is inappropriate and possibly dangerous to encourage children to engage in trauma-focused therapy when they are still at risk.

1. Exposure techniques

For children, gradual exposure techniques are recommended. These techniques gradually expose a child to thoughts, memories and other cues or reminders of the traumatic experience. When children can tolerate the memories without significant emotional distress they are less likely to resort to avoidant behaviors. The goal is that when children face trauma-related memories or cues, more adaptive responses like feelings of control, mastery, pride and courage will gradually replace fearful/anxious responses. There are a variety of different exposure techniques used to elicit children's participation and provide them with a sense of control. It is important, regardless of the exposure technique used, that a therapist clearly presents to the child the rationale behind exposure. No matter how well a therapist prefaces the exposure procedure, resistance by children may be an initial reaction to this therapeutic approach because significant emotional and physical discomfort may be experienced. For this reason it is important to inform caregivers and children that some increased symptoms are common responses at first. In order to attain relief in the long run, some level of anxiety or distress may need to be endured while confronting fears. Preparing caregivers for children's possible negative reactions to therapy will increase cooperation and compliance.

Gradual exposure techniques are primarily designed to be useful when post-traumatic stress symptoms are present. Children who do not exhibit fear or anxiety may not need extensive focus on the traumatic experience itself. Emotions such as embarrassment, shame, or sadness associated with recalling the event may be reasonable reactions or may be better addressed through a focus on attributions and perceptions about the event. A child's symptoms may worsen if a therapist insists upon constantly talking about the traumatic memories or events. There is currently no evidence that talking about the details of what happened is necessary to recovery in children.

In sum, a child's capacity to talk about the trauma without experiencing significant distress or use of avoidance coping is an indication of successful emotional processing. However, a child's unwillingness to talk about it may not be because of post-traumatic stress reaction but instead a legitimate response(e.g., tired of talking about it, embarrassed). In these situations, various indirect methods of addressing trauma-related issues like art, book making and play techniques may be more useful. Mediums such as clay or PLAY-DOH can also facilitate children in depicting different aspects of the traumatic event.

2. Strategies for gradual exposure

The process of gradual exposure begins by confronting the least anxiety provoking stimuli first and works its way through more distressing stimuli (e.g. the child might identify hearing the word "rape" as upsetting, but less so than remembering what actually happened). Talking, writing, speaking into a tape-recorder, responding to "mock interview," or drawing a picture with explanation can be used to accomplish exposure. Role-playing, puppet play, and doll-play can be helpful especially with young children. Some children may choose to create books, poems or songs about their traumatic experiences.

Direct Exposure: This method is appropriate for an older child with good visualization skills. The child is asked to recall specific sensory details of traumatic event, focusing on visual memories. Fantasy is discouraged when recalling the account. This approach should not be confused with hypnotic suggestion or guided imagery. For example, a therapist asks the child to close her eyes (if comfortable) and recall a scene of the traumatic event as if she were there. The therapist poses some specific questions to help the child stay focused like, "describe the room you were in, the time of day, or what the child smells, hears, feels, and thinks at the time." Too many questions may interfere with the child's visualization. The therapist should only ask as many questions as they feel necessary to help the child visualize the scene. The session should not end until the child's anxiety level has decreased or coping techniques have been used to help the child regain a sense of calmness.

In Vivo Exposure: this technique is most used in the later stages of the exposure therapy. The child is helped to identify situations for in vivo practice of exposure to fear inducing stimuli. This should occur in a situation where there is no actual danger or risk thus enabling the child to experience mastery and competence (e.g., confronting fear of the dark by turning off the light during the session, sleeping alone in her room, walking to school).

8.3.3 Exploring and correcting inaccuate attributions

Most interventions for traumatized children also involve the evaluation of cognitive assumptions children may have made relating the traumatic experience. Children make sense of their experiences in the world by developing belief systems. Like adults, most children have a generally positive view of themselves, other people and the world. Being the victim of a traumatic stress situation can conflict with those beliefs. In order to resolve the conflict, children may change their ideas and thoughts about themselves and others or develop inaccurate, distorted and confused beliefs about the trauma[]. Examples of faulty attributions are "Nothing is safe anymore", "It was all my fault", "I must be a bad person for this to have happened." For some children, unfortunately, a traumatic event can serve to confirm already existing negative perceptions. When treating children with PTSD, it is important to explore and correct these distorted thought patterns related to the trauma. The maladaptive assumptions or beliefs must first be identified. This means it is important initially to allow children to express beliefs even though they may be inaccurate (e.g. self blame-"I asked for it because I went to his house" - or thinking that drinking caused the offender to abuse). Then through various therapeutic exercised, like role playing, telling stories, and providing corrective feedback, these negative or inaccurate thoughts can be disputed. The therapist helps the child generate positive thoughts to replace negative distorted ones instead of just telling children what they should think. With younger children, play therapy using toys and dolls, art materials, and games may be a more effective approach to explore their inaccurate attributions.

1. Strategies for correcting cognitive distortions

Cognitive coping triangle: The therapist facilitates discussion with the child about the interrelationship among thoughts, feelings, and behavior starting with a general discussion and moving toward trauma-specific examples. Using examples from every-day life is a useful way of conveying these connections and then relating them to post-traumatic symptoms. For example, the child is presented with a negative and a positive scenario involving peers. For each situation, the child is asked what his/her thoughts, feelings and behaviors would be. The child practices identifying the emotions generated by different thoughts and then identifying thoughts underlying emotions. The therapist helps the child work through examples modeling the process and pointing out how different thoughts about the same situation can result in very different feelings and behaviors. This process may be difficult. Visual aids like pencil and paper, a chalkboard, or a dry-erase board are used to help work through fictitious examples until the child understands the problem triangle concept.

Disputing negative/unproductive-thoughts: The therapist explains that changing distressing thoughts and emotions is a skill that can be gradually acquired through practice. The therapist stresses that negative thoughts are not necessarily valid or permanent. The therapist presents fictitious examples through storytelling in which the child practices substituting positive replacement thoughts for negative unhealthy ones. For example, the therapist may use the "Best Friend Role Play" in which the child role plays with the therapist, (or puppet, empty chair, etc.) imagining that their best friend is having negative thoughts and their job is to convince the best friend that these thoughts are NOT true. It is important to distinguish between the personal thoughts and feelings of the therapist and the role that they are playing during these exercises. For younger children, the use of a puppet reinforces the idea that they are engaged in a game and distinguishes the character's beliefs in the role-play from the therapist's beliefs.

Generating positive self statements: The therapist teaches the child a series of positive self-statements that can replace negative dysfunctional thoughts. Children's self-statements are made to fit their individual difficulties. For example, a child with low self esteem and poor self-image may be encouraged to say, "I am just as good as other kids" or generate reasons why they are special. A withdrawn and/or fearful child may be taught to say, "It's fun trying new things or I am very brave sometimes."

2. Pharmacotherapy

Preliminary studies have shown that some children with PTSD present with physiologic abnormalities much like those seen in adults with PTSD. Even though randomized trials have not yet been conducted, preliminary reports have prompted clinicians to sue a variety of medication with children suffering from PTSD symptoms and associated symptoms of depression or panic. The psychopharmacological agents that have been recommended include propranol, carbamazapine, clonidine, and antidepressants. Most often these medications are not considered the primary intervention but prescribed in conjunction with psychotherapy. Research on psychopharmacological treatments for children with PTSD have revealed that certain psychotropic medications have significantly reduced reexperiencing symptoms like nightmares and other PTSD related symptoms in uncontrolled clinical trials[]. As a general practice, "medication should be selected on the basis of established practice in treating the co-morbid condition (e.g., antidepressants for children with prominent depressive symptoms)". Due to their favorable side effect profile and effectiveness in treating both depressive and anxiety disorders, serotonin reuptake inhibitors (SSRIs) are often he first psychotropic medications selected for treating pediatric PTSD. Imipramine also is often chosen to treat children suffering from co-morbid panic symptoms.

Helping Children Cope With Trauma

After any disaster, children are most afraid that the event will recur, that they or someone they love will be hurt or killed, and that they may be separated from those they love and will be left alone. Suggested strategies for helping children cope with trauma include the following[]:

- Children younger than 6 years of age should not be exposed to TV videotape coverage of the attacks (or any television coverage of war or prolonged violence), and the viewing time allowed for older children should be limited.
- Encourage children to express their feelings about what has happened. Parents should share their feelings with them. Regressive behaviors (e.g., thumb sucking, night awakenings, and bed-wetting) may occur in response to traumatic events. Parents should know not to punish or scold their child for these types of behaviors, but instead to try to help the child put their feelings into words.
- Children need to be frequently reassured that they are safe and that they are loved.
- Parents should be encouraged to be honest with their children about what has occurred and to provide facts about what has happened. Children usually know when something is being "sugar-coated."
- Encourage parents to try to return the child and the family to a normal routine as soon as possible. This will help provide a sense of security and safety.
- Encourage parents to spend extra time with the child, especially doing something fun or relaxing for both of them.
- Remember the importance of touch. A hug can reassure children that they are loved.
- Each family should review safety procedures so children will be prepared the next time an emergency situation occurs.
- Encourage parents to talk with teachers, baby-sitters, and day care providers and others who may be with the child so that they will understand how the child has been affected.
- Watch for signs of repetitive play in which children re-enact all or part of the disaster. Although excessive reenactment of a traumatic experience may be a warning sign, this behavior is an appropriate form of expression of emotions.
- Encourage children who are not able to articulate their feelings to express themselves through coloring, drawing, and painting.
- Remind parents to praise and recognize responsible behavior and reassure children that their feelings are normal in response to an abnormal situation.

9. Conclusion

Events that are threatening to life or bodily integrity will produce traumatic stress in its victim. This is a normal, adaptive response of the mind and body to protect the individual by preparing him to respond to the the threat by fighting or fleeing. If the fight or flight is successful, the traumatic stress will usually be released or dissipated allowing the victim to return to a normal level of functioning. PTSD develops: when fight or flight is not possible;

the threat persists over a long period of time; and/or the threat is so extreme that the instinctive response of the victim is to freeze. There is a mistaken assumption that anyone experiencing a traumatic event will have PTSD. This is far from true. Studies vary, but confirm that only a fraction of those facing trauma will develop PTSD (Elliott 1997, Kulka et al 1990, Breslau et al 1991). What distinguishes those who do not is still a hot topic of discussion, but there are many clues. Factors mediating traumatic stress appear to include: preparation for expected stress (when possible), successful fight or flight responses, prior experience, internal resources, support from family, community, and social networks, debriefing, emotional release, and psychotherapy.
Severe trauma during childhood can have a devastating effect on the development of the brain and all functions mediated by this complex organ - emotional, cognitive, behavioural and physiological. Intense emotional reactions in the face of these events are expected and normal, and the range of feelings experienced may be quite broad. Every conflict forces one to live through some terrible experiences. Indeed, millions of people have been present at events far beyond the worst nightmares. Trauma researchers believe that it is the repression of memories and feelings that is at the heart of trauma suffering in both the short and long term. Time does not heal trauma. Every culture has its own way of dealing with traumatic experiences. And much also depends on the family circumstances, as well as on their age and the nature of their exposure to traumatic events. In all cultures, one of the most important factors is the cohesion of the family and community, and the degree of nurture and support that one receives through events in which they had defied death.
Identification of a portion of those suffering from PTSD will be straightforward. But others may be difficult to spot owing to complicated life or defensive systems. Evaluation of the state of the autonomic nervous system will assist in the diagnosis of PTSD and in setting treatment objectives where appropriate. Preexisting negative appraisals, impaired retrieval of autobiographical memories, and decrements in verbal memory may represent trait-like cognitive phenomena that denote greater vulnerability to PTSD following trauma and predict symptom course. These areas of research have tremendous potential for contributing to prevention and treatment of PTSD, as would additional research examining relationships between cognitive phenomena that may shed light on underlying mechanisms. In addition, future research delineating cognitive difficulties in PTSD in the absence of comorbid depression would further elucidate factors contributing to and resulting from different posttraumatic sequelae. Regarding cognitive phenomena specific to trauma memories, current research examining PTSD-related intrusions provides further evidence that they are not qualitatively distinct from other intrusive cognitions.

10. References

[1] American Psychiatric Association. Diagnostic and statistical manual of mental disorders: DSM-IV. Washington, DC: American Psychiatric Association, 1994.
[2] American Psychiatric Association. Diagnostic and Statistical Manual of Mental Disorders: DSM-III-R. Washington, DC: American Psychiatric Association, 1987.
[3] World Health Organisation: The ICD – 10 Classification of Mental Behavioural Disorders – Diagnostic Criteria for Research. Geneva: World Health Organisation; 1992.
[4] Blake DD, Weathers FW, Nagy LM, et al. "The development of a Clinician-Administered PTSD Scale". J Trauma Stress 1995; 8 (1): 75–90.

[5] Brewin CR. Brief screening instrument for post traumatic stress disorder. British Journal of Psychiatry 2002; 181, 158 - 162.
[6] Foa EB, Cashman L, Jaycox L & Perry K. The validation of a self-report measure of posttraumatic stress disorder : the Posttraumatic Diagnostic Scale. Psychological Assessment 1997; 9, 445–451.
[7] Yehuda R, Halligan SL, Golier JA, Grossman R, Bierer LM. "Effects of trauma exposure on the cortisol response to dexamethasone administration in PTSD and major depressive disorder".Psychoneuroendocrinology 2004;29(3):389–404.
[8] Amaya-Jackson L. Posttraumatic stress disorder in children and adolescents. In Kaplan and Sadock's Comprehensive Textbook of Psychiatry. 7th edition. Edited by Sadock B, Sadock V. Philadelphia: Lippincott Williams and Wilkins; 2000: 2763-2769.
[9] Kelleher I, Harley M, Lynch F, Arseneault L, Fitzpatrick C, Cannon M. "Associations between childhood trauma, bullying and psychotic symptoms among a school-based adolescent sample". Br J Psychiatry 008; 193 (5):378–2.
[10] Binder EB, Bradley RG, Liu W, et al. "Association of FKBP5 polymorphisms and childhood abuse with risk of posttraumatic stress disorder symptoms in adults". JAMA 2008; 299 (11): 1291–305.
[11] Kaminer T. "Distorted maternal mental representations and atypical behavior in a clinical sample of violence-exposed mothers and their toddlers". J Trauma Dissociation 2008; 9 (2): 123–47.
[12] Koenen KC, Moffitt TE, Poulton R, Martin J, Caspi A . "Early childhood factors associated with the development of post-traumatic stress disorder: results from a longitudinal birth cohort". Psychol Med 2007; 37 (2): 181–92.
[13] Clarke C. Childhood and Adulthood Psychological Ill Health as Predictors of Midlife and Anxiety disorders. Archives of General Psychiatry 2007; 64: 668-678.
[14] Lapp KG, Bosworth HB, Strauss JL, et al. "Lifetime sexual and physical victimization among male veterans with combat-related post-traumatic stress disorder". Mil Med 2005; 170 (9): 787–90.
[15] Otte C, Neylan TC, Pole N, et al. "Association between childhood trauma and catecholamine response to psychological stress in police academy recruits". Biol. Psychiatry 2005; 57(1): 27–32.
[16] Brewin CR, Andrews B, Valentine JD. "Meta-analysis of risk factors for posttraumatic stress disorder in trauma-exposed adults". J Consult Clin Psychol 2000; 68 (5): 748–66.
[17] Dubner AE, Motta RW. "Sexually and physically abused foster care children and posttraumatic stress disorder.". Journal of consulting and clinical psychology 1999; 67 (3): 367–73.
[18] Foy DW, Madvig BT, Pynoos RS, Camilleri AJ: Etiologic factors in the development of posttraumatic stress disorder in children and adolescents. Journal of School Psychology 1996, 34:133-145.
[19] Wolfe DA, Sas L, Wekerle C: Factors associated with the development of post traumatic stress disorder among child victims of sexual abuse.Child Abuse Negl 1994, 18:37-50.

[20] Lonigan CJ, Shanon MP, Taylor CM, Finch AJ, Sallee FR: Children exposed to disasters: II. Risk factors for the development of post-traumatic symptomatology. J Am Acad Child Adolesc Psychiatry 1994, 33(1):94-105.

[21] Schnurr PP, Lunney CA, Sengupta A. "Risk factors for the development versus maintenance of posttraumatic stress disorder". J Trauma Stress 2004; 17 (2): 85–95.

[22] Ozer EJ, Best SR, Lipsey TL, Weiss DS. "Predictors of posttraumatic stress disorder and symptoms in adults: a meta-analysis". Psychol Bull 2003; 129 (1): 52–73.

[23] True WR, Rice J, Eisen SA, et al. "A twin study of genetic and environmental contributions to liability for posttraumatic stress symptoms". Arch. Gen. Psychiatry 2003; 50 (4): 257–64.

[24] Binder EB, Bradley RG, Liu W, et al. "Association of FKBP5 polymorphisms and childhood abuse with risk of posttraumatic stress disorder symptoms in adults". JAMA 2008; 299 (11): 1291–305.

[25] Ressler KJ, Merer KB, Bradley B, et al. "Post-traumatic stress disdorder is associated with PACAP and PAC1 receptor". Nature 2011; 470(7335):492–497.

[26] Yehuda R, Halligan SL, Grossman R, Golier JA, Wong C. "The cortisol and glucocorticoid receptor response to low dose dexamethasone administration in aging combat veterans and holocaust survivors with and without posttraumatic stress disorder". Biol Psychiatry 2002; 52 (5): 393–403.

[27] Bohnen N, Nicolson N, Sulon J, Jolles J. "Coping style, trait anxiety and cortisol reactivity during mental stress". J Psychosom Res 1991; 35 (2-3): 141–7.

[28] Geracioti TD Jr, Baker DG, Ekhator NN, West SA, Hill KK, Bruce AB, Schmidt D, Rounds-Kugler B, Yehuda R, Keck PE Jr, Kasckow JW. "CSF norepinephrine concentrations in posttraumatic stress disorder". Am J Psychiatry 2001; 158 (8): 1227–1230.

[29] Sautter FJ, Bissette G, Wiley J, et al. "Corticotropin-releasing factor in posttraumatic stress disorder (PTSD) with secondary psychotic symptoms, nonpsychotic PTSD, and healthy control subjects". Biol. Psychiatry 2003; 54 (12): 1382–8.

[30] Kloet CS, Vermetten E, Geuze E, et al. "Elevated plasma corticotrophin-releasing hormone levels in veterans with posttraumatic stress disorder". Prog. Brain Res. 2008; 167: 287–91.

[31] Yehuda R. "Biology of posttraumatic stress disorder". J Clin Psychiatry2001. 62 Suppl 17:41–6.

[32] Lindley SE, Carlson EB, Benoit M. "Basal and dexamethasone suppressed salivary cortisol concentrations in a community sample of patients with posttraumatic stress disorder".Biol. Psychiatry 2004; 55 (9): 940–5.

[33] Jatzko A, Rothenhöfer S, Schmitt A, Gaser C, Demiracka T, Weber-Fahr W, Wessa M, Magnotta V, Braus DF. et al. "Hippocampal volume in chronic posttraumatic stress disorder (PTSD): MRI study using two different evaluation methods". Journal of Affective Disorders 2006; 94 (1-3): 121–126.

[34] Milad MR, Pitman RK, Ellis CB, Gold AL, Shin LM, Lasko NB, Zeidan MA, Handwerger K, Orr SP et al. "Neurobiological basis of failure to recall extinction memory in posttraumatic stress disorder". Biol Psychiatry 2009;66(12):1075–82.

[35] Mason JW, Giller EL, Kosten TR, Harkness L. "Elevation of urinary norepinephrine/cortisol ratio in posttraumatic stress disorder". J Nerv Ment Dis 1988; 176 (8): 498–502.
[36] Kaplan, HI; Sadock, BJ, Grebb, JA. Kaplan and Sadock's synopsis of psychiatry: Behavioral sciences, clinical psychiatry, 7th ed 1994. Baltimore: Williams & Williams: 606–609.
[37] Breslau N, Kessler RC. "The stressor criterion in DSM-IV posttraumatic stress disorder: an empirical investigation". Biol. Psychiatry 2001; 50 (9): 699–704.
[38] Ursano RJ, Bell C, Eth S, et al. "Practice guideline for the treatment of patients with acute stress disorder and posttraumatic stress disorder". Am J Psychiatry 2004; 161 (11 Suppl): 3–31.
[39] Cahill SP. & Foa EB. A glass half empty or half full? Where we are and directions for future research in the treatment of PTSD. In S. Taylor (ed.), Advances in the Treatment of Posttraumatic Stress Disorder: Cognitive-behavioral perspectives (pp. 267-313) New York: Springer 2004.
[40] Committee on Treatment of Posttraumatic Stress Disorder, Institute of Medicine: Treatment of Posttraumatic Stress Disorder: An Assessment of the Evidence. Washington, D.C.: National Academies Press, 2008.
[41] Joseph, JS; Gray, M.J. BAO "Exposure Therapy for Posttraumatic Stress Disorder".Journal of Behavior Analysis of Offender and Victim: Treatment and Prevention 2008; 1 (4): 69–80.
[42] Seidler GH, Wagner FE. "Comparing the efficacy of EMDR and trauma-focused cognitive-behavioral therapy in the treatment of PTSD: a meta-analytic study". Psychol Med 2006; 36 (11): 1515–22.
[43] Berger W, Mendlowicz MV, Marques-Portella C, Kinrys G, Fontenelle LF, Marmar CR, Figueira I. "Pharmacologic alternatives to antidepressants in posttraumatic stress disorder: a systematic review.". Prog Neuropsychopharmacol Biol Psychiatry 2009; 33 (2): 169–80.
[44] Cooper J, Carty J, Creamer M. "Pharmacotherapy for posttraumatic stress disorder: empirical review and clinical recommendations". Aust N Z J Psychiatry 2005; 39 (8): 674–82.
[45] Brady K, Pearlstein T, Asnis GM, et al. "Efficacy and safety of sertraline treatment of posttraumatic stress disorder: a randomized controlled trial". JAMA 2000; 283 (14): 1837–44.
[46] Hertzberg MA, Butterfield MI, Feldman ME, et al. "A preliminary study of lamotrigine for the treatment of posttraumatic stress disorder". Biol. Psychiatry 1999; 45 (9): 1226–9.
[47] Pitman RK, Sanders KM, Zusman RM, et al. "Pilot study of secondary prevention of posttraumatic stress disorder with propranolol". Biol. Psychiatry 2002; 51 (2): 189–92.
[48] Sapolsky RM, Romero LM, Munck AU. "How do glucocorticoids influence stress responses? Integrating permissive, suppressive, stimulatory, and preparative actions." Endocr Review 2000; 21 (1): 55–89.
[49] Khoshnu E. "Clonidine for Treatment of PTSD". Clinical Psychiatry News 2006; 34(10):22

[50] American Academy of Child and Adolescent Psychiatry: Practice parameters for the assessment and treatment of post-traumatic stress disorder in children and adolescents. Judith A Cohen, principal author. J Am Acad Child Adolesc Psychiatry 1998, 37(10 Suppl):4S-26S.
[51] Goenjian AK, Karayan I, Pynoos RS, Minassian D, Najarian LM, Steinberg AM, Fairbanks LA: Outcome of psychotherapy among early adolescents after trauma. Am J Psychiatry 1997, 154(4): 536-542.

2

Combat Related Posttraumatic Stress Disorder – History, Prevalence, Etiology, Treatment, and Comorbidity

Jenny A. Bannister, James J. Mahoney III and Tam K. Dao
University of Houston,
USA

1. Introduction

This chapter seeks to provide a better understanding of combat related posttraumatic stress disorder. Some of the information presented in this chapter may apply broadly to all populations affected by posttraumatic stress disorder, but should not be used as a primary reference for the disorder as a whole. This chapter will first provide a brief history of the diagnosis and discuss the current diagnostic criteria including potential changes that have been suggested for the Diagnostic Statistical Manual – V. Next, the chapter will present the prevalence of posttraumatic stress disorder and explain potential gender differences in soldiers affected by the disorder. Theories of how an individual obtains posttraumatic stress disorder will be discussed and current and novel treatments will be explained. A brief discussion on traumatic brain injury will also be presented, as it is a common comorbidity of combat related posttraumatic stress disorder.

2. History of posttraumatic stress disorder

Posttraumatic stress disorder was not officially recognized as psychological disorder until the Diagnostic Statistical Manual -III, which was published in 1980 (American Psychiatric Association, 1980; Lasiuk & Hegadoren, 2006). Posttraumatic stress disorder was known by an array of different labels previous to 1980, such as combat neurosis, railway spine, shell shock, soldier's heart, and stress response syndrome. Although it has been speculated that posttraumatic stress disorder has existed in all trauma stricken populations throughout history,the occurrence has been documented primarily in soldiers who experienced combat related trauma. (Jones et. al., 2003; Lasiuk & Hegadoren, 2006). One exception to this pattern is the historical concept of hysteria. Hysteria has also received much attention, but the symptoms associated with this term have evolved throughout history and therefore the term can only be loosely associated with posttraumatic stress disorder. This chapter will make reference to numerous historical figures that noted the similarities of hysteria to the symptoms that they were observing, but it must be noted that this concept is loosely defined.

2.1 Railway spine, soldiers' heart, and hysteria

Another non-military population who exhibited posttraumatic stress disorder-like symptoms is seen in the documentation of a phenomenon called *railway spine*. This is the first time where a cluster of symptoms that resembled posttraumatic stress disorder was documented on within medical literature. Railway spine was observed in London in the late 1700's in railway passengers and workers who were in train crashes (Lasiuk & Hegadoren, 2006; Micale, 1990; Ray, 2008). They experienced the physical effects of the crash such as whiplash, but more importantly they were said to have born the psychological effects of the trauma from the crash. Some of the symptoms of railway spine that resemble posttraumatic stress disorder include: nightmares about the crash, avoiding trains as a means of transportation, and difficulty sleeping. At the time, these symptoms were seen by some as being consistent with hysteria, which was believed to more commonly occur in females.

Since some of the individuals who suffered from railway spine believed that the railway companies should be legally liable for their passenger's and worker's well being, there was much debate as to whether a train crash could cause chronic psychological impairment (Lasiuk & Hegadoren, 2006). At the time, some believed that the individuals were faking their symptoms in order to receive financial gain from the railway companies. Others believed that the symptoms were legitimate which inspired a debate about what caused the symptoms. An English surgeon named John Eric Erichsen believed that hysteria should not be associated with railway spine and that its cause was rooted in an organic illness (Lasiuk & Hegadoren, 2006; Erichsen, 1866 & 1886 as cited in van der Kolk, 2007). Another English surgeon, Herbert Page opposed Erichsen's belief and argued that fear could be a sufficient cause for the symptoms (Lasiuk & Hegadoren, 2006; van der Kolk, 2007). Herman Oppenheim argued that railway spine could result from slight molecular changes in the central nervous system and renamed it *traumatic neurosis* (Lasiuk & Hegadoren, 2006; Oppenheim, 1889 as cited in van der Kolk, 2007). This is the first time where the title of the disorder implies that trauma is implicated in the development of the disorder. Kraepelin later used the term traumatic neurosis in reference to a reaction that was seen in those who survived through accidents or other disasters (Kraepelin, 1899 as cited in Ray, 2008).

During the late 1800's, many theorists became interested in the etiology of hysteria. Individuals such as Charcot and Janet contended that an individual must experience trauma in order to develop hysteria-like symptoms (Ray, 2008). Both individuals also agreed that hysteria was not solely a female disorder and pointed out many male populations that experienced symptoms that mimicked hysteria. One population Janet highlighted was males who had suffered from railway spine. Janet developed the term *neurasthenia* which encompassed a number of reactions to emotional trauma (Ray, 2008). Neurasthenia included symptoms such as headaches, fatigue, sleep issues, and emotional and somatization disorders.

Disorders that are similar to combat related posttraumatic stress disorder emerged once again during the Boer, Crimean, and American Civil Wars (Ray, 2008). Terms such as *soldiers' heart* and *DaCosta syndrome* were developed to describe symptoms that were frequently seen in soldiers after being exposed to combat situations. Some of the symptoms associated with soldiers' heart included "extreme fatigue, tremors, dyspnea, palpitations, [and] sweating" (Ray, 2008, p. 218). The central focus when providing a soldiers' heart diagnosis was the abnormality of the soldier's heartbeat. Little attention was paid to their emotional response to the trauma. Since soldiers were expected to be courageous, when a soldier showed any kind of fatigue they were only briefly sent to the back of the battle lines, so that they could recoup (Ray, 2008). After they received some time in the back, they were believed to have recovered and were sent back to the front lines. As a result, soldiers were likely exposed to multiple traumas during war.

2.2 World War I – Shell shock

A British military psychologist named Charles Samuel Myers was the first to use the term *shell shock* in medical literature (Myers, 1915 as cited in van der Kolk, 2007). Previous to his writings, the term was used in reference to British soldiers during World War I who had been exposed to a detonation or explosion, but had not sustained a visible head injury (Jones et. al., 2007). Some of the symptoms soldiers exhibited included tremors, dizziness, increased sensitivity to noise, headaches, difficulty concentrating, and amnesia (Turner, 1915 as cited in Jones et. al., 2007). Frederick Mott, a British neuropathologist suggested that shell shock impacted the tissue in the brain and spinal chord and could be fatal in extreme cases (Mott, 1917 as cited in Jones et. al., 2007). He also believed that some of the symptoms could be attributed to the gases that soldiers were exposed to during an explosion and that the gases could cause damage to the central nervous system (Mott, 1919 as cited in Jones et. al., 2007). Myers later conducted research on shell shock and suggested that the disorder may also result from psychological distress. He believed this because many of the soldiers who showed symptoms that were consistent with those of shell shock, had not been anywhere near an explosion (van der Kolk, 2007). The British Army was compelled to accept Myers' hypothesis because it enabled them to force soldiers to return to combat since the problem was psychological and they were not physically injured (Jones et. al., 2007). Subsequently the army declared two subtypes of shell shock, those who had been exposed to an explosion and those who were said to suffer from "nervousness" due to their anxiety about combat (Sloggett, 1916 as cited in Jones et. al., 2007).

By 1917, shell shock was said to have accounted for one in seven discharges from the British Army (Salmon, 1917 as cited in Jones et. al., 2007). Many doctors at the time believed that having shell shock was synonymous to being a coward (van der Kolk, 2007). Since numerous soldiers sought pensions for the effects of shell shock, the British Army became much more conservative with the diagnosis (Jones et. al., 2007). They intended that only the soldiers who were actually exposed to an explosion receive the diagnosis. Consequently, soldiers who were still serving and were dismissed from their duties for symptoms that resembled shell shock were said to be "not yet diagnosed nervous" (Jones et. al., 2007). Of those soldiers, individuals that did not have visible wounds and did not recover from their symptoms were labelled as "neurasthenic" (Jones et. al., 2007). Soon after the United State entered World War I, similar symptoms were observed in American soldiers.

Following World War I, many psychiatrists attempted to translate the clinical skills they gained during the war to working with the general public. Although the majority of psychiatrists were unsuccessful in impacting the field, Abram Kardiner was able to incite some changes (van der Kolk, 2007). Kardiner was one of Freud's students, and after treating veterans of World War I he tried to develop a theory on *war neurosis* that fit with psychoanalysis. Many of the symptoms that he made note of to characterize war neurosis are still highly relevant to the diagnostic criteria that we use for posttraumatic stress disorder today. Kardiner documented on symptoms that he labelled as "physioneurosis," which is nearly synonymous to the current symptom of physiological hyper-arousal (Kardiner, 1941 as cited in van der Kolk, 2007). He also made note of many of the re-experiencing and numbing or avoidant symptoms of posttraumatic stress disorder. Some of the symptoms which he acknowledged include irritability or proneness to anger, becoming withdrawn or detached, and individuals feeling as if they were re-experiencing the trauma when triggered by a neutral stimuli (Kardiner, 1941 as cited in van der Kolk, 2007). He continued to conceptualize war neurosis based upon psychoanalytic theories and he believed that those with the disorder were fixated on the trauma (Kardiner, 1941 as cited in van der Kolk, 2007). Despite psychiatrists working extensively with those with shell shock, much of the public was still sceptical of the diagnosis and believed soldiers were malingering (Ray, 2008).

2.3 World War II – Combat neurosis

During World War II, numerous names developed for what was previously labeled as shell shock even though each label was describing a very similar set of symptoms (Ray, 2008). Although having numerous names for one disorder could potentially result in confusion, it was seen as positive growth because it showed that multiple clinicians and researchers were coming to the same conclusion, that combat neurosis was a valid diagnosis. A new population of individuals suffering from similar symptoms - those who had survived the Nazi concentration camps, also expanded the professional understanding of combat neurosis. Observing concentration camp survivors brought Harry Abram to expand the concept of combat neurosis to a number of other trauma stricken populations which included: those under stress and those experiencing a life-threatening illness or an emergency situation (Abram, 1970 as cited in Ray, 2008). In his description, he suggested that the syndrome was comprised of both physical and psychological factors. An equal integration of both components was a novel argument because all past theories had put the primary emphasis on only one aspect without realizing the interplay between both components (Ray, 2008).

2.4 The diagnostic statistical manual, vietnam war, and posttraumatic stress disorder

In 1952, the first Diagnostic Statistical Manual included a diagnosis known as *stress response syndrome* (American Psychiatric Association, 1952 as cited in Lamprecht & Sack, 2002). The diagnosis was conceptualized as transient personality characteristic and was considered to be a normal reaction to extreme stress. Furthermore, with treatment, the symptoms were believed to subside once the ego regained balance (Lamprecht & Sack, 2002). The belief that people commonly recover from the syndrome was maintained despite multiple case examples to the contrary. The second Diagnostic Statistical Manual retained a very similar definition of stress response syndrome despite evidence demonstrating the need for adjustments (American Psychological Association, 1968 as cited in Lamprecht & Sack, 2002). It became a common belief among professionals that everyone had a breaking point, and that stress response syndrome was a normal response to an extreme stressor (Lamprecht & Sack, 2002).

Eventually the prevalence of soldiers who suffered from the chronic effects of stress response syndrome following the Vietnam War became undeniable. Vietnam veterans lobbied for compensation from the government for the trauma that they suffered (Lasiuk & Hegadoren, 2006). This forced the American Psychiatric Association to reconsider their conceptualization of the disorder, and in 1980 the term *posttraumatic stress disorder* was officially adopted into the Diagnostic Statistical Manual - III (American Psychiatric Association, 1980). In this version, posttraumatic stress disorder was defined by its' overt symptoms so that the characterization was not biased to a particular theory (Ray, 2008). In the revision of the Diagnostic Statistical Manual – III, they further refined the criteria for posttraumatic stress disorder. A distinction was made between common life stressors and a traumatic event, which was considered outside of the realm of normal human experience. Posttraumatic stress disorder was defined as having experienced a traumatic event, causing marked distress and fear, helplessness, or horror (American Psychiatric Association, 1987 as cited in Lasiuk & Hegadoren, 2006). Civilian populations such as those who suffered child abuse, sexual abuse, and intimate partner violence were also included under the diagnosis. Extreme changes were made in the diagnostic criteria of posttraumatic stress disorder in the Diagnostic Statistical Manual - IV, which closely resembles the diagnostic criteria that we follow today in the revised version.

2.5 Current definition of posttraumatic stress disorder

Posttraumatic stress disorder is currently defined by the Diagnostic Statistical Manual - IV Text Revision as an anxiety disorder resulting from exposure to a traumatic event involving personal or secondary threat to life or wellbeing and causing intense fear, helplessness, or horror (American Psychiatric Association, 2000). Posttraumatic stress disorder is characterized by physiological hyper-arousal, avoidance of stimuli that would provoke anxiety or general emotional numbing, and recurrence of psychologically re-experiencing aspects of the trauma.

2.6 Potential diagnostic statistical manual – v changes

It is apparent that the definition of posttraumatic stress disorder has evolved throughout the years. As such, it is to be expected that the Diagnostic Statistical Manual criteria will continue to be adapted as we learn more about the disorder. Some of the proposed changes for the Diagnostic Statistical Manual - V criteria will be presented in this section. Currently, none of the changes presented here have been officially accepted. Upon publication of the Diagnostic Statistical Manual - V, readers should reevaluate the proposed changes that have been presented in this chapter. It is not expected that the current proposal will severely alter the prevalence rates of posttraumatic stress disorder or severely impact how clinicians evaluate or treat the disorder (Frueh et. al., 2010).

The Diagnostic Statistical Manuel - IV - TR criterion for posttraumatic stress disorder specifies three symptom clusters: re-experiencing, avoidance or emotional numbing, and hyperarousal (American Psychiatric Association, 2000). These symptoms arise from primary or secondary exposure to a traumatic event that evokes feelings of extreme horror, fear, or helplessness. Re-experiencing is described as having nightmares, intrusive memories, feeling as if the event were reoccurring, and experiencing psychological and/or physiological distress when encountering internal or external reminders of the trauma. Avoidance or emotional numbing is defined as trying to avoid thoughts or feelings about the trauma, trying to avoid people or places that serve as reminders of the trauma, impaired memory for the trauma, feeling detached from others, having a sense of a foreshortened future, restricted affect, and anhedonia. Finally, hyper-arousal is defined as difficulty sleeping, irritability or anger, difficulty concentrating, hyper-vigilance, and exhibiting an exaggerated startle response. In order to receive a diagnosis of posttraumatic stress disorder, an individual must exhibit one re-experiencing symptom, three avoidance or emotional numbing symptoms, and two hyper-arousal symptoms. The symptoms must be present for over one month following the traumatic event and must cause impaired functioning or distress. If the symptoms have been apparent for less than three months the posttraumatic stress disorder is labeled as acute, but if present for over three months, the label is then changed to chronic posttraumatic stress disorder. The criteria for the Diagnostic Statistical Manual - IV - TR (current edition) and the proposed changes for the Diagnostic Statistical Manual - V can be seen in the table below (figure 1).

Diagnostic Statistical Manual - V	Diagnostic Statistical Manual - IV - TR
A. The person was exposed to one or more of the following event(s): death or threatened death, actual or threatened serious injury, or actual or threatened sexual violation, in one or more of the following ways:	A. The person has been exposed to a traumatic event in which both of the following were present:
1. Experiencing the event(s) him/herself	1. The person experienced, witnessed, or was confronted with an event or events that involved actual or threatened death or serious injury, or a threat to the physical integrity of self or others
2. Witnessing, in person, the event(s) as they occurred to others	2. The person's response involved intense fear, helplessness, or horror. **Note:** In children, this may be expressed instead by disorganized or agitated behavior
3. Learning that the event(s) occurred to a close relative or close friend; in such cases, the actual or threatened death must have been violent or accidental	
4. Experiencing repeated or extreme exposure to aversive details of the event(s) (e.g., first responders collecting body parts; police officers repeatedly exposed to details of child abuse); this does not apply to exposure through electronic media, television, movies, or pictures, unless this exposure is work related.	
B. Intrusion symptoms that are associated with the traumatic event(s) (that began after the traumatic event(s)), as evidenced by 1 or more of the following:	B. The traumatic event is persistently reexperienced in one (or more) of the following ways:
1. Spontaneous or cued recurrent, involuntary, and intrusive distressing memories of the traumatic event(s). **Note:** In children, repetitive play may occur in which themes or aspects of the traumatic event(s) are expressed.	1. Recurrent and intrusive distressing recollections of the event, including images, thoughts, or perceptions. **Note:** In young children, repetitive play may occur in which themes or aspects of the trauma are expressed.
2. Recurrent distressing dreams in which the content and/or affect of the dream is related to the event(s). **Note:** In children, there may be frightening dreams without recognizable content.	2. Recurrent distressing dreams of the event. **Note:** In children, there may be frightening dreams without recognizable content.
3. Dissociative reactions (e.g., flashbacks) in which the individual feels or acts as if the	3. Acting or feeling as if the traumatic event were recurring (includes a sense of reliving

traumatic event(s) were recurring (Such reactions may occur on a continuum, with the most extreme expression being a complete loss of awareness of present surroundings.) **Note:** In children, trauma-specific reenactment may occur in play.	the experience, illusions, hallucinations, and dissociative flashback episodes, including those that occur on awakening or when intoxicated). **Note:** In young children, trauma-specific reenactment may occur.
4. Intense or prolonged psychological distress at exposure to internal or external cues that symbolize or resemble an aspect of the traumatic event(s)	4. Intense psychological distress at exposure to internal or external cues that symbolize or resemble an aspect of the traumatic event
5. Marked physiological reactions to reminders of the traumatic event(s)	5. Physiological reactivity on exposure to internal or external cues that symbolize or resemble an aspect of the traumatic event
C. Persistent avoidance of stimuli associated with the traumatic event(s) (that began after the traumatic event(s)), as evidenced by efforts to avoid 1 or more of the following:	C. Persistent avoidance of stimuli associated with the trauma and numbing of general responsiveness (not present before the trauma), as indicated by three (or more) of the following:
1. Avoids internal reminders (thoughts, feelings, or physical sensations) that arouse recollections of the traumatic event(s)	1. Efforts to avoid thoughts, feelings, or conversations associated with the trauma
2. Avoids external reminders (people, places, conversations, activities, objects, situations) that arouse recollections of the traumatic event(s).	2. Efforts to avoid activities, places, or people that arouse recollections of the trauma
D. Negative alterations in cognitions and mood that are associated with the traumatic event(s) (that began or worsened after the traumatic event(s)), as evidenced by 3 or more of the following: **Note**: In children, as evidenced by 2 or more of the following:	
1. Inability to remember an important aspect of the traumatic event(s) (typically dissociative amnesia; not due to head injury, alcohol, or drugs).	3. Inability to recall an important aspect of the trauma
2. Persistent and exaggerated negative expectations about one's self, others, or the world (e.g., "I am bad," "no one can be trusted," "I've lost my soul forever," "my whole nervous system is permanently ruined," "the world is completely dangerous").	4. Markedly diminished interest or participation in significant activities
3. Persistent distorted blame of self or others about the cause or consequences of the traumatic event(s)	5. Feeling of detachment or estrangement from others

4. Pervasive negative emotional state -- for example: fear, horror, anger, guilt, or shame	6. Restricted range of affect (e.g., unable to have loving feelings)
5. Markedly diminished interest or participation in significant activities.	7. Sense of a foreshortened future (e.g., does not expect to have a career, marriage, children, or a normal life span)
6. Feeling of detachment or estrangement from others.	
7. Persistent inability to experience positive emotions (e.g., unable to have loving feelings, psychic numbing)	
E. Alterations in arousal and reactivity that are associated with the traumatic event(s) (that began or worsened after the traumatic event(s)), as evidenced by 3 or more of the following: **Note**: In children, as evidenced by 2 or more of the following:	D. Persistent symptoms of increased arousal (not present before the trauma), as indicated by two (or more) of the following:
1. Irritable or aggressive behavior	1. Difficulty falling or staying asleep
2. Reckless or self-destructive behavior	2. Irritability or outbursts of anger
3. Hypervigilance	3. Difficulty concentrating
4. Exaggerated startle response	4. Hypervigilance
5. Problems with concentration	5. Exaggerated startle response
6. Sleep disturbance -- for example, difficulty falling or staying asleep, or restless sleep.	
F. Duration of the disturbance (symptoms in Criteria B, C, D and E) is more than one month.	E. Duration of the disturbance (symptoms in Criteria B, C, and D) is more than 1 month.
G. The disturbance causes clinically significant distress or impairment in social, occupational, or other important areas of functioning.	F. The disturbance causes clinically significant distress or impairment in social, occupational, or other important areas of functioning.
H. The disturbance is not due to the direct physiological effects of a substance (e.g., medication or alcohol) or a general medical condition (e.g., traumatic brain injury, coma).	
Specify if: *With Delayed Onset:* if diagnostic threshold is not exceeded until 6 months or more after the event(s) (although onset of some symptoms may occur sooner than this).	*Specify* if: *Acute*: if duration of symptoms is less than 3 months *Chronic*: if duration of symptoms is 3 months or more *With Delayed Onset:* if onset of symptoms is at least 6 months after the stressor

Fig. 1. DSM IV - TR (American Psychiatric Association, 2000) and Proposed Criteria for the DSM - V (American Psychiatric Association, 2010 obtained 7.6.11 from http://www.dsm5.org/ProposedRevisions/Pages/proposedrevision.aspx?rid=165#)

The first change that was proposed is meant to give more clarity to what qualifies as a traumatic event. The current criteria states that the person must have both "experienced, witnessed, or [been] confronted with an event or events that involved actual or threatened death or serious injury, or a threat to the physical integrity of self or others" and "the person's response involved intense fear, helplessness, or horror" (American Psychiatric Association, 2000). In order to provide more clarity, it has been proposed that a traumatic event should be constituted by actual or threatened: death, serious injury, or sexual violation. In addition, the individual must have either personally experienced the traumatic event, witnessed it in person, heard about it happening to a close friend or relative (where death or threatened death must be either violent or accidental), or have had one extreme exposure or repeated exposures to the unpleasant details of the event (American Psychiatric Association, 2010). A criticism of the current definition is that it could be interpreted that witnessing the media's portrayal of a situation would qualify as a traumatic event, although not all clinicians would endorse this interpretation. Therefore, it has been proposed that the Diagnostic Statistical Manuel - V prohibit that a media portrayal of an event qualify as traumatic unless the exposure is work related. The new definition also removes the qualifier that the person must react with intense fear, helplessness, or horror. One of the arguments for this change is that the current definition does not allow for individual differences in how people respond to trauma. A contrary argument for removing this qualifier is that most people respond to trauma in a manner that is consistent with the criteria.

In a worldwide sample of 28,490 participants who experienced a potentially traumatic event, only 1.4% of participants did not respond to the event with intense fear, helplessness or horror while meeting all other criteria for a diagnosis of posttraumatic stress disorder (Karam et. al., 2010). This study bolsters the argument that those who experience an event which results in them meeting all of the other criteria for the disorder will almost always respond with intense horror, fear, or helplessness. For that reason, including this additional criterion does not provide meaningful information. Those that do not meet the criteria may for some reason respond differently to trauma, but this would be something to be explored in therapy rather than addressed in their diagnosis. In addition, excluding this qualifier from the diagnostic criteria could reduce the amount of time it takes to assess for posttraumatic stress disorder.

Another relatively significant change that is being proposed is to use four symptom clusters to diagnose posttraumatic stress disorder instead of three. More specifically, it has been proposed to divide the avoidance and emotional numbing cluster into an avoidance cluster and separate cluster focusing on distorted thinking and negative emotions. In order to obtain a diagnosis using the divided clusters, an individual would need to avoid either internal or external reminders of the trauma. In addition they would need three symptoms from the distorted thinking and negative emotions cluster. This cluster would be comprised of seven symptoms which include: inability to remember the trauma, anhedonia, feeling detached, restricted affect, pervasive experience of negative emotions, distorted blame of self or others for the trauma, and persistent distorted negative thoughts about one's self, others, and the world. This change is being proposed because factor analysis has suggested that the current model does not account for all of the dimensions of posttraumatic stress disorder (Frueh et. al., 2010). More specifically this means that avoidance and emotional numbing are distinct concepts.

A problem that has emerged with the current diagnosis is that posttraumatic stress disorder symptoms overlap with many of the symptoms from major depressive disorder. Some of the symptoms that overlap include: anhedonia, sleep problems, irritability, and difficulty concentrating. Due to the high comorbidity of posttraumatic stress disorder and major depressive disorder, some have suspected that counting the same symptom for both disorders accounts for much of the comorbidity rather than the disorders actually co-occurring. A study by Elhia and colleagues (2008) demonstrated that the disorders do in fact frequently co-occur because when the symptoms that overlap with depression and anxiety are removed, the lifetime prevalence figure for posttraumatic stress disorder only decreases from 6.81% to 6.42%. Removing the overlapping symptoms would cause some individuals to reach a subclinical level, but it is valuable to be aware that the disorders are in fact distinct. Concern about the overlap in criteria may become even more common because the proposed changes will make posttraumatic stress disorder less distinct from depression. It is being proposed that the current symptoms, which are very similar to those of major depressive disorder remain in the criteria. Furthermore, it has been proposed that additional criterion that also overlaps with major depressive disorder be added to the diagnostic criteria. It is unclear how this change will impact future comorbidity of posttraumatic stress disorder and major depressive disorder, which may prove to be a problem.

Additional changes that are being proposed include adding an extra symptom to the hyper-arousal cluster and removing the distinction between chronic and acute posttraumatic stress disorder (American Psychological Association, 2010). Other minor changes are being proposed to reword some of the criteria in order to provide clarity. Removing the distinction between chronic and acute posttraumatic stress disorder is being proposed because there is not enough evidence to show that they are separate concepts rather than just two separate time points on the same continuum. The additional symptom that may be added to the hyper-arousal cluster includes engaging self-destructive or reckless behavior. With the additional symptom, individuals would need to have three of six symptoms from the hyper-arousal cluster to receive a diagnosis.

We have clearly come a long way in our understanding of posttraumatic stress disorder throughout the years, but there are many aspects of the disorder that we still do not understand. As we learn more about the mechanisms of the disorder, the definition will continue to be adapted within the Diagnostic Statistical Manual.

3. Epidemiology

Prevalence estimates of posttraumatic stress disorder are important because they can be used to determine how to allocate resources for those affected by the disorder (Ramchand et. al., 2010). This section will present prevalence figures for the general population and the figures for the current and past wars, and will conclude with a discussion about the gender differences in posttraumatic stress disorder.

3.1 Prevalence

In the general population of the United States, posttraumatic stress disorder has been found to have a lifetime prevalence of 6.8 percent (Kessler et. al., 2005). Posttraumatic stress disorder is frequently seen in military personnel due to their elevated potential for exposure to trauma during combat. In the current war in Iraq and Afghanistan, the prevalence of posttraumatic stress disorder in soldiers post-deployment is believed to be between 10.3% and 17% (Sundin et. al., 2009). The prevalence of posttraumatic stress disorder for Vietnam Veterans ranges from 8.5% to 19.3% and between 1.9% and 24% for soldiers in the Persian

Gulf War (Sundin et. al., 2009). Prevalence figures vary widely in the military based upon a number of variables, such as how posttraumatic stress disorder was assessed, how much time has elapsed since the trauma, the level of combat exposure, the number of completed tours, gender, and the unit the individual was assigned to during deployment.

3.2 Gender differences in posttraumatic stress disorder in the military

Posttraumatic stress disorder has been known to develop following a broad range of traumatic situations. Due to this chapter's focus on combat related posttraumatic stress disorder this section will only present traumatic situations that are commonly experienced by those in the military. Some of the more common trauma experiences include: combat situations where the soldier felt as though their life was in danger or witnessed the death or threatened death of another person, seeing dead bodies or mutilated body parts during an assignment, or sexual assault while in the military. Men experience posttraumatic stress disorder as a result of combat situations more frequently than women because women are not permitted to have infantry positions. Women are more likely to experience sexual assault, which is unfortunately a frequent occurrence in the military (Williams & Bernstein, 2011).

3.2.1 Men in the military

Men in the military are vulnerable to an array of traumatic situations during combat. Individual differences exist and dictate whether a person has a heightened likelihood for developing posttraumatic stress disorder and how severe the stressor must be in order for them to develop the disorder (See the section on etiology for a more in depth discussion of individual differences in vulnerability for developing posttraumatic stress disorder). Some soldiers may be traumatized by just hearing the sounds of explosions due to a fear of being harmed by an explosive device. More resilient soldiers may obtain posttraumatic stress disorder from being involved an automobile accident while deployed or by being exposed to an explosive device that detonated near them or injured someone around them. Furthermore, they could be traumatized from engaging in hand-to-hand combat or in a firefight with the enemy. Finally, soldiers could be traumatized while retrieving severely injured soldiers or collecting bodies or body parts of soldiers who were killed in combat. Women in the military are also at risk for being involved in the previously mentioned traumatic situations, but have a decreased likelihood because their job assignments are intended to keep them away from direct combat. The list of potential combat scenarios provided is not meant to be all-inclusive, as there are a number of unpredictable situations in war that can cause a soldier to develop posttraumatic stress disorder.

3.2.2 Women in the military

Women in the military are thought to have an increased probability of experiencing a traumatic event during their service because of their ability to be sexually assaulted. Female soldiers and male soldiers placed in non-combat positions experience the same level of risk for encountering a traumatic event during deployment. Female soldiers are additionally at risk for being sexually assaulted by other soldiers (Williams & Bernstein, 2011). Although men are also sexually assaulted while in the military, women are more frequently assaulted. Lipari and Lancaster (2003) found that in active duty personnel, 3% of women have been sexually assaulted while in the military as compared to 1% of men. Furthermore, Sadler and colleagues (2003) found in a sample of 558 female veterans, 28% had experienced a rape or an attempted rape while in the military, 8% experienced some form of sexual coercion, and 27% experienced unwanted sexual attention. The Department of Defense (2004) found that 71% of the women seeking treatment for posttraumatic stress disorder, who had served in the Vietnam War and subsequent wars, had been raped while in the military. Some of the risk factors that increase a female soldier's chance of being sexually assaulted include being between the ages of 17 to 24 years old, using alcohol, and past history of sexual assault (Williams & Bernstein, 2011).

4. Etiology

This section will discuss the mechanisms through which an individual develops posttraumatic stress disorder. Brief attention will be given to the nature of traumatic stressors and the linear progression from acute stress disorder to posttraumatic stress disorder. The primary emphasis of this section will be on the psychological and biological theories regarding what makes a person vulnerable to posttraumatic stress disorder and what maintains it after symptoms arise.

4.1 From trauma to acute stress disorder to posttraumatic stress disorder

Not every individual who experiences a traumatic event will subsequently develop posttraumatic stress disorder. As the name implies, posttraumatic stress disorder results from an experience with a traumatic stressor. Some of the stressors that can cause the disorder include: natural disasters, combat, sexual assault, physical assault, abuse or neglect as a child, car accidents, surgery, and witnessing something life threatening happen to a loved one. A person can develop posttraumatic stress disorder from a single stressor or may encounter multiple traumatic situations. An individual who encounters multiple events may either develop the disorder after the first event and the subsequent events then exacerbate their symptoms or they may develop the disorder only after experiencing multiple traumatic events.

As previously discussed, an individual must experience their symptoms for at least a month in order to receive a posttraumatic stress disorder diagnosis (American Psychiatric Association, 2000). Individuals with symptoms lasting less than a month are given an acute stress disorder diagnosis. Although everyone who has posttraumatic stress disorder has also had acute stress disorder, not everyone who experiences acute stress disorder will go on to develop posttraumatic stress disorder. The diathesis-stress model helps explain why some individuals do not develop the disorder after a traumatic experience (Elwood et. al., 2009). This model refers to the interaction between a person's environment (the severity of the stressors that they encounter) and their biological and psychological predispositions, which can create vulnerability for developing the disorder. Those with high diathesis only require a minimal stressor in order to develop the disorder, whereas someone with no diathesis may never develop the disorder even when presented with an extreme stressor. The next section will present the psychological and biological theories on the characteristics that may act as a diathesis for developing posttraumatic stress disorder.

4.2 Theories

It is important to understand some of the basic theories on posttraumatic stress disorder in order to appreciate how these theories have then been integrated into the current theories that are far more complex. This section will provide a brief introduction to stress response theory, theory of shattered assumption, conditioning theory, and information-processing theory. This section will be followed by a discussion about some of the current psychological theories, including: emotional processing theory, dual representation theory, and Ehlers and Clark's (2000) cognitive theory on posttraumatic stress disorder. This section will conclude with a discussion on the biological correlates of posttraumatic stress disorder. It must be noted that the majority of the research on the biological aspects of posttraumatic stress disorder comes from correlational studies. Inferences cannot be made as to whether the biological abnormalities existed before the trauma and acted as a vulnerability for acquiring the disorder or developed after being exposed to the trauma.

4.2.1 Basic psychological theories

Stress response theory posits that a person develops posttraumatic stress disorder when they are unable to reconcile their beliefs about the world with what happened during the trauma (Horowitz 1976 & 1986 as cited in Brewin & Holmes, 2003). People have an internal working model of how the world operates and a traumatic experience often violates some of those core beliefs. When the individual is unable to logically integrate what happened to them within their world-view, defense mechanisms become activated to repress the trauma. The defense mechanisms at play are said to mimic many of the avoidance and numbing symptoms of posttraumatic stress disorder. Since a drive to reconcile the trauma with one's world-view still unconsciously exists, the person will experience intrusive reminders of the trauma to force them to cope with what happened. The individual will continue to experience these symptoms until they resolve the discrepancy, which is said to explain why some suffer from chronic posttraumatic stress disorder. Clearly, this theory is highly rooted in psychodynamic principles. Although it does not explain the full range of symptoms in those with posttraumatic stress disorder, stress response theory provided a framework for the theories that followed it.

The *theory of shattered assumptions* is very similar to stress response theory in that it places an emphasis on the individual's assumptions about the world. According to this theory, the assumptions that are said to be the most important to how a person responds to trauma include believing that: the world is a good place, what happens within the world makes sense, and that they are generally a good person and worthy of having good things happen to them (Janoff-Bulman, 1992). One of the initial assumptions of this theory was that those with the most positive beliefs about the world would also be the most severely impacted by trauma. Since this belief was disproved by the fact that previous trauma serves as a risk factor for developing posttraumatic stress disorder, the theory was revised to say that those who have previously been exposed to trauma have already had their view of the world shattered. Having this negative outlook makes them vulnerable for developing posttraumatic stress disorder in the future. Similar to stress response theory, this theory provides an incomplete rationale for all of the symptoms associated with posttraumatic stress disorder.

The *conditioning theory* of posttraumatic stress disorder is based upon Mowrer's two-factor learning theory (1960 as cited in Brewin & Holmes, 2003). The process of fear acquisition occurs when a traumatic experience is paired with a neutral stimulus, resulting in a fear response to the previously neutral stimuli. Once the neutral stimulus becomes a conditioned stimulus, the person begins to generalize their fear to other situations (Keane, et. al., 1985 as cited in Brewin & Holmes, 2003). Using a behavioral framework, individuals with posttraumatic stress disorder should habituate to their feared stimuli due to the re-experiencing symptoms of the disorder. Individuals with the disorder do not habituate because once they begin re-experiencing the trauma they then engage activities that are consistent with the avoidance or numbing symptoms of the disorder. Since their distress subsides, they are then reinforced to continue engaging in avoidance and numbing tactics to cope with the trauma. Although this theory is highly useful for explaining posttraumatic stress disorder in many ways, it has been criticized because it is missing the cognitive component of the disorder. The cognitive component is important because it is often necessary for explaining individual differences in acquisition of the disorder (Brewin & Holmes, 2003).

Information processing theory integrates the cognitive components of the disorder into conditioning theory (Lang et. al., 1979 as cited in Brewin & Holmes, 2003). The general assumption of this theory is that when a person has a traumatic experience, the memory is stored differently than those from normal experiences. Posttraumatic stress disorder is then the result of a memory not being processed correctly. Information processing theory focuses solely on the cognitive components of the trauma and does not broadly integrate the social and personal context of the event. The memory of the trauma is comprised of: the surroundings during the trauma, other concrete aspects of the event, the person's physical and emotional reactions, and their assessment of the event. The consolidation of the experience, including all of the aforementioned components into a memory is called a *fear network*. Subsequently, when an individual is exposed to something that resembles an aspect of the fear network, the entire network then gets activated which triggers the same emotional response that was experienced during the trauma. An example of the fear network is a soldier ducking to the ground in fear when he hears a balloon pop because he was traumatized after witnessing an explosion while in combat. In this example, a loud noise, feeling fearful, and ducking to the ground, all are a part of the soldier's fear network. Simply hearing a sound that was similar to an explosion was sufficient to trigger the entire fear network.

Edna Foa added to this theory by explaining that what separates posttraumatic stress disorder from other anxiety disorders is that a traumatic event causes the person to question their basic assumptions about their personal safety in a global manner (Foa et. al., 1989 as cited in Brewin & Holmes, 2003). Since their assumptions about safety have been violated, their threshold to activate the fear network is low. In addition, because the individual does not feel safe, they are much more aware of their surroundings causing a reciprocal relationship between the decreased threshold and their sense of safety. An individual can reintegrate the different components of their fear network back into a normal memory if they are exposed to those components in a way that teaches them to that they are not actually in danger. This concept will re-visited and elaborated upon in the therapy section regarding Prolonged Exposure.

4.2.2 Contemporary psychological theories

Emotional processing theory is based on information processing theory, but takes into consideration individual perceptions before, during, and after the trauma (Foa & Riggs, 1993 as cited in Brewin & Holmes; Foa & Rothbaum, 1998 as cited in Brewin & Holmes). Furthermore, this theory proposes that those with more rigid views before the trauma will have worse outcomes following the experience. Having an extremely positive view or extremely negative view pre-trauma is considered a risk factor for developing posttraumatic stress disorder. Positive views include believing the world is very safe or the person thinking they are completely capable of dealing with stress. Negative views would include believing that the world is a bad place or that bad things always happen to them. When an individual with rigid negative views of the world encounters a traumatic situation, it confirms that their views of the world were accurate. Therefore, an individual's outlook before experiencing trauma can impact how they perceive the event while it is happening and how they reflect on what happened. This theory is clinically relevant because if during treatment, an individual can be repeatedly re-exposed to the traumatic experience they can habituate to the feared stimulus and may reevaluate and hopefully reconsider how they reflect on the trauma. A client can be re-exposed to the trauma in session by either asking the client to imagine the experience or ask them to have real life encounters with innocuous situations that remind them of the memory.

Dual representation theory, as its name implies, makes the assumption that people store memories in two distinct ways (Bewin et. al., 1996 as cited in Brewin & Holmes, 2003). More specifically, memories tied to emotionally traumatic situations are stored differently than those from every day occurrences. Memories can either be stored as *verbally accessible* or *situationally accessible*. A verbally accessible memory is one that can be intentionally retrieved. A situationally accessible memory cannot be recalled at will, and can only be triggered by perceptual reminders of the trauma, such as sights, sounds, or physiological responses. When a memory becomes pathological, it is because it has become dissociated from being verbally accessible and is only situationally accessible. In addition, only primary emotions are stored in situationally accessible memories, such as fear, hopelessness, or horror. In order to transform a traumatic memory into a normal one, the individual must learn to express the traumatic situation verbally as though it were regarding a daily occurrence. This changes the emotions associated with the situational memory from negative emotions to positive ones due to the continued pairing of positive emotions with the memory.

Ehlers and Clark (2000) proposed a *cognitive model of posttraumatic stress disorder*, which highlights the discrepancy of the disorder from other anxiety disorders. This is because individuals who develop posttraumatic stress disorder perceive a current threat from a past event instead of a future event. Furthermore, this theory suggests that what distinguishes those who develop posttraumatic stress disorder from those who experience trauma but do not develop the disorder, is whether they equate experiencing past trauma to also having an increased susceptibility to future danger.

Ehlers and Clark's model proposes that there are multiple negative appraisals that people can make after experiencing a traumatic event that may lead to the belief that there is also a current threat for danger. The content of the appraisals include an individual's beliefs regarding: the fact that the event occurred and that it happened to them, their behavior and emotions during the trauma, the meaning of initial occurrence of posttraumatic stress disorder symptoms and the chronic symptoms (such as re-experiencing, emotional numbing, and concentration problems), the positive and negative reactions of other's to the trauma, and the physical or global consequences of the trauma (Ehlers & Clark, 2000, pg. 322). Many of the negative appraisals that can lead to posttraumatic stress disorder contain themes about the individual assuming personal responsibility for the trauma, believing that others perceive the event as their fault, and assuming that their cognitive and emotional responses to the trauma are going to be permanent. Since individuals with posttraumatic stress disorder often assume personal responsibility for the trauma by attributing it's occurrence a personal deficiency, they also overestimate the likelihood of something dangerous happening again.

Ehlers and Clark's theory adopts some of the same concepts from dual representation theory and posits that a pathological memory contains only the sensory and emotional aspects of the event. Since the individual has not integrated the memory into their autobiographical memory, they are unable to provide all of the details of the event on cue. Remembering the details of the event may buffer from having unwanted recollections by providing context for memory. The chronological details of the event are also important because those with posttraumatic stress disorder may not be consciously aware of all of the precursors of the event, but can still be triggered by a stimulus that preceded the trauma. These individuals may also show biased attention for the negative aspects of what occurred before, during, and after the trauma. Furthermore, they often engage in behaviors that cause or exacerbate their symptoms, such as avoiding reminders of the trauma. Their avoidance often causes intrusive recollections, fails to give them the opportunity to disprove their beliefs about the trauma, and inhibits them from creating an autobiographical memory of the event. This theory provides the most integrative and detailed explanation of posttraumatic stress disorder and clearly incorporates many of the theories that preceded it. The theory's multifaceted explanation of posttraumatic stress disorder provides clinicians with a complex framework for viewing their clients. Due to the complexity of the theory, clinicians can choose which aspects are the most relevant to the cognitive distortions that they are seeing in their client.

4.2.3 Biological theories

In recent years, researchers have extended the biological theories on depression to posttraumatic stress disorder due to the comorbidity of both disorders. Kilpatrick and colleagues (2007) were one of the first research teams to generalize the genetic research on the serotonin transporter gene (5-HTTLPR) from depression to posttraumatic stress disorder. Previous research established that those with two short 5-HTTLPR alleles had a higher risk of developing depression than those with two long alleles or a combination of a short and long allele (Lesch et. al., 1996). The environment also plays a huge role in whether someone develops depression despite the genetic component of the disorder. Using this framework, Kilpatrick and colleagues (2007) investigated whether having two short 5-HTTLPR alleles increased the likelihood of developing posttraumatic stress disorder in participants who were exposed to hurricane Rita, which hit Florida in 2004. They found that low social support and high hurricane exposure proved to be risk factors for developing posttraumatic stress disorder. In addition individuals who had high levels of hurricane exposure, low levels of social support, and had two short alleles had a 4.5 times greater chance of developing posttraumatic stress disorder than the rest of the sample.

Research has also looked at monozygotic twins to examine the biological differences in a twin with posttraumatic stress disorder compared to their twin who does not have posttraumatic stress disorder. Pitman and colleagues (2006) examined twin pairs, where one twin obtained posttraumatic stress disorder through involvement in the Vietnam War and the other twin did not experience combat exposure or develop posttraumatic stress disorder. They found that the twin with posttraumatic stress disorder demonstrated higher heart rate reactivity to a startling noise than his brother. This response is thought to be in part the result of hyperactivity in the amygdala. They also discovered that high-risk twin pairs often had some level of neurological dysfunction. The study inferred that this preexisting dysfunction might act as a vulnerability for developing posttraumatic stress disorder. When examining hippocampal volume using magnetic resonance imaging, they found that the twins with more severe posttraumatic stress disorder had a smaller hippocampus than average, but their twin brothers also had reduced hippocampal volume. Since this is a correlational study, the authors caution that more research is needed to draw causal inferences from these results.

Stress hormones such as cortisol have also been examined and found to correlate with posttraumatic stress disorder. A meta-analysis by de Kloet and colleagues (2006) concluded that those with posttraumatic stress disorder have lower baseline levels of cortisol than those without the disorder. Conversely, when exposed to a stressor, those with the posttraumatic stress disorder show an elevated cortisol response in comparison to those without the disorder. Although many theories have been proposed as to why this relationship exists, there is no conclusive evidence explaining why people with posttraumatic stress disorder have deceased baseline levels of cortisol, yet have an exaggerated stress response.

In accordance with the diathesis-stress model, both the psychological and biological theories on posttraumatic stress disorder should be taken into consideration because diathesis is comprised of both components. All of the contemporary theories on posttraumatic stress disorder are valuable to help conceptualize the disorder and no one theory has become dominant within the research. Each theory can be applied based on its relevance to a particular client.

5. Treatment

A number of treatments have been shown to be effective in treating posttraumatic stress disorder. Many of the treatments that are used for the disorder are rooted in cognitive behavioral therapy. This section will focus primarily on the treatments that have proven effective with those suffering from combat related posttraumatic stress disorder. The Veterans Administration in particular, has endorsed both cognitive processing therapy and prolonged exposure therapy (Karlin et. al., 2010). This section will also address a few of the more novel treatments for posttraumatic stress disorder such as the use of virtual reality and biofeedback. Some clinicians and researchers have recently incorporated virtual reality technology into prolonged exposure therapy. In addition, with the use of biofeedback, veterans can be taught to monitor their own physiological reactions, which are often elevated due to the hyper-arousal component of posttraumatic stress disorder.

5.1 Popular treatments for combat related posttraumatic stress disorder

This section will discuss cognitive processing therapy, prolonged exposure therapy, and the medications that can be used for individuals with posttraumatic stress disorder. An array of therapies exists for treating posttraumatic stress disorder and what is covered below should not be considered an all-inclusive list of the effective treatments.

5.1.1 Cognitive processing therapy

Cognitive Processing Therapy places an emphasis on the meaning that an individual assigns to their traumatic experience (Karlin et. al., 2010; Resick & Schnicke, 1992). The treatment is divided into three phases and is typically administered over the course of 12 sessions. In addition, the treatment can be used in individual or group therapy. The three phases are comprised of: education, processing, and challenging. During the education phase, clients learn about the symptoms of posttraumatic stress disorder, how treatment will work, and is taught about the interaction between thoughts and feelings. They are also asked to consider how the event has impacted their outlook on the world. More specifically they are asked to examine the changes that may have occurred in their beliefs about themselves, others, and how the world operates. During the processing phase, the client is asked to either write about or discuss the traumatic event and work to identify thinking patterns that may be hindering their recovery. In the final phase of therapy, the challenging phase, the therapist works with the client to help them reframe their distorted beliefs about themselves, others, and the world. In doing this, the client develops a more balanced view of their environment.

5.1.2 Prolonged exposure therapy

Prolonged exposure therapy was designed specifically for individuals with posttraumatic stress disorder. The length of treatment typically ranges from 8 to 15 sessions, although it was initially designed to be 10 sessions (Foa & Kozak, 1986; Foa et al., 2007). This treatment draws from cognitive behavioral theories and it operates on the assumption that exposure to a feared stimulus will eventually extinguish the fear. During the first and second session, the primary focus is to provide psycho-education regarding the techniques that will be used, explain the rationale for using those techniques, and discuss the ways that people typically react to a traumatic event. Subsequent sessions will be dedicated to either imagery exposure or in vivo exposure. *In vivo exposure* is where the client goes out into the real world and encounters the feared object or situation in person with the goal of habituation. The in vivo scenarios that are used during treatment are low risk and are often commonplace experiences. These scenarios are appropriate for treatment because individuals with posttraumatic stress disorder will often avoid an array of low threat situations because they trigger unpleasant memories. *Imagery exposure* involves the person imagining the feared situation. More specifically, the client is prompted to talk about the most disturbing aspects of their trauma with the therapist. This gives them the ability to reprocess what actually happened and the opportunity to reorganize how they reflect on the traumatic event. The length of treatment depends on the client and is terminated when they no longer have symptoms that inhibit them from engaging in every day activities.

5.1.3 Medication

Due to the biological component of posttraumatic stress disorder, individuals who suffer from the disorder can also receive antidepressants to help ameliorate their symptoms. Medication can be used in conjunction with psychotherapy or can be used alone. Although a number of medications are currently being investigated for the treatment of posttraumatic stress disorder, the Food and Drug Administration has only approved two medications (Friedman & Davidson, 2007). Both of the medications that they approved, Sertraline and Paroxetine, are selective serotonin reuptake inhibitors. As we learn more about the biological mechanisms of the disorder the medications that are recommended for posttraumatic stress disorder will continue to change.

5.2 New treatments

Although there is limited research on novel treatments for posttraumatic stress disorder, some treatments are showing promising results. Two of those treatments include heart rate variability biofeedback training and virtual reality exposure therapy.

5.2.1 Heart rate variability biofeedback training

As mentioned in previous sections, hyper-arousal is one of the symptoms found in those with posttraumatic stress disorder. Persistent hyperarousal has been linked to physiological abnormalities such as increased blood pressure, exaggerated heart rate response to stressors, and an elevated resting heart rate (Cohen et al. 1997; Pitman et al. 1987). This has led researchers to speculate that posttraumatic stress disorder may alter sympathetic nervous system reactivity. In addition, researcher found that between 80% to 100% of individuals with posttraumatic stress disorder can be distinguished from those without by looking solely at their physiological reactivity (Orr & Roth 2000), which can be indicative of autonomic nervous system dysfunction. Heart rate variability can be used as an indicator of how the autonomic nervous system is functioning (Appelhans & Luecken 2006). Those with posttraumatic stress disorder typically have low heart rate variability (Tan et. al. 2011). Heart rate variability is the mean value of heart rate fluctuations over a period of time and is reflective of the interplay between the sympathetic and parasympathetic nervous system (Akselrod et al. 1981; Cohen et al. 1999). Research has established that by breathing at an ideal resonance frequency (approximately 5.5 breaths per minute), an individual can increase their heart rate variability (Vaschillo et al. 2002). Ideal resonance frequency varies by person.

Clients undergoing *heart rate variability training* are asked to first meet with the therapist to determine what breathing rate will produce their greatest heart rate variability (Lehrer et. al., 2000; Tan et. al., 2011). Clients are then instructed to practice breathing at this rate at home. They may either practice with a CD that guides them through the breathing techniques or they may be given a machine that notifies them when they are not breathing at their ideal rate. In a pilot study by Tan and colleagues (2011), participants who underwent eight, 30 minute training sessions experienced a significant reduction in posttraumatic stress disorder symptoms from pretest to posttest.

5.2.2 Virtual reality exposure therapy

Virtual reality exposure therapy has been used to treat soldiers that served in Vietnam, Operation Enduring Freedom, and Operation Iraqi Freedom. Computer programs were developed for both populations containing scenes that look similar to the surroundings veterans would have experienced during combat. The Vietnam virtual reality environment contains a scene with a virtual jungle and includes sounds of the jungle, gunfire, and nearby helicopters and has a separate scene within a helicopter (Gerardi et. al., 2010). *Virtual Iraq* was developed for veterans of current war. (A. A. Rizzo, et al., 2008). Virtual Iraq contains scenes of a Middle Eastern themed city, where the person is able to travel through the city by foot or in a truck. This environment can be adapted based on the client's therapeutic needs. In addition to the virtual reality scene, the individual is also presented with auditory, tactile, and olfactory stimulation. The client sits on a platform equipped with subwoofers, and the therapist controls which sounds the client hears. Furthermore, the platform vibrates in coordination with the virtual reality environment. The clinician also controls the smells that are emitted from the "olfaction box" which includes various scents such as: burning rubber, body odor, and gasoline. Since all of these stimuli are presented simultaneously, it increases the reality of the virtual environment (A. A. Rizzo, et al., 2010).

Individuals undergoing treatment with the Virtual Iraq technology typically come in twice a week for 90 minutes over the course of five weeks (A. A. Rizzo, et al., 2008). The initial sessions are dedicated to identifying the details of the traumatic event and teaching the client stress management techniques such as deep breathing. They are also taught how to use the technology and to rate their distress so that it can be used as a reference throughout treatment. In a study on the efficacy of this treatment modality, Reger and Gahm (2008) found that patient's PTSD Checklist score decreased by approximately 50% post-treatment and they also showed a significant functional improvement. A major criticism of this type of therapy is the cost of the technology. Although this complaint is justified, virtual reality may prove to be a very valuable tool for clinicians that can afford to use it.

6. Traumatic brain injury

Posttraumatic stress disorder has been regarded as a signature wound of the wars in Iraq and Afghanistan. Of equal importance is the fact that *traumatic brain injuries* have been called the other signature wound of the wars (National Council on Disability, 2009). This section will focus on traumatic brain injuries due the pronounced overlap of this medical condition with posttraumatic stress disorder in soldiers returning from Iraq and Afghanistan. During the current wars, it is believed that up to 23% of soldiers have obtained a traumatic brain injury during deployment (Terrio et. al., 2009). In addition, 5% to 7% of Operation Enduring Freedom and Operation Iraqi Freedom soldiers are thought to have a probable comorbidity of posttraumatic stress disorder and a traumatic brain injury (Carlson et. al., 2011). Hoge and colleagues (2008), using a sample of 2525 Army infantry soldiers, found that 43.9% of soldiers who had lost consciousness after experiencing trauma to the head met criteria for posttraumatic stress disorder three to four months after returning from Iraq. Furthermore, 27.3% of the soldiers who solely experienced altered consciousness following a trauma to the head also met criteria for posttraumatic stress disorder. Although this is a biased sample due to an Army infantry soldiers' disproportionately high levels of combat exposure, it highlights the clear overlap of posttraumatic stress disorder and traumatic brain injury.

A traumatic brain injury diagnosis is given when an individual experiences an external disturbance to the head, resulting in trauma to the brain, and causing a lack of consciousness or diminished cognitive capacity (Department of Defense, 2007). A traumatic brain injury diagnosis is categorized in terms of severity and labeled as mild, moderate, or severe. In 2009, of the soldiers diagnosed with a traumatic brain injury, 78.4% of the cases were classified as a mild traumatic brain injury (Levin, 2010). A mild traumatic brain injury is defined as experiencing trauma to the head that causes a loss of consciousness for less than 30 minutes and an alteration of consciousness or mental state and posttraumatic amnesia for less than 24 hours (Department of Defense, 2007).

Soldiers in the current war frequently come into contact with explosive devices and as a result can obtain a traumatic brain injury in three ways (Department of Defense, 2007). A soldier is said to have a *primary blast injury* when they were close enough to an explosion to experience the extreme changes in atmospheric pressure, otherwise known as a "blast wave." A blast wave can easily permeate a combat helmet and can ultimately cause trauma to the brain. A *secondary blast injury* can be obtained when a fragment from the explosion hits the soldier on the head hard enough to cause brain injury symptoms. This type of injury can be external but may also permeate the skull. Lastly, a soldier is said to have obtained a *tertiary blast injury* when an explosion causes the soldier to either be knocked to the floor or into something resulting in trauma to the head.

Despite the high comorbidity, researchers continue to struggle to detangle the overlap of symptoms between posttraumatic stress disorder and traumatic brain injury. The residual symptoms that one experiences as a result of a traumatic brain injury are called *postconcussive symptoms*. Many of the symptoms associated with posttraumatic stress disorder overlap with postconcussive symptomology, which include irritability, memory deficits, sleep problems, and difficulty focusing attention (Kennedy & Moore, 2010). Some of the symptoms that can often be unique to a traumatic brain injury diagnosis include balance problems, dizziness, and headaches (Kennedy & Moore, 2010).

Brenner and colleagues (2010) examined the unique contribution of posttraumatic stress disorder and traumatic brain injury to a sample of injured Army personnels' endorsement of postconcussive symptoms (headache, dizziness, memory problems, balance problems, irritability). They concluded that soldiers with either posttraumatic stress disorder or a traumatic brain injury endorsed more postconcussive symptoms than those without a diagnosis. Those with both posttraumatic stress disorder and a traumatic brain injury endorsed more symptom prevalence than those with a single diagnosis. Although it is noteworthy that a comorbid posttraumatic stress disorder and traumatic brain injury diagnosis can increase postconcussive symptomology, it is also important to recognize that the co-occurrence of either disorder can reciprocally exacerbate the other (King 2008).

Researchers have speculated that standard treatments for posttraumatic stress disorder could be less effective when a comorbid traumatic brain injury diagnosis exists (Bryant, 2001; Carlson et. al., 2011). This is solely speculation because there has been limited research to explore the efficacy of current treatments for those with this comorbidity. King (2008) suggests that early education about postconcussive symptomology and an explanation of the reciprocal relationship of the co-occurrence of posttraumatic stress disorder and traumatic brain injury can aid in proper detection and treatment. It is important for further research to explore the effectiveness of treatment for those with a comorbid diagnosis due to the high prevalence of soldiers who suffer from the co-occurring disorders. In addition, it is important for clinicians to be aware that the presence of a mild traumatic brain injury in a patient with posttraumatic stress disorder may make recovery from the posttraumatic stress disorder more challenging (Chard et. al., 2011).

7. Conclusion

Throughout the years we have gained a far better understanding of posttraumatic stress disorder. We have refined our diagnostic criteria for the disorder and developed more complex theories for understanding its' etiology. With the high prevalence of soldiers who are affected by posttraumatic stress disorder, it is important that we continue to refine our understanding of the disorder so that can continue to improve the therapeutic techniques we are using to treat veterans who suffer from the disorder. Future research should examine how the common comorbidities of posttraumatic stress disorder, such as traumatic brain injuries impact treatment outcomes.

8. References

Akselrod, S.; Gordon, D.; Ubel, F.; Shannon, D.; Barger, A. C. & Cohen, R. J. (1981). Power spectral analysis of heart rate fluctuation: A quantitative probe of beat-to-beat cardiovascular control. *Science*, Vol. 213, pp. (220–222)

Appelhans, B.M. & Luecken, L. J. (2006). Heart rate variability as an index of regulated emotional responding. *Review of General Psychology*, Vol. 10, pp. (229–240)

American Psychiatric Association. (1980). *Diagnostic Criteria from DSM-III*, The Association, 0890420467, Washington, DC, USA

American Psychiatric Association. (2000). *Diagnostic Criteria from DSM-IV-TR*, The Association, 0890420262, Washington, DC, USA

American Psychiatric Association. (2010). G 05 Posttraumatic Stress Disorder, In: *American Psychiatric Association DSM-5 Development*, 11.06.2011, Available from: <http://www.dsm5.org/ProposedRevisions/Pages/proposedrevision.aspx?rid=165#>

Brenner, L.A.; Ivins, B.J.; Schwab, K.; Warden, D.; Nelson, L.A.; Jaffee, M. & Terrio, H. (2010) Traumatic Brain Injury, Posttraumatic Stress Disorder, and Postconcussive Symptom Reporting Among Troops Returning From Iraq. *Journal of Head Trauma Rehabilitation*, Vol. 25, No. 5, (September-October 2010), pp. (307-312)

Brewin, C.R. & Holmes, E.A. (2003) Psychological theories of posttraumatic stress disorder. *Clinical Psychology Review*, Vol. 23, No. 3, (May 2003), pp. (339-376)

Bryant, R. A. (2008). Disentangling mild traumatic brain injury and stress reactions. *New England Journal of Medicine*, Vol. 358, No. 5, (January 2008), pp. (525-527)

Carlson, K.F.; Kehle, S.M.; Meis, L.A.; Greer, N.; MacDonald, R.; Rutks, I.; Sayer, N.A.; Dobscha, S.K. & Wilt, T.J. (2011). Prevalence, Assessment, and Treatment of Mild Traumatic Brain Injury and Posttraumatic Stress Disorder: A Systematic Review of the Evidence. *Journal of Head Trauma Rehabilitation*, Vol. 26, No. 2, (March-April 2011), pp. (103-115)

Chard, K.M.; Schumm, J.A.; McIlvain, S.M.; Bailey, G.W. & Parkinson, R.B. (2011) Exploring the Efficacy of a Residential

Treatment Program Incorporating Cognitive Processing Therapy-Cognitive for Veterans With PTSD and Traumatic Brain Injury. *Journal of Traumatic Stress*, Vol. 24, No. 3, (June 2011), pp. (347-351)

Cohen, H.; Kotler, M.; Matar, M.; Kaplan, Z.; Miodownik, H. & Cassuto, Y. (1997). Power spectral analysis of heart rate variability in posttraumatic stress disorder patients. *Biological Psychiatry*, Vol. 41, No. 5, (March 1997) pp. (627–629)

Cohen, H.; Matar, M.; Kaplan, Z. & Kotler, M. (1999). Power spectral analysis of heart rate variability in psychiatry. *Psychotherapy and Psychosomatics*, Vol. 68, pp. (59–66)

de Kloet, C.S.; Vermetten, E.; Geuze, E.; Kavelaars, A.; Heijnen, C.J. & Westenberg H.G.M. (2006). Assessment of HPA-axis function in posttraumatic stress disorder:

Pharmacological and non-pharmacological challenge tests, a review. *Journal of Psychiatric Research,* Vol. 40, No. 6, (September 2006), pp. (550-567)

Department of Defense (2004). Care for victims of sexual assault task force report. Washington, DC: *Department of Defense.*

Department of Defense. (2007). Mild Traumatic Brain Injury Pocket Guide, In: *Department of Defense,* 21.05.2011, Available from: <http://www.dcoe.health.mil/ForHealthPros/traumatic brain injuryInformation.aspx >

Ehlers, A. & Clark, D.M. (2000). A cognitive model of posttraumatic stress disorder. *Behavioral Research and Therapy,* Vol. 38, No. 4, (April 2000), pp. (319-345)

Elhai JD, Grubaugh AL, Kashdan TB, Frueh BC. (2008). Empirical examination of a proposed refinement to DSM-IV posttraumatic stress disorder symptom criteria using the National Comorbidity Survey Replication data. *Journal of Clinical Psychiatry,* Vol. 69, No. 4, (April 2008), pp. (597-602)

Elwood, L.S.; Hahn, K.S.; Olatunji, B.O. & Williams, N.L. (2009). Cognitive vulnerabilities to the development of PTSD: A review of four vulnerabilities and the proposal of an integrative vulnerability model. *Clinical Psychology Review,* Vol. 29, No. 1, (February 2009), pp. (87-100)

Foa, E.B.; Hembree, E.A. & Rothbaum, B.O. (2007). Prolonged Exposure therapy for PTSD: Emotional processing of traumatic experiences: Therapist guide. Oxford: Oxford University Press.

Foa, E.B. & Kozak, M.J. (1986). Emotional processing of fear: Exposure to corrective information. *Psychological Bulletin,* Vol. 99, pp. (20–35)

Friedman, M.J. & Davidson, J.R.T. (2007). Pharmacotherapy for PTSD, In: *Handbook of Posttraumatic Stress Disorder: Science and practice,* Friedman M. J., Keane T. M. and Resick P. A., pp. (376-405), Guilford Press, 978-1-59385-473-7, New York, NY, USA

Frueh, B.C.; Elhai, J.D. & Acierno, R. (2010). The Future of Posttraumatic Stress Disorder in the DSM. *Psychological Injury and Law,* Vol. 3, No. 4, (December 2010), pp. (260-270)

Gerardi, M.; Cukor, J.; Difede, J.; Rizzo, A. & Rothbaum, B.O. (2010). Virtual reality exposure therapy for post-traumatic stress disorder and other anxiety disorders. *Current Psychiatry Reports,* Vol. 12, No. 4, (August 2010), pp. (298-305)

Hoge, C.W.; McGurk, D.; Thomas, J.L.; Cox, A.L.; Engel, C.C. & Castro, C.A. (2008). Mild Traumatic Brain Injury in U.S. Soldiers Returning from Iraq. *The New England Journal of Medicine,* Vol. 358, No. 5, (January 2008), pp. (453-463)

Janoff-Bulman, R. (1992). *Shattered assumptions: Towards a new psychology of trauma,* Free Press, 978-0029160152, New York City, NY

Jones, E.; Fear, N.T. & Wessely, S. (2007). Shell Shock and Mild Traumatic Brain Injury: A Historical Review. *American Journal of Psychiatry,* Vol. 164, No. 11, (November 2007), pp. (1641-1645)

Jones, E.; Vermaas, R.H.; McCartney, H.; Beech, C.; Palmer, I.; Hyams, K. & Wessely, S. (2003). Flashbacks and post-traumatic stress disorder: the genesis of a 20th-century diagnosis, *British Journal of Psychiatry,* , Vol. 182, (2003), pp. (158-163)

Karam, E.G.; Andrews, G.; Bromet, E.; Petukhova, M.; Ruscio, A.M.; Salamoun, M.; Sampson, N.; Stein, D.J.; Alonso, J.;

Andrade, L.H.; Angermeyer, M.; Demyttenaere, K.; de Girolamo, G.; de Graaf, R.; Florescu, S.; Gureje, O.; Kaminer, D.; Kotov, R.; Lee, S.; Lépine, J.P.; Medina-Mora, M.E.; Oakley Browne, M.A.; Posada-Villa, J.; Sagar, R.; Shalev, A.Y.; Takeshima, T.;

Tomov, T. & Kessler, R.C. (2010). The Role of Criterion A2 in the DSM-IV Diagnosis of Posttraumatic Stress Disorder. *Biological Psychiatry*, Vol. 68, No. 5, (September 2010), pp. (465-473)

Karlin, B.E.; Ruzek, J.I.; Chard, K.M.; Eftekhari, A.; Monson, C.M.; Hembree, E.A.; Resick, P.A. & Foa, E.B. (2010). Dissemination of Evidence-Based Psychological Treatments for Posttraumatic Stress Disorder in the Veterans Health Administration. *Journal of Traumatic Stress*, Vol. 23, No. 6, (December 2010), pp. (663–673)

Kennedy, D.C. & Moore, D.J. (2010). *Military neuropsychology*, Springer Publishing Company, 9780826104496, New York, NY, USA

Kessler, R.C.; Berglund, P.; Demler, O.; Jin, R.; Merikangas, K.R. & Walters, E.E. (2005). Lifetime Prevalence and Age-of-Onset Distributions of DSM-IV Disorders in the National Comorbidity Survey Replication. *Archives of General Psychiatry*, Vol. 62, (June 2005), pp. (593-602)

Kilpatrick, D.G.; Koenen, K.C.; Ruggiero, K.J.; Acierno, R.; Galea, S.; Resnick, H.S.; Roitzsch, J.; Boyle, J. & Glernter, J. (2007). The serotonin transporter genotype and social support and moderation of posttraumatic stress disorder and depression in hurricane-exposed adults. *American Journal of Psychiatry*, Vol. 164, No. 11, (November 2007), pp. (1693-1699)

King, N.S. (2008). PTSD and Traumatic Brain Injury- Folklore and Fact? *Brain Injury*, Vol. 22, No. 1, (January 2008), pp. (1-5), 0269–9052 print/ISSN 1362–301X

Lamprecht, F. & Sack, M. (2002). Posttraumatic Stress Disorder Revisited. *Psychosomatic Medicine*, Vol. 64, No. 2, (March 2002), pp. (222-237)

Lasiuk, G.C. & Hegadoren, K.M. (2006). Posttraumatic Stress Disorder Part I: Historical Development of the Concept. *Perspectives in Psychiatric Care*, Vol. 42, No. 1, (February 2006), pp. (13-20)

Lehrer, P.; Vaschillo, E. & Vaschillo, B. (2000). Resonant frequency biofeedback training to increase cardiac variability: Rational and manual for training. *Applied Psychophysiology and Biofeedback*, Vol. 25, pp. (177–191)

Lesch, K.P.; Bengel, D.; Heils, A.; Sabol, S.Z.; Greenberg, B.D.; Petri, S.; Banjamin, J.; Muller, C.R.; Hamer, D.H. & Murphy, D.L. (1996). Association of anxiety-related traits with a polymorphism in the serotonin transporter gene regulatory region. *Science*, Vol. 5292, No. 274, (November 1996) pp. (1527– 1531)

Levin, A. (2010). Blast-Affected Troops to Get Mandatory Traumatic Brain Injury Evaluations. *Psychiatric News*, Vol. 45, No. 19, (October 2010), pp. (6), Available from: < http://pn.psychiatryonline.org/content/45/19/6.1.short?rss=1 >

Lipari, R.N. & Lancaster, A.R. (2003). Armed forces 2002 sexual harassment survey (DMDC Report No. 2003- 026, November 2003). Arlington, Virginia: Defense Manpower Data Center.

Micale, M.S. (1990). Charcot and the Idea of Hysteria in the Male: Gender, Mental Science, And Medical Diagnosis In Late Nineteenth-Century France. *Medical History*, Vol. 34, No. 4, (October 1990), pp. (363-411)

National Council on Disability (U.S.). (2009). *Invisible Wounds: Serving Service Members and Veterans with Posttraumatic Stress Disorder and Traumatic Brain Injury*. Washington, D.C.: National Council on Disability.

Pitman, R.K.; Gilbertson, M.W.; Gurvits, T.V.; May, F.S.; Lasko, N.B.; Metzger, L.J.; Shenton, M.E.; Yehuda, R. & Orr, S.P. (2006). Clarifying the Origin of Biological

Abnormalities in Posttraumatic Stress Disorder Through the Study of Identical Twins Discordant for Combat Exposure. *Annals New York Academy of Sciences,* Vol. 1071, (July 2006), pp. (242-254)

Pitman, R.K.; Orr, S.P.; Forgue, D.F.; de Jong, J.B. & Claiborn, J. M. (1987). Psychophysiologic assessment of posttraumatic stress disorder imagery in Vietnam combat veterans. *Archives of General Psychiatry,* Vol. 44, No. 11, (November 1987) pp. (970–975)

Ramchand, R.; Schell, T.L.; Karney, B.R.; Osilla, K.C.; Burns, R.M. & Caldarone, L.B. (2010). Disparate Prevalence Estimates of PTSD Among Service Members Who Served in Iraq and Afghanistan: Possible Explanations. *Journal of Traumatic Stress,* Vol. 23, No. 1, (February 2010), pp. (59-68)

Ray, S.L. (2008). Evolution of Posttraumatic Stress Disorder and Future Directions. *Archives of Psychiatric Nursing,* Vol. 22, No. 4, (August 2008), pp. (217-225)

Reger, G.M. & Gahm, G.A. (2008). Virtual reality exposure therapy for active duty soldiers. *Journal of Clinical Psychology,* Vol. 64, No. 8, (August 2008), pp. (940-946)

Resick, P.A. & Schnicke, M.K. (1992). Cognitive Processing Therapy for Sexual Assault Victims. *Journal of Consulting and Clinical Psychology,* Vol. 60, No. 5, (October 1992), pp. (748-756)

Rizzo, A.A.; Graap, K.; Perlman, K.; McLay, R.N.; Rothbaum, B.O. & Reger, G. (2008). Virtual Iraq: initial results from a VR exposure therapy application for combat-related PTSD. *Study of Health Technology and Informatics,* Vol. 132, (2008), pp. (420-425)

Rizzo, A.S.; Difede, J.; Rothbaum, B.O.; Reger, G.; Spitalnick, J. & Cukor, J. (2010). Development and early evaluation of the Virtual Iraq/Afghanistan exposure therapy system for combat-related PTSD. *Annals of the New York Academy of Sciences,* Vo. 1208, (October 2010), pp. (114-125)

Sadler, A.G.; Booth, B.M.; Cook, B.L. & Doebbeling, B.N. (2003). Factors associated with women's risk of rape in the military environment. *American Journal of Industrial Medicine,* Vol. 43, No. 3 (March 2003), pp. (262–273)

Sundin, J.; Fear, N.T.; Iversen, A.; Rona, R.J. & Wessely, S. (2009). PTSD after deployment to Iraq: conflicting rates, conflicting claims. *Psychological Medicine,* Vol. 40, No. 3, (March 2010), pp. (367-382)

Tan, G.; Dao, T.K.; Farmer, L.; Sutherland, R.J. & Gevirtz, R. (2011). Heart Rate Variability (HRV) and Posttraumatic Stress Disorder (PTSD): A Pilot Study. *Applied Psychophysiology and Biofeedback,* Vol. 36, No. 1, (March 2011), pp. (27-35)

Terrio, H.; Brenner, L. A.; Ivins, B. J.; Cho, J. M.; Helmick, K.; Schwab, K.; Scally, K.; Bretthauer, R. & Warden, D. (2009). Traumatic Brain Injury Screening: Preliminary Findings in a US Army Brigade Combat Team. *Journal of Head Trauma Rehabilitation,* Vol. 24, No. 1, (January-February 2009), pp. (14-23)

Van der Kolk, B.A. (2007). The History of Trauma in Psychiatry, In: *Handbook of Posttraumatic Stress Disorder: Science and practice,* Friedman M. J., Keane T. M. and Resick P. A., pp. (19-36), Guilford Press, 978-1-59385-473-7, New York, NY, USA

Vaschillo, E.; Lehrer, P.; Rishe, N. & Konstantinov, M. (2002). Heart rate variability biofeedback as a method for assessing baroreflex function: A preliminary study of resonance in the cardiovascular system. *Applied Psychophysiology and Biofeedback,* Vol. 27, pp. (1–27)

Williams, I. & Bernstein, K. (2011). Military Sexual Trauma Among U.S. Female Veterans. *Archives of Psychiatric Nursing*, Vol. 25, No. 2 (April 2011), pp. (138–147)

3

Psychiatric Management of Military-Related PTSD: Focus on Psychopharmacology

Don J. Richardson[1,2,3], Jitender Sareen[4,5] and Murray B. Stein[6]
[1]*Parkwood Operational Stress Injury Clinic, St. Joseph's Health Care- London, Ontario,*
[2]*Department of Psychiatry, University of Western Ontario, London, Ontario,*
[3]*Centre for National Operational Stress Injury, Veterans Affairs Canada,*
[4]*Operational Stress Injury Clinic, Deer Lodge, Winnipeg, Manitoba,*
[5]*Professor of Psychiatry, Psychology and Community Health Sciences, University of Manitoba, Winnipeg, Manitoba,*
[6]*Professor of Psychiatry and Family & Preventive Medicine, University of California San Diego,*
[1,2,3,4,5]*Canada*
[6]*USA*

1. Introduction

Military-related posttraumatic stress disorder (PTSD) occurs in a significant minority of veterans and often presents with complex psychiatric co-morbidity (Kessler et al., 1995, Keane and Kaloupek, 1997, Keane and Wolfe, 1990, Forbes et al., 2003, Kulka et al., 1990, Sareen et al., 2004). Twelve month and lifetime prevalence rates of PTSD in the Canadian Regular Forces has been reported as 2.8% and 7.2% respectively (Statistics Canada, 2002). In Canadian veterans pensioned with a medical condition, the 1 month prevalence was 10.3% (Richardson et al., 2006). Other military samples have shown 6 month and lifetime prevalence rates of 11.6 and 20.0% respectively (O'Toole et al., 1996). The large variation in PTSD rates might be a function of the time elapsed between the end of a mission and the start of the mental health evaluation, the nature and frequency of potentially traumatic events within each mission and differences in measurement used i.e. self-report screening tools vs. diagnostic interview.

Patients with PTSD often present first to their primary care clinician with mental health issues, (Del Piccolo et al., 1998) and as such demonstrate increased healthcare service use and costs (Kulka et al., 1990, Ronis et al., 1996, Marshall et al., 1998, Hankin et al., 1999, Kessler et al., 1999, Switzer et al., 1999, Elhai and Ford, 2005, Elhai et al., 2005, Gavrilovic et al., 2005, Richardson et al., 2006). Studies indicate that military-related PTSD is more prone to somatisation (McFarlane et al., 1994) and is associated with more physical health problems (Boscarino, 1997, Boscarino and Chang, 1999, Schnurr and Jankowski, 1999, Schnurr et al., 2000, Sledjeski et al., 2008, Jakupcak et al., 2008, Sareen et al., 2007, Elhai et al., 2007). Evidence also shows that PTSD is often associated with significant comorbidity including major depression, substance abuse, suicidality, (Kessler et al., 1995, Keane and

Kaloupek, 1997, Keane and Wolfe, 1990, Forbes et al., 2003, Kulka et al., 1990, Gradus et al., 2010, Nepon et al., 2010, Sareen et al., 2005) and chronic disability contributing to impaired quality of life (Mills et al., 2006, Richardson et al., 2008 , Richardson et al., 2010).

Military personnel are more likely to be exposed to trauma than the general public (Breslau et al., 1991). Potentially traumatic events can include combat, imprisonment, torture, witnessing atrocities, comrades being wounded or killed, or rescue missions following natural disasters. Peacekeeping missions to Bosnia, Somalia and Rwanda have also involved complex rules of engagement that prevented immediate and active intervention, with a resultant sense of intense vulnerability to attack (Litz et al., 1997b, Litz et al., 1997a, American Psychiatric Association, 2004, Litz, 1996). However military members can also be exposed to non-military specific trauma including rape, motor vehicle accidents, assault and natural disasters.

Risk factors for the development of PTSD have been extensively studied in the military and veteran population. Pre-trauma risk factors for PTSD include a family and/or personal history of psychiatric illness, past trauma including history of childhood abuse (Brewin et al., 2000, Ozer et al., 2003a, Sandweiss et al., 2011). Women are twice as likely to develop PTSD, although men are more likely to be exposed to a traumatic events (Kessler et al., 1995, Breslau et al., 1998). In the military, men still vastly outnumber women, especially in trades that involved combat. Other proposed pre-trauma risk factors from community studies include: younger age, single marital status and lower socioeconomic status (Breslau et al., 2006, Richardson et al., 2007).

Suggested peri-traumatic risk factors include: trauma severity and life threat, (Brewin et al., 2000, Hoge et al., 2004a, Richardson et al., 2007) bodily injury (Koren et al., 2005) and the number of operational deployments (Richardson et al., 2007, Statistics Canada, 2002). The dose-response effect between number of operational deployments was confirmed in a recent re-analysis of PTSD's prevalence among U.S. male Vietnam veterans (Dohrenwend et al., 2006)and in American soldiers deployed in Afghanistan (Hoge et al., 2004b). The emotional response at the time of the trauma, such as feeling unable to control a situation and peritraumatic dissociation, (Brewin et al., 2000, Yehuda, 1999, Ozer et al., 2003a) has also been identified as significant peri-traumatic risk factors. Although more recent studies have cast some doubt on the vailidity of the importance of peri-traumatic dissociation (Candel et al., 2003). More recent studies have demonstrated that pain control in trauma care was significantly associated with a lower risk of PTSD after injury (Holbrook et al., 2010), and both increase heart rate at the time of the trauma (Bryant et al., 2011) and intensive care admission following traumatic injury (O'Donnell et al., 2010) were associated with increased risk of developing PTSD.

Post-traumatic risk factors may include: lack of access to treatment, stigmatization, ongoing life stressors and lack of social support (Brewin et al., 2000, Ozer et al., 2003b, Yehuda et al., 1998). Access to treatment is important, as there is a significant association between soldiers diagnosed with a psychiatric conditions and high attrition rates from the military (Hoge et al., 2002). Deployed members are frequently exposed to long separations from their families and friends and ongoing financial strain might add to the distress a deployed member might face after they return home. Shame and guilt are also posttraumatic risk factors (Yehuda et al., 1998) that military members frequently often face.

Formal psychometric instruments have been developed to assess deployment risk and resiliency factors in relation to mental health outcomes, such as the Deployment Risk and Resilience Inventory (King et al., 2006).

Military members face barriers to rapid, effective treatment for mental illness (Hoge et al., 2004b). Military culture, fear of stigmatization and concerns of career debasement can deter help-seeking , particularly at an early stage when symptoms may be more likely to respond to treatment (Hoge et al., 2002, Elhai et al., 2005, Gavrilovic et al., 2005, McFall et al., 2000, Hoge et al., 2004b). Such delays in accessing treatment may further contribute to the functional impairment often associated with PTSD.

Military-related PTSD responds to both psychotherapeutic and psychopharmacological treatments. (Foa, 2006, Benedek et al., 2009). However, psychotherapy meta-analysis showed that military-related PTSD has the lowest effect size when compared to civilian PTSD (Bradley et al., 2005). Treatment response for PTSD related to a car accident, sexual assault or other more-typically civilian trauma, might not garner the same response for a military-related PTSD. Recent psychotherapy studies have been more encouraging, demonstrating effectiveness in randomized controlled trials including cognitive behavioral psychotherapy, prolonged exposure and cognitive processing therapy (Monson et al., 2006, Nacasch et al., 2010, Tuerk et al., 2011, Morland et al., 2010).

Pharmacological treatment has also demonstrated poor response in military-related PTSD (Schoenfeld et al., 2004a, Shalev et al., 1996, Friedman, 1997). Factors such as chronicity, high comorbidity rates (Friedman, 1997, Shalev et al., 1996, Forbes et al., 2003) and anger that is often present in military-related PTSD (Forbes et al., 2005) have been identified as predictors of poor response. Prior trauma history and past history of psychiatric illness has also been identified as important predictors of treatment outcome (Hourani and Yuan, 1999). Military specific factors, such as the nature of deployment, which often involves months of persistent hyperarousal and hypervigilance in unfamiliar surroundings away from their social support, have also been demonstrated as being a negative predictor in veterans with combat exposure (Foa et al., 2009, King et al., 1995, Creamer and Forbes, 2004). Although a recent Cochrane review demonstrated the effectiveness of pharmacological interventions for PTSD, especially serotonin specific reuptake inhibitors (SSRIs) (Stein et al., 2006), the American Psychiatric Association PTSD Treatment Guideline update concluded that there was insufficient evidence demonstrating the benefit of an SSRI in the veteran population (Benedek et al., 2009).

Due to the complex nature of the clinical presentation of PTSD, from the continuum of adjustment disorders and subthreshold PTSD to 'full-blown' PTSD, this paper aims to confine itself to a general overview of the psychiatric management of military-related PTSD. Despite the challenges researchers face in conducting studies on the effectiveness of military-related PTSD treatment (Institute of Medicine (IOM), 2008), if evidence-based practices are utilized using established guidelines (American Psychiatric Association, 2004, Australian Centre for Post Traumatic Mental Health and National Health and Medical Research Council, 2007) remission can be achieved in 30%–50% of cases of PTSD (Friedman, 2006).

2. Psychiatric management

2.1 Assessment

The presentations of military-related PTSD is often complex. Military members and veterans may initially present indirectly with an emotional, behavioural or addiction concern or an unrelated, less stigmatizing somatic problem such as a physical complaint (Australian Centre for Post Traumatic Mental Health and National Health and Medical Research

Council, 2007). The psychiatric assessment should detail the presenting symptoms and elicit a trauma history, including childhood and adolescent trauma, and exposure to military trauma (combat or peacekeeping operations) (Friedman, 2006). The details of the traumatic event should be limited to information that clarifies the diagnosis as the recounting of an extremely traumatic event is often highly triggering and can lead to significant symptom exacerbation.

Clinically, PTSD presents as four symptom clusters: reexperiencing the traumatic events, avoidance of reminders and emotional numbing (which are grouped together as one symptoms cluster in DSM-IV but are seen as distinct and will likely be denoted as such in DSM-5), and hyperarousal symptoms (American Psychiatric Association, 2004, American Psychiatric Association, 2001). Military members with PTSD relive their trauma in intrusive recollections during the day, including flashbacks, or at night as bad dreams or nightmares. Many complain of both physical and emotional symptoms of anxiety when exposed to reminders of their traumatic event. They may avoid reminders of the trauma and describe emotional numbness or an inability to experience a normal range of emotions with family or friends. They may complain of hyperarousal symptoms such as insomnia, irritability, frequent anger outburst, poor concentration and hypervigilance. According to DSM-IV-TR, acute PTSD has a duration of between 1 and 3 months, whilst chronic PTSD has a duration of more than three months (American Psychiatric Association, 2001).

The clinician can screen for PTSD using available short screening instruments such as the four-item yes/no screening instrument—the Primary Care PTSD Screen—designed for use by primary care practitioners. It has a sensitivity of 78% and specificity of 87% for PTSD in patients who endorse three or more items, (Friedman, 2006) figure 1. Patients who screen positive should be assessed for PTSD using the DSM IV diagnostic criteria, figure 2, or using more elaborative screening instruments such as the Clinician Administered PTSD Scale (CAPS)(Blake et al., 1995) or a self-rating scale such as the PTSD Checklist (Military Version) (Weathers et al., 1993). Veterans may also present with some symptoms of PTSD without meeting the full diagnostic criteria (Zlotnick et al., 2002, Schützwohl and Maercker, 1999, Stein et al., 1997, Charney et al., 1986, Weiss et al., 1992). Even if the full criteria are not met, studies indicate that these individuals may experience significant functional impairment (Olfson et al., 2001). In a study of Canadian veterans, Asmundson and colleagues (Asmundson et al., 2002) demonstrated increased psychopathology in veterans with sub-threshold PTSD when compared to the non-deployed, non-traumatized veterans.

Assessing suicide risk is also critical. The presence of PTSD symptoms increases the possibility of suicidal ideation (Marshall et al., 2001). PTSD often presents with co-morbidities such as depression and addictions (Kessler et al., 1995, Forbes et al., 2003). Studies have estimated that more than 50% of PTSD patients have symptoms of a major depressive disorder (Kessler et al., 1995), but in the veteran population, possibly due to delayed treatment, the percentage may be much higher (Keane and Wolfe, 1990, Southwick et al., 1991, Forbes et al., 2003). Co-morbid depression also significantly increases suicide risk (Kaufman and Charney, 2000). Issues of aggression and anger are also well documented in war veterans, (Lewis, 1990, Forbes et al., 2003, Forbes et al., 2004, Biddle et al., 2002) and during the initial PTSD assessment, male military members may report violent thoughts and aggressive behavior, including homicidal thoughts. Assessing comorbidity, suicidal or homicidal ideations and social support is important in order to determine the need for inpatient treatment or referral for specialist care (American Psychiatric Association, 2004).

In your life, have you ever had any experience that was so frightening, horrible, or upsetting that, ***in the past month,*** you...

1. Have had nightmares about it or thought about it when you did not want to? *Yes/No*
2. Tried hard not to think about it or went out of your way to avoid situations that reminded you of it? *Yes/No*
3. Were constantly on guard, watchful, or easily startled? *Yes/No*
4. Felt numb or detached from others, activities, or your surroundings? *Yes/No*

Screen is positive if patient answers "yes" to any three items.

Fig. 1. Primary Care PTSD Screen

Enquiry should also be made into family functioning, the health of spouse and children, social functioning and vocational issues (American Psychiatric Association, 2004). Family, friends and peers can also provide valuable collateral information as to the current and past functioning of the military member or veteran and eliciting their support at the initial assessment can assist with the treatment process.

2.2 Treatment

Once a firm diagnosis has been established, psychoeducation in group format or individually regarding diagnosis and treatment is critical for both patient and family (American Psychiatric Association, 2004, Turnbull and McFarland, 1996, Van Der Kolk et al., 1996a, Foa et al., 2000). Patient education is a fundamental component of the treatment of as PTSD. Providing psychoeducation can enhance patient satisfaction and improve treatment compliance (Gray et al., 2004). Effective treatment requires that patients understand the treatment plans and return for follow-up assessment and treatment (American Psychiatric Association, 2004). Veterans need information soon after the initial assessment of the different stages of treatment for PTSD (Herman, 1992). The initial phase of treatment focuses on symptom stabilization and the treatment of co-morbid conditions such as depression, addictions and anxiety disorders. Educating patients regarding the phases of treatment reassures those frightened by the notion of psychiatric medication and psychotherapy as well as to set appropriate expectations for treatment. Some patients expect they will be forced to talk about feared traumatic events from the outset and are relieved to know that trauma work comes after their anxiety and distress are more manageable. While symptoms might initially be overwhelming and require pharmacological intervention, early work on mastering anxiety and anger using psychological tools, provides a sense of self- control. Safety in therapy is paramount and only after acute symptoms, particularly suicidality and homicidality, are addressed should the exploration of traumatic events be approached. Once symptoms stabilize, patients are more able to engage in psychotherapy (Van Der Kolk et al., 1996b).

The person has been exposed to a traumatic event in which both of the following were present:

1. the person experienced, witnessed, or was confronted with an event or events that involved actual or threatened death or serious injury, or a threat to the physical integrity of self or others
2. the person's response involved intense fear, helplessness, or horror. Note: In children, this may be expressed instead by disorganized or agitated behavior

B. The traumatic event is persistently reexperienced in one (or more) of the following ways:

1. recurrent and intrusive distressing recollections of the event, including images, thoughts, or perceptions. Note: In young children, repetitive play may occur in which themes or aspects of the trauma are expressed.
2. recurrent distressing dreams of the event. Note: In children, there may be frightening dreams without recognizable content.
3. acting or feeling as if the traumatic event were recurring (includes a sense of reliving the experience, illusions, hallucinations, and dissociative flashback episodes, including those that occur on awakening or when intoxicated). Note: In young children, trauma-specific reenactment may occur.
4. intense psychological distress at exposure to internal or external cues that symbolize or resemble an aspect of the traumatic event
5. physiological reactivity on exposure to internal or external cues that symbolize or resemble an aspect of the traumatic event

C. Persistent avoidance of stimuli associated with the trauma and numbing of general responsiveness (not present before the trauma), as indicated by three (or more) of the following:

1. efforts to avoid thoughts, feelings, or conversations associated with the trauma
2. efforts to avoid activities, places, or people that arouse recollections of the trauma
3. inability to recall an important aspect of the trauma
4. markedly diminished interest or participation in significant activities
5. feeling of detachment or estrangement from others
6. restricted range of affect (e.g., unable to have loving feelings)
7. sense of a foreshortened future (e.g., does not expect to have a career, marriage, children, or a normal life span)

D. Persistent symptoms of increased arousal (not present before the trauma), as indicated by two (or more) of the following:

1. difficulty falling or staying asleep
2. irritability or outbursts of anger
3. difficulty concentrating
4. hypervigilance
5. exaggerated startle response

E. Duration of the disturbance (symptoms in Criteria B, C, and D) is more than 1 month.

F. The disturbance causes clinically significant distress or impairment in social, occupational, or other important areas of functioning. Specify if:

Acute: if duration of symptoms is less than 3 months

Chronic: if duration of symptoms is 3 months or more

Specify if:

With Delayed Onset: if onset of symptoms is at least 6 months after the stressor.

---

a Reprinted from *Diagnostic and Statistical Manual of Mental Disorders, 4th Edition, Text Revision*. Washington, DC, American Psychiatric Association, 2000. Copyright © 2000. American Psychiatric Association.

Fig. 2. DSM-IV-TR Diagnostic Criteria for Posttraumatic Stress Disorder (DSM-IV-TR code 309.81)[a]

2.2.1 Psychotherapy

The therapeutic relationship focuses on the "therapeutic use of self", the interpersonal process and the authentic relationship between clinician and client (Carper, 1978). Developing a trusting therapeutic relationship is a challenge and one of paramount importance. Establishing trust in therapy takes time, and so it is often helpful to set the timeframe for therapy soon after the initial assessment. Patients need to be reassured that their clinician does not expect that trust will develop immediately, but requires time to develop. Genuineness and empathy are essential in order to develop an authentic, trusting therapeutic relationship with a veteran. Because of their initial paucity of basic trust, especially of individuals in authority (Glover, 1988.), younger veterans seeking help will often challenge their clinician to determine if the clinician is indeed "genuine."

It is crucial to find a therapist with experience in treating PTSD and knowledgeable on military culture. Both prolonged exposure and cognitive behavioral psychotherapy (CBT) are considered first-line treatment for PTSD. In prolonged exposure, the patient reiterates the trauma during planned treatment sessions, including every sensory experience associated with it, until the memory no longer provokes significant anxiety. With CBT, both the conditioned fear and cognitive distortions associated with PTSD are addressed. Common cognitive distortions include perceiving the world as dangerous, seeing oneself as powerless or inadequate, or feeling guilty for outcomes that could not have been prevented (Friedman, 2006). Most clinical guidelines have also accepted that Eye Movement Desensitization and Reprocessing (EMDR) is an evidence-based treatment for PTSD (American Psychiatric Association, 2004, Friedman, 2006). In EMDR, patients are instructed to imagine painful traumatic memories and associated negative cognitions such as guilt and shame while visually focusing on the rapid movement of the clinician's finger (Friedman, 2006). However dismantling studies have demonstrated that the "eye movement" component is not necessary for the treatment response and that the theoretical bases for its method of action has yet to be determined (Davidson and Parker, 2001). Regardless of the treatment modality, stabilization is critical as the potential danger of initiating "trauma-focused psychotherapy" prior to stabilization may exacerbate pre-existing co-morbid symptoms of depression and substance abuse.

Group based psychotherapy is also commonly used, focusing on psychoeducation, anger, depression, substance use, social and vocational skills, relaxation training as well as other facets of PTSD (American Psychiatric Association, 2004, Foy et al., 2000).

2.2.2 Pharmacological management

As demonstrated in Table 1, a number of medications have been used to treat PTSD. Selective Serotonin Reuptake Inhibitors (SSRIs) have the most empirical evidence for efficacy in the treatment of all three PTSD symptom clusters and are usually considered as a first-line treatment for PTSD (American Psychiatric Association, 2004, National Institute for Clinical Excellence, 2005, Schoenfeld et al., 2004b). SSRIs are also effective agents for the treatment of co-morbid mood and anxiety disorders commonly associated with PTSD. Both paroxetine and sertraline have received FDA approval for the treatment of PTSD in the United States (American Psychiatric Association, 2004). In Canada, only paroxetine has Health Canada approval for the treatment of PTSD.

Second-generation, dual acting antidepressants such as venlafaxine and mirtazepine, are widely used in treating major depression and other anxiety disorders but have less

Class and drug	Adult (mg/Day)[a]	Common side effects
Antidepressant- SSRIs[b]		
Citalopram	20–60	Anxiety, fatigue, nausea, dry mouth, sexual dysfunction
Escitalopram	10-30	Nausea, fatigue, dry mouth, sexual dysfunction
Fluvoxamine	100–250	Anxiety, Nausea, headache, sedation , insomnia, sexual dysfunction
Fluoxetine	20–80	Nausea, insomnia, tremor, sexual dysfunction
Paroxetine	20–60	Anxiety, Nausea, drowsiness, insomnia, sexual dysfunction
Sertraline	50–200	Nausea, insomnia, loose stools, sexual dysfunction
Dual acting antidepressant		
Bupropion (SR or XL)	150–300	Agitation, tremor, dizziness, insomnia, excessive sweating, hypertension
Mirtazapine	15–45	Sedation, increased appetite, weight gain, dry mouth
Venlafaxine	75–375	Nausea, Nervousness, insomnia, somnolence, dizziness, anorexia, sexual dysfunction, hypertension
Adrenergic inhibitors		
Prazosin	2–10	Dizziness, headache, drowsiness, fatigue, risk of syncope
Mood Stabilizers		
Carbamazepine	400–1,000	Dizziness, drowsiness, nausea; risk of aplastic anemia, agranulocytosis
Gabapentin	300–3000	Drowsiness, dizziness, ataxia, fatigue
Lamotrigine	25–400	Dizziness, ataxia, drowsiness, headache; risk of skin rash, Stevens-Johnson syndrome syndrome (rare)
Topiramate	50–400	Drowsiness, dizziness, ataxia, confusion
Valproate	250–2,000	Nausea, gastrointestinal problems, weight change, sedation, tremor, hepatic failure, teratogenic
Antipsychotics		
Aripiprazole	5-10	Restlessness or need to move (akathisia), insomnia, fatigue, blurred vision, constipation.
Olanzapine	5–10	Drowsiness, dizziness, weight gain, dry mouth, akathesia, parkinsonism events; risk of new-onset diabetes mellitus
Quetiapine	50–300	Somnolence, dizziness, postural hypotension
Risperidone	0.5–4	Extrapyramidal symptoms, agitation, anxiety, insomnia, rhinitis

[a] Dosage recommandations repesent clinical consensus.

[b] Selective serotonin reuptake inhibitors

Table 1. Dosage and common side effects of drugs used to treat PTSD (adapted from Current Concepts in Pharmacotherapy for PTSD, Schoenfeld et al., 2004)

empirical data demonstrating their efficacy for the specific treatment of PTSD (Hopwood et al., 2000, Smajkic et al., 2001, Davidson et al., 2003, Chung et al., 2004, Connor et al., 1999). They are often considered as a second-line treatment in patients who have failed to respond to a trial of an SSRI. However, since SSRIs have not demonstrated their efficacy in the treatment of Vietnam or combat-related PTSD thus far, (Schoenfeld et al., 2004b, Friedman et al., 2007) second generation antidepressants may be considered as first-line treatment. The tricyclic antidepressants (TCAs) and monoamine oxidase inhibitors (MAOIs) have some limited data to support their use in the treatment of combat-related PTSD; (Kosten et al., 1991, Davidson et al., 1990) however, they are not commonly used because of their side effect profile and toxicity.

Benzodiazepines are not recommended as monotherapy for the treatment of PTSD, (Friedman, 2006, Braun et al., 1990, Gelpin et al., 1996) but are sometimes used as adjuncts in treating anxiety or insomnia (American Psychiatric Association, 2004). There is a risk of rebound insomnia when a benzodiazepine, used as a hypnotic, is discontinued especially after long-term use (Cooper et al., 2005). The use of benzodiazepines among patients with military-related PTSD who have comorbid substance abuse should be avoided.

2.3 Combining treatment resistant PTSD

In the veteran population, response to treatment might be significantly affected by the severity and chronicity of PTSD (Friedman et al., 2000). Although there is no treatment algorithm for reference, patients who demonstrate a partial response (25-50% improvement) after 8 to 12 weeks of treatment with the first antidepressant trial, augmentation or combination strategies could be considered. Of note though, optimization of monotherapy is critical and close monitoring of potential side effects, especially in the early stages of combination pharmacotherapy, is essential when considering augmentation or combination strategies (Cooper et al., 2005). Common combination treatments include adding mirtazapine or bupropion to an SSRI or venlafaxine. Other augmenting agents for PTSD include atypical antipsychotics and anticonvulsants, although the patient should be fully informed about potential benefits and side effects

The utility of atypical antipsychotics such as risperidone, olanzapine and aripiprazole for the treatment of PTSD in combination with an antidepressant has been demonstrated in numerous studies, including randomized controlled trials (Richardson et al., 2011, Stein et al., 2002, Bartzokis et al., 2001, Hamner et al., 2003, Monnelly et al., 2003). However, a recent study with military-related PTSD did not find that risperidone significantly decreased PTSD symptoms when compared to placebo (Krystal et al., 2011). These agents have been particularly beneficial in managing hyperarousal symptoms such as hypervigilance and irritability as well as for severe dissociation symptoms (Schoenfeld et al., 2004b). There is no established role for the use of conventional antipsychotics in the treatment of PTSD.

Anticonvulsants such as carbamazapine, valproate, topiramate, lamotrigine are increasingly used in combination with antidepressants to treat symptoms of depression, mood instability and impulsivity observed in PTSD (Lipper et al., 1986, Keck et al., 1992, Fesler, 1991, Berlant and Van Kammen, 2002, Hertzberg et al., 1999, Hamner et al., 2001). These agents are generally reserved as third line agents and used in combination with first or second line agents, due to the paucity of evidence for their efficacy.

Antiadrenergic agents such as propranolol and prazosin may have a role as a preventive strategy in the acute traumatic stress reaction (Friedman et al., 1993, Cooper et al., 2005,

Vaiva et al., 2003) or in combination with antidepressants to treat excessive hyperarousal or hyperactive symptoms (Friedman, 2006).

For significant symptoms of insomnia that persist with the use of therapeutic doses of antidepressants, a trial of low-dose mirtazepine (15 mg) or trazodone (50- 100 mg) may be helpful. Alternative non-benzodiazepine hypnotics include zopiclone and zaleplon. Zaleplon may be helpful for patients presenting with middle insomnia resulting from nightmares. Its rapid onset of action and very short half-life (approximately one hour) permits patients to take it in the middle of the night (Samuels, 2005). There is evidence demonstrating the benefits of using prazosin, an adrenergic inhibitors to reduce nightmares in combat veterans (Raskind et al., 2002, Raskind et al., 2003; Miller, 2008; Peterson et al., 2011).

2.3.1 Combining psychotherapy and pharmacotherapy

In clinical practice, despite limited empirical evidence, most veterans with PTSD receive psychotherapy in combination with pharmacotherapy either concurrently (at the same time) or sequentially (one modality after another) (Alderman et al., 2009). There is limited research using combination treatment for PTSD (Canadian Psychiatric Association, 2006, Marshall and Cloitre, 2000). A recent Cochrane systematic review of four clinical trials using SSRI with PE/CBT concluded that not enough evidence is available to support or refute the effectiveness of combined psychological & pharmacotherapy" (Hetrick et al., 2010). Many patients receive psychotherapy and pharmacotherapy either at the same time or one after another. Even though this is generally considered standard clinical practice in our specialty clinics, there is very limited research demonstrating the benefit of combination treatment. A recent Cochrane review published this year, found only four published trials of combination treatment and concluded that there was not sufficient evidence at this time to either support or refute the effectiveness of combined psychological and pharmacotherapy (Hetrick et al., 2010). One study demonstrated the benefits of psychotherapy augmentation in patients who have had a partial response to pharmacotherapy (Rothbaum et al., 2006).

3. Special treatment consideration

3.1 Treatment adherence

Medication compliance is crucial for treatment to be effective. Medication non-compliance may be related to the psychological meaning of taking medication (Fenton and McGlashan, 2000). Veterans may believe that taking medication means they are weak or defective, or they fear that they will become addicted to the medication, (National Institute for Clinical Excellence, 2005) that it will change their personality or lead to job loss. These false beliefs or fears about medications should be explored and confronted prior to starting medication. Providing a safe environment and a positive doctor-patient interaction will help develop trust and may make the veteran more accepting of treatment, improving medication compliance (Weiden and Rao, 2005, Kluft, 2002). Engaging and educating all care providers is essential so the veteran feels safe and comfortable with treatment. Peer social support programs, such as Operational Stress Injury Social Support Program (OSISS) in Canada, may play a valuable role in encouraging medication and treatment compliance. Family involvement may also assist treatment adherence, although this requires further study (Phillips et al., 2001). Education about the potential risk of increased suicidal thoughts associated with antidepressant medication, particularly at the time of initiation of treatment, should be discussed and reviewed with the patient (National Institute for Clinical Excellence, 2005).

Patients may wish to discontinue their medication once they start to feel better or can no longer tolerate side effects such as weight gain or sexual dysfunction. However, studies have demonstrated the benefits of continuing medication at least up to one year (Richardson et al., 2011). There are no published guidelines on the length of time that patients suffering from anxiety disorders should continue taking their medication; however, existing guidelines for major depression suggest that the medication should be continued for at least six months after symptom remission has been reached (Canadian Psychiatric Association, 2001).

3.2 Dosing considerations

Since veterans with PTSD often present with marked anxiety, they may be very sensitive to the potential heightened anxiety sometimes seen early in treatment with antidepressants. Patients benefit from a "start low, go slow" approach to medication titration, such as starting at ¼ to ½ the usual starting dose and then gradually increasing to a therapeutic level (Cooper et al., 2005, American Psychiatric Association, 1998). While the initiation of medication might be slow and cautious, ultimately the dose should be titrated to full symptom remission at maximum tolerated doses.

4. Conclusion

The presentation of military-related PTSD is often complex. The primary care clinician should consider early referral for specialist military psychological and psychiatric care. Understanding military culture and the nature of military deployments helps the clinician appreciate the challenges veterans' face, which is essential to establishing a trusting therapeutic alliance. Treatment often involves a combination of medications making compliance more challenging. Although remission is not always possible, pharmacological interventions assist with symptom reduction and improve functioning and quality of life. Pharmacological interventions also assist with stabilization and facilitate psychotherapeutic interventions such as trauma-focused psychotherapy.

The treatment of veterans with PTSD often involves a multidisciplinary team of health professionals and it is important that the physician maintain a close interagency liaison with a view to 'shared care'.

5. Acknowledgment

Preparation of this article was supported by a Canadian Institutes of Health Research New Investigator award (#152348), and the Manitoba Health Research Council Chair award to (Dr. Sareen). The views expressed in this manuscript are those of the authors and do not necessarily represent the views of the Veterans Affairs Canada.

6. References

Alderman, C. P., McCarthy, L. C. & Marwood, A. C. 2009. Pharmacotherapy for Post-traumatic Stress Disorder. *Expert Review of Clinical Pharmacology*, 2, 77-86.

American Psychiatric Association 1998. Practice Guideline for the Treatment of Patients with Panic Disorder. *American Journal of Psychiatry*.

American Psychiatric Association 2001. *Diagnostic and statistical manual of mental disorders,* Washington, DC, Author.

American Psychiatric Association 2004. Practice Guidelines for the Treatment of Patients with Acute Stress Disorder and Posttraumatic Stress Disorder. *American Journal of Psychiatry,* 161, 1-57.

Asmundson, G. J. G., Stein, M. B. & McCreary, D. R. 2002. Posttraumatic stress disorder symptoms influence health status of deployed peacekeepers and nondeployed military personnel. *Journal of Nervous and Mental Disease,* 190, 807-815.

Australian Centre for Post Traumatic Mental Health and National Health and Medical Research Council 2007. Australian Guidelines for the Treatment of Adults with Acute Stress Disorder and Post Traumatic Stress Disorder. Melbourne.

Bartzokis, G., Freeman, T. & Roca, V. 2001. Risperidone for patients with chronic combat-related posttraumatic stress disorder. *154th Annual Meeting of the APA.* New Orleans, La.

Benedek, D. M., Friedman, M. J., Zatzick, D. & Ursano, R. J. 2009. Guideline Watch: Practice Guideline for the Treatment of Patients With Acute Stress Disorder and Posttraumatic Stress Disorder. Available: http://www.psychiatryonline.com/content.aspx?aid=156498.

Berlant, J. & Van Kammen, D. 2002. Open-label topiramate as primary or adjunctive therapy in chronic civilian posttraumatic stress disorder: a preliminary report. *Journal of Clinical Psychiatry,* 63, 15–20.

Biddle, D., Elliott, P., Creamer, M., Forbes, D. & Devilly, G. 2002. Self-reported problems: a comparison between PTSD diagnosed veterans, their spouses, and clinicians. *Behaviour Research and Therapy,* 40, 853–865.

Blake, D. D., Weathers, F. W., Nagy, L. M., Kaloupek, D. G., Gusman, F. D., Charney, D. S. & Keane, T. M. 1995. The development of a clinician-administered PTSD scale. *Journal of Traumatic Stress,* 8, 75-90.

Boscarino, J. A. 1997. Diseases among men 20 years after exposure to severe stress: Implications for clinical research and medical care. *Psychosomomatic Medicine,* 59, 605-614.

Boscarino, J. A. & Chang, J. 1999. Electrocardiogram abnormalities among men with stress-related psychiatric disorders: Implications for coronary heart disease and clinical research. *Annals of Behavioral Medicine,* 21, 227-234.

Bradley, R., Greene, J., Russ, E., Dutra, L. & Westen, D. 2005. A Multidimensional Meta-Analysis of Psychotherapy for PTSD. *American Journal of Psychiatry,* 162, 214-227.

Braun, P., Greenberg, D., Dasberg, H. & Lerer, B. 1990. Core symptoms of posttraumatic stress disorder unimproved by alprazolam treatment. *Journal of Clinical Psychiatry,* 51, 236-238.

Breslau, N., Davis, G. C., Andreski, P. & Peterson, E. 1991. Traumatic events and posttraumatic stress disorder in an urban population of young adults. *Archives of General Psychiatry,* 48, 216-222.

Breslau, N., Kessler, R., Chilcoat, H., Schultz, L., Davis, G. & Andreski, P. 1998. Trauma and posttraumatic stress disorder in the community: the 1996 Detroit Area Survey of Trauma. *Arch Gen Psychiatry,* 55, 626-632.

Breslau, N., Lucia, V. C. & Alvarado, G. F. 2006. Intelligence and Other Predisposing Factors in Exposure to Trauma and Posttraumatic Stress Disorder: A Follow-up Study at Age 17 Years *Arch Gen Psychiatry,* 63, 1238-1245.

Brewin, C. R., Andrews, B. & Valentine, J. D. 2000. Meta-analysis of risk factors for posttraumatic stress disorder in trauma-exposed adults. *Journal of Consulting and Clinical Psychology,* 68, 748-766.

Bryant, R., Creamer, M., O'Donnell, M., Silove, D. & McFarlane, A. 2011. Heart rate after trauma and the specificity of fear circuitry disorders. *Psychological Medicine,* Jun 15 [Epub ahead of print], 1-8.

Canadian Psychiatric Association 2001. Clinical Practice Guidelines for the Treatment of Depressive Disorder. *The Canadian Journal of Psychiatry.*

Canadian Psychiatric Association 2006. Posttraumatic Stress Disorder. In Clinical Practice Guiselines Management of Anxiety Disorders. *Canadian Journal of Psychiatry,* 51,Suppl 2, 57-63.

Candel, I., Merkelbach, H. & Kuijpers, M. 2003. Dissociative experiences are related to commissions in emotional memory. *Behaviour Research and Therapy,* 41, 719-725.

Carper, B. 1978. Fundamental patterns of knowing in nursing,. *Advances in Nursing Science,* 13-23.

Charney, D. S., Price, L. H. & Heninger, G. R. 1986. Desipramine-yohimbine combination treatment of refractory depression. Implications for the beta-adrenergic receptor hypothesis of antidepressant action. *Archives of General Psychiatry,* 43, 1155-61.

Chung, M., Min, K., Jun, Y., Kim, S., Kim, W. & Jun, E. 2004. Efficacy and tolerability of mirtazapine and sertraline in Korean veterans with posttraumatic stress disorder: a randomized open label trial. . *Human Psychopharmacology* 19 489-94.

Connor, K., Davidson, J., Weisler, R. & Ahearn, E. 1999. A pilot study of mirtazapine in post-traumatic stress disorder. *International Clinical Psychopharmacology,* 14, 29-31.

Cooper, J., Carty, J. & Creamer, M. 2005. Pharmacotherapy for posttraumatic stress disorder: empirical review and clinical recommendations. *Australian and New Zealand Journal of Psychiatry,* 39, 674-682(9).

Creamer, M. & Forbes, D. 2004. Treatment of Posttraumatic Stress Disorder in Military and Veteran Populations *Psychotherapy: Theory, Research, Practice, Training,* 41, 388-398.

Davidson, J., Kudler, H., Smith, R., Mahorney, S., Lipper, S., Hammett, E., Saunders, W. & Cavenar, J. J. 1990. Treatment of posttraumatic stress disorder with amitriptyline and placebo. *Arch Gen Psychiatry,* 47, 259-266.

Davidson, J., Weisler, R., Butterfield, M., Casat, C., Connor, K., Barnett, S. & Van Meter, S. 2003. Mirtazapine vs placebo in posttraumatic stress disorder: a pilot trial. *Biological Psychiatry,* 53, 188–191.

Davidson, P. R. & Parker, K. C. 2001. Eye movement desensitization and reprocessing (EMDR): a meta-analysis. *Journal of Consulting and Clinical Psychology,* 69, 305-316.

Del Piccolo, L., Saltini, A. & Zimmerman, C. 1998. Which patients talk about stressful events and social problems to the general practitioner? *Psychological Medicine,* 28, 1289-1299.

Dohrenwend, B., Turner, J., Turse, N., Adams, B., Koenen, K. & Marshall, R. 2006. The psychological risks of Vietnam for U.S. veterans: a revisit with new data and methods. *science,* 313, 979-82.

Elhai, J. D. & Ford, J. D. 2005. Recent psychiatric disorders, trauma exposure, and posttraumatic stress disorder as predictors of mental health service utilization in a nationally representative U.S. sample. *Manuscript Submitted for Publication.*

Elhai, J. D., North, T. C. & Frueh, B. C. 2005. Health service use predictors among trauma survivors: A critical review. *Psychological Services,* 2, 3-19.

Elhai, J. D., Richardson, J. D. & Pedlar, D. 2007 Predictors of general medical and psychological treatment use among a national Canadian sample of United Nations peacekeeping veterans. *Journal of Anxiety Disorders,* 21, 580-589.

Fenton, W. & McGlashan, T. 2000. Schizophrenia: Individual Therapy. *In:* Saddock B & V, S. (eds.) *Comprehensive Textbook of Psychiatry.* Lippincott Williams and Wilkins.

Fesler, F. 1991. Valproate in combat-related posttraumatic stress disorder. *Journal of Clinical Psychiatry,* 52, 361–364.

Foa, E., Keane, T., Friedman, L. & Cohen, J. A. 2009. Introduction. *In:* Foa E, Keane T, Friedman LM & Judith, C. (eds.) *Effective Treatments for PTSD.* New York: The Guilford press.

Foa, E., Keane, T. & Friedman, M. 2000. *Effective Treatments for PTSD,* New York, Guilford.

Foa, E. B. 2006. Psychosocial therapy for posttraumatic stress disorder. *Journal of Clinical Psychiatry,* 67, 40-45.

Forbes, D., Bennett, N., Biddle, D., Crompton, D., McHugh, T., Elliott, P. & Creamer, M. 2005. Clinical Presentations and Treatment Outcomes of Peacekeeper Veterans With PTSD: Preliminary Findings. *American Journal of Psychiatry,* 162, 2188-2190.

Forbes, D., Creamer, M., Hawthorne, G., Allen, N. & McHugh, T. 2003. Comorbidity as a predictor of symptom change after treatment in combat-related posttraumatic stress disorder. *Journal of Nervous and Mental Disease,* 191, 93-99.

Forbes, D., Hawthorne, G., Elliott, P., McHugh, A. F., Biddle, D., Creamer, M. & Novaco, R. W. 2004. A concise measure of anger in combat-related posttraumatic stress disorder. *Journal of Traumatic Stress,* 17, 249-256.

Foy, W. F., Glynn, S. M., Schnurr, P., Jankowski, M., Wattenberg, M., Weiss, D., Marmar, C. & Gusman, F. 2000. Group Therapy. *In:* Foa, E., Keane, T. & Friedman, M. (eds.) *Effective treatments for PTSD.* New York: The Guildford Press.

Friedman, M. 1997. Drug treatment for PTSD: answers and questions. *Ann NY Acad Sci,* 359-371.

Friedman, M., Charney, D. & Southwick, S. 1993. Pharmacotherapy for recently evacuated military casualties. *Military Medicine,* 158, 493–497.

Friedman, M., Davidson, J., Mellman, T. & Southwick, S. 2000. Pharmacotherapy. *In:* Foa E, Keane T & M, F. (eds.) *Effective Treatments for PTSD.* The Guilford Press.

Friedman, M., Marmar, C., Baker, D., Sikes, C. & Farfel, G. 2007. Randomized, double-blind comparison of sertraline and placebo for posttraumatic stress disorder in a Department of Veterans Affairs setting. *Journal of Clinical Psychiatry,* 68, 711-20.

Friedman, M. J. 2006. Posttraumatic Stress Disorder Among Military Returnees From Afghanistan and Iraq. *American Journal of Psychiatry,* 163, 586-593.

Gavrilovic, J. J., Schutzwohl, M., Fazel, M. & Priebe, S. 2005. Who seeks treatment after a traumatic event and who does not? A review of findings on mental health service utilization. *Journal of Traumatic Stress,* 18, 595-605.

Gelpin, E., Bonne, O., Peri, T., Brandes, D. & Shalev, A. 1996. Treatment of recent trauma survivors with benzodiazepines: a prospective study. *Journal of Clinical Psychiatry,* 57, 390-394.

Glover, H. 1988. Four syndromes of post-traumatic stress disorder: stressors and conflicts of the traumatized with special focus on the Vietnam combat veteran. Journal of Traumatic Stress. *Journal of Traumatic Stress,* 1(1), 57-78.

Gradus, J. L., Qin, P., Lincoln, A. K., Miller, M., Lawler, E., Sorensen, H. T. & Lash, T. L. 2010. Posttraumatic Stress Disorder and Completed Suicide. *American Journal of Epidemiology,* 171, 721-727.

Gray, M. J., Elhai, J. D. & Frueh, B. C. 2004. Enhancing patient satisfaction and increasing treatment compliance: Patient education as a fundamental component of PTSD treatment. *Psychiatric Quarterly,* 75, 321-332.

Hamner, M., Brodrick, P. & Labbate, L. 2001. Gabapentin in PTSD: a retrospective, clinical series of adjunctive therapy. *Annals of Clinical Psychiatry,* 13, 141–146.

Hamner, M., Faldowski, R., Ulmer, H., Frueh, B., Huber, M. & Arana, G. 2003. Adjunctive risperidone treatment in post-traumatic stress disorder: a preliminary controlled trial of effects on comorbid psychotic symptoms. *International Clinical Psychopharmacology,* 18, 1-8.

Hankin, C. S., Spiro, A., Miller, D. R. & Kazis, L. 1999. Mental disorders and mental health treatment among U.S. Department of Veterans Affairs outpatients: The Veterans Health Study. *American Journal of Psychiatry,* 156, 1924-1930.

Hertzberg, M., Butterfield, M., Feldman, M., Beckham, J., Sutherland, S., Connor, K. & Davidson, J. 1999. A preliminary study of lamotrigine for the treatment of posttraumatic stress disorder. *Biological Psychiatry,* 45, 1226–1229.

Hetrick, S., Purcell, R., Garner, B. & Parslow, R. 2010. Combined pharmacotherapy and psychological therapies for post traumatic stress disorder (PTSD). *Cochrane Database of Systematic Reviews 2010.*

Hoge, C. W., Castro, C. A., Messer, S. C., McGurk, D., Cotting, D. I. & Koffman, R. L. 2004a. Combat Duty in Iraq and Afghanistan, Mental Health Problems, and Barriers to Care.

Hoge, C. W., Castro, C. A., Messer, S. C., McGurk, D., Cotting, D. I. & Koffman, R. L. 2004b. Combat duty in Iraq and Afghanistan: Mental health problems and barriers to care. *New England Journal of Medicine,* 351, 13-22.

Hoge, C. W., Lesikar, S. E., Guevara, R., Lange, J., Brundage, J. F., Engel, C. C., Messer, S. C. & Orman, D. T. 2002. Mental disorders among U.S. Military personnel in the 1990s: Association with high levels of health care utilization and early military attrition. *American Journal of Psychiatry,* 159, 1576-1583.

Holbrook, T. L., Galarneau, M. R., Dye, J. L., Quinn, K. & Dougherty, A. L. 2010. Morphine Use after Combat Injury in Iraq and Post-Traumatic Stress Disorder. *New England Journal of Medicine,* 362, 110-117.

Hopwood, M., Morris, P. L. P., Debenham, P., Bonwick, R., Parkin, I., Ignatiadis, S., Norman, T. & Burrows, G. D. 2000. An Open Label Trial of Venlafaxine in War Veterans with Chronic Post Traumatic Stress Disorder. *Australian and New Zealand Journal of Psychiatry,* 34.

Hourani, L. L. & Yuan, H. 1999. The mental health status of women in the Navy and Marine Corps: Preliminary findings from the Perceptions of Wellness and Readiness Assessment. *Military Medicine,* 164, 174–181.

Institute of Medicine (IOM) 2008. *Treatment of PTSD: an Assessment of the Evidence,* Washington, DC, National Academies Press

Jakupcak, M., Luterek, J., Hunt, S., Conybeare, D. & McFall, M. 2008. Posttraumatic stress and its relationship to physical health functioning in a sample of Iraq and Afghanistan war veterans seeking postdeployment VA health care. *Journal of Nervous and Mental Disease,* 196, 425-428.

Kaufman, J. & Charney, D. 2000. Comorbidity of mood and anxiety disorders. *Depression Anxiety,* 12, 69–76.

Keane, T. M. & Kaloupek, D. G. 1997. Comorbid psychiatric disorders in PTSD: Implications for research. *Annual New York Academy of Sciences,* 21, 24-34.

Keane, T. M. & Wolfe, J. 1990. Comorbidity in post-traumatic stress disorder: An analysis of community and clinical studies. *Journal of Applied Social Psychology,* 20, 1776-1788.

Keck, P. J., McElroy, S. & Friedman, L. 1992. Valproate and carbamazepine in the treatment of panic and posttraumatic stress disorders, withdrawal states, and behavioral dyscontrol syndromes. *Journal of Clinical psychopharmacology,* 12, 36S–41S.

Kessler, R. C., Sonnega, A., Bromet, E., Hughes, M. & Nelson, C. B. 1995. Posttraumatic stress disorder in the National Comorbidity Survey. *Archives of General Psychiatry,* 52, 1048-1060.

Kessler, R. C., Zhao, S., Katz, S. J., Kouzis, A. C., Frank, R. G., Edlund, M. J. & Leaf, P. 1999. Past-year use of outpatient services for psychiatric problems in the National Comorbidity Survey. *American Journal of Psychiatry,* 156, 115-123.

King, D. W., King, L. A., Gudanowski, D. M. & Vreven, D. L. 1995. Alternative representations of war zone stressors: Relationships to posttraumatic stress disorder in male and female Vietnam veterans. *Journal of Abnormal Psychology and Aging,* 104, 184–196.

King, L. A., King, D. W., Vogt, D. S., Knight, J. & Samper, R. E. 2006. Deployment Risk and Resilience Inventory: A Collection of Measures for Studying Deployment-Related Experiences of Military Personnel and Veterans. *Military Psychology,* 18, 89-120.

Kluft, R. P. 2002. Negotiating the Therapeutic Alliance: A Relational Treatment Guide. *American Journal of Psychiatry,* 159, 885-.

Koren, D., Norman, D., Cohen, A., Berman, J. & Klein, E. M. 2005. Increased PTSD Risk With Combat-Related Injury: A Matched Comparison Study of Injured and Uninjured Soldiers Experiencing the Same Combat Events. *American Journal of Psychiatry,* 162, 276-28.

Kosten, T., Frank, J., Dan, E., McDougle, C. & Giller, E. J. 1991. Pharmacotherapy for post-traumatic stress disorder using phenelzine or imipramine. *Journal of Nervous and Mental Disease,* 179, 366-370.

Krystal, J. H., Rosenheck, R. A., Cramer, J. A., Vessicchio, J. C., Jones, K. M., Vertrees, J. E., . . . Stock, C. (2011). Adjunctive Risperidone Treatment for Antidepressant-Resistant Symptoms of Chronic Military Service–Related PTSD. JAMA: The Journal of the American Medical Association, 306(5), 493-502. doi: 10.1001/jama.2011.1080

Kulka, R. A., Schlenger, W. E., Fairbank, J. A., Hough, R. L., Jordan, B. K., Marmar, C. R. & Weiss, D. S. 1990. *Trauma and the Vietnam War generation: Report of findings from the National Vietnam Veterans Readjustment Study,* New York, Brunner/Mazel.

Lewis, D. 1990. Neuropsychiatric and experiential correlates of violent juvenile delinquency. *Neuropsychological Review,* 1, 125-36.

Lipper, S., Davidson, J., Grady, T., Edinger, J., Hammett, E., Mahorney, S. & Cavenar, J. J. 1986. Preliminary study of carbamazepine in post-traumatic stress disorder. *Psychosomatics,* 27, 849–854.

Litz, B. T. 1996. The Psychological Demands of Peacekeeping. *PTSD Clinical Quarterly,* 6, 1-8.

Litz, B. T., King, L. A., King, D. W., Orsillo, S. M. & Friedman, M. 1997a. Warriors as Peacekeepers: Features of the Somalia Experience and PTSD. *Journal of Consulting and Clinical Psychology,* 65, 1001-1010.

Litz, B. T., Orsillo, S. M., Friedman, M., Erhlich, P. & Batres, A. 1997b. Posttraumatic Stress Disorder Associated with Peacekeeping Duty in Somalia for U.S. Military Personnel. *American Journal of Psychiatry,* 154, 178-184.

Londborg, P., Hegel, M., Goldstein, S., Goldstein, D., Himmelhoch, J., Maddock, R., Patterson, W., Rausch, J. & Farfel, G. 2001. Sertraline treatment of post-traumatic stress disorder: results of 24 weeks of open label continuation treatments. *Journal of Clinical Psychiatry,* 62, 325–331.

Marshall, R. & Cloitre, M. 2000. Maximizing treatment outcome in PTSD by combining psychotherapy with pharmacotherapy. *Current Psychiatry Reports*, 335–340.

Marshall, R. D., Olfson, M., Hellman, F., Blanco, C., Guardino, M. & Struening, E. L. 2001. Comorbidity, Impairment, and Suicidality in Subthreshold PTSD. *Am J Psychiatry,* 158, 1467-1473.

Marshall, R. P., Jorm, A. F., Grayson, D. A. & O'Toole, B. I. 1998. Posttraumatic stress disorder and other predictors of health care consumption by Vietnam veterans. *Psychiatric Services,* 49, 1609-1611.

McFall, M., Malte, C., Fontana, A. & Rosenheck, R. A. 2000. Effects of an outreach intervention on use of mental health services by veterans with posttraumatic stress disorder. *Psychiatric Services,* 51, 369-374.

McFarlane, A., Atchison, M., Rafalowicz, E. & Papay, P. 1994. Physical symptoms in post-traumatic stress disorder. *Journal of Psychosomatic Research,* 38, 715-726.

Miller, L. J. 2008. Prazosin for the Treatment of Posttraumatic Stress Disorder Sleep Disturbances. Pharmacotherapy, 28, 656-666.

Mills, K. L., Teesson, M., Ross, J. & Peters, L. 2006. Trauma, PTSD, and Substance Use Disorders: Findings From the Australian National Survey of Mental Health and Well-Being. *American Journal of Psychiatry,* 163, 652-658.

Monnelly, E., Ciraulo, D., Knapp, C. & Keane, T. 2003. Low-dose risperidone as adjunctive therapy for irritable aggression in posttraumatic stress disorder. *Journal of Clinical psychopharmacology,* 23, 193–196.

Monson, C. M., Schnurr, P. P., Resick, P. A., Friedman, M. J., Young-Xu, Y. & Stevens, S. P. 2006. Cognitive processing therapy for veterans with military-related posttraumatic stress disorder. *Journal of Consulting and Clinical Psychology,* 74, 898-907.

Morland, L. A., Greene, C. J., Rosen, C. S., Foy, D., Reilly, P., Shore, J., He, Q. & Frueh, C. B. 2010. Telemedicine for Anger Management Therapy in a Rural Population of Combat Veterans With Posttraumatic Stress Disorder: A Randomized Noninferiority Trial. *Journal of Clinical Psychiatry,* 71, 855–863.

Nacasch, N., Foa, E., Huppert, J., Tzur, D., Fostick, L., Dinstein, Y., Polliack, M. & Zohar, J. 2010. Prolonged exposure therapy for combat- and terror-related posttraumatic stress disorder: a randomized control comparison with treatment as usual. *Journal of Clinical Psychiatry,* 16, Epub ahead of print.

National Institute for Clinical Excellence 2005. Post-traumatic stress disorder (PTSD): The management of PTSD in adults and children in primary and secondary care. *In:* London (ed.). National Institute for Clinical Excellence.

Nepon, J., Belik, S.-L., Bolton, J. & Sareen, J. 2010. The relationship between anxiety disorders and suicide attempts: findings from the National Epidemiologic Survey on Alcohol and Related Conditions. *Depression and Anxiety,* 27, 791–798.

O'Donnell, M., Creamer, M., Holmes, A., Ellen, S., McFarlane, A., Judson, R., Silove, D. & Bryant, R. 2010. Posttraumatic stress disorder after injury: does admission to intensive care unit increase risk? *Journal of Trauma,* 69, 627-32.

O'Toole, B. I., Marshall, R. P., Grayson, D. A., Schureck, R. J., Dobson, M., French, M., Pulvertaft, B., Meldrum, L., Bolton, J. & Vennard, J. 1996. The Australian Vietnam veterans health study: III. Psychological health of Australian Vietnam veterans and its relationship to combat. *International Journal of Epidemiology,* 25, 331-340.

Olfson, R., Hellman, M., Blanco, F., Guardino, C. & Struening, M. 2001. Comorbidity, impairment, and suicidality in subthreshold PTSD. *American Journal of Psychiatry,* 1467-1473.

OSISS Operational Stress Injury Social Support Program. *In:* National Defense & Canada, V. A. (eds.).

Ozer, E. J., Best, S. R., Lipsey, T. L. & Weiss, D. S. 2003a. Predictors of posttraumatic stress disorder and symptoms in adults: A meta-analysis. *Psychological Bulletin,* 129, 52-73.

Ozer, E. J., Best, S. R., Lipsey, T. L. & Weiss, D. S. 2003b. Predictors of posttraumatic stress disorder and symptoms in adults: A meta-analysis. *Psychological Bulletin,* 129, 52-73.

Peterson, A., Luethcke, C., Borah, E., Borah, A. & Young-McCaughan, S. 2011. Assessment and Treatment of Combat-Related PTSD in Returning War Veterans. Journal of Clinical Psychology in Medical Settings, 18, 164-175.

Phillips, S., Burns, B., Edgar, E., Mueser, K., Linkins, K., Rosenheck, R., Drake, R. & McDonel Herr, E. 2001. Moving assertive community treatment into standard practice. *Psychiatr Serv,* 52, 771-779.

Raskind, M., Thompson, C., Petrie, E., Dobie, D., Rein, R., Hoff, D., McFall, M. & Peskind, E. 2002. Prazosin reduces nightmares in combat veterans with posttraumatic stress disorder. *Journal of Clinical Psychiatry,* 63, 565-568.

Raskind, M. A., Peskind, E. R., Kanter, E. D., Petrie, E. C., Radant, A., Thompson, C. E., Dobie, D. J., Hoff, D., Rein, R. J., Straits-Troster, K., Thomas, R. G. & McFall, M. M. 2003. Reduction of nightmares and other PTSD symptoms in combat veterans by prazosin: A placebo-controlled study. . *American Journal of Psychiatry,* 160, 371-373.

Richardson, J., Long, M. E., Pedlar, D. & Elhai, J. D. 2010 Posttraumatic Stress Disorder and Health Related Quality of Life (HRQol) in Pension-Seeking Canadian WW II and Korean Veterans *Journal of Clinical Psychiatry,* 1099-1101.

Richardson, J. D., Elhai, J. D. & Sareen, J. 2011. Predictors of Treatment Response in Canadian Combat and Peacekeeping Veterans with Military-Related PTSD. *Journal of Nervous and Mental Disease,* 199, 639-645.

Richardson, J. D., Elhai, J. & Pedlar, D. 2006. Association of PTSD and Depression with Medical and Specialist Care Utilization in Modern Peacekeeping Veterans in CanadaWith Health-Related Disabilities. *Journal of Clinical Psychiatry,* 67, 1240-1245.

Richardson, J. D., Long, M. E., Pedlar, D. & Elhai, J. D. 2008 Posttraumatic Stress Disorder and Health Related Quality Of Life (HRQol) among a Sample of Treatment- and Pension-Seeking deployed Canadian Forces Peacekeeping Veterans. *Canadian Journal of Psychiatry,* 53, 594-600.

Richardson, J. D., Naifeh, J. A. & Elhai, J. 2007. Posttraumatic Stress Disorder and Associated Risk Factors in Canadian Peacekeeping Veterans With Health-Related Disabilities. *Canadian Journal of Psychiatry,* 52, 510-518.

Richardson, J. D., Fikretoglu, D., Liuf, A., & Mcintosh, D. (2011). Aripiprazole Augmentation in the Treatment of Military-Related PTSD with Major Depression: a retrospective chart review. BMC Psychiatry, 11(86), 1-7.

Ronis, D. L., Bates, E. W., Garfein, A. J., Buit, B. K. & et al. 1996. Longitudinal patterns of care for patients with posttraumatic stress disorder. *Journal of Traumatic Stress*, 9, 763-781.

Rothbaum, B., Cahill, S., Foa, E., Davidson, J., Compton, J., Connor, K., Astin, M. & Hahn, C. 2006. Augmentation of sertraline with prolonged exposure in the treatment of posttraumatic stress disorder. *journal of Traumatic Stress*, 19.

Samuels, C. H. 2005. Bedtime Blues: Managing Primary Insomnia. *The Canadian Journal of CME*, 67-69.

Sandweiss, D. A., Slymen, D. J., LeardMann, C. A., Smith, B., White, M. R., Boyko, E. J., Hooper, T. I., Gackstetter, G. D., Amoroso, P. J., Smith, T. C. & for the Millennium Cohort Study Team 2011. Preinjury Psychiatric Status, Injury Severity, and Postdeployment Posttraumatic Stress Disorder. *Archives of General Psychiatry*, 68, 496-504.

Sareen, J., Cox, B., Clara, I. & Asmundson, G. 2005. The relationship between anxiety disorders and physical disorders in the U.S. National Comorbidity Survey. *Depression and Anxiety*, 21, 193-202.

Sareen, J., Cox, B. J., Stein, M. B., Afifi, T. O., Fleet, C. & Asmundson, G. J. G. 2007. Physical and mental comorbidity, disability, and suicidal behavior associated with posttraumatic stress disorder in a large community sample. *Psychosomatic Medicine*, 69, 242-248.

Sareen, J., Stein, M., Cox, B. & Hassard, S. 2004. Understanding comorbidity of anxiety disorders and antisocial behavior: Findings from two large community surveys. *Journal of Nervous and Mental Disease*, 192, 178-86.

Schnurr, P. P. & Jankowski, M. K. 1999. Physical health and post-traumatic stress disorder: Review and synthesis. *Seminar in Clinical Neuropsychiatry*, 4, 295-304.

Schnurr, P. P., Spiro, A. & Paris, A. H. 2000. Physician-diagnosed medical disorders in relation to PTSD symptoms in older male military veterans. *Health Psychology*, 19, 91-97.

Schoenfeld, F. B., Marmar, C. R. & Neylan, T. C. 2004a. Current Concepts in Pharmacotherapy for Posttraumatic Stress Disorder. *Psychiatric Services*, 55, 519-531.

Schoenfeld, F. B., Marmar, C. R. & Neylan, T. C. 2004b. Current Concepts in Pharmacotherapy for Posttraumatic Stress Disorder. *Psychiatr Serv*, 55, 519-531.

Schützwohl, M. & Maercker, A. 1999. Effects of varying diagnostic criteria for posttraumatic stress disorder are endorsing the concept for partial PTSD. *Journal of Traumatic Stress*, 12, 155-165.

Shalev, A., Bonne & Eth, S. 1996. Treatment of posttraumatic stress disorder: A review. . *Psychosomatic Medicine*, 165-182.

Sledjeski, E. M., Speisman, B. & Dierker, L. C. 2008. Does number of lifetime traumas explain the relationship between PTSD and chronic medical conditions? Answers from the National Comorbidity Survey-Replication (NCS-R). *Journal of Behavioral Medicine*, 31, 341-349.

Smajkic, A., Weine, S., Djuric-Bijedic, Z., Boskailo, E., Lewis, J. & Pavkovic, I. 2001. Sertraline, paroxetine, and venlafaxine in refugee posttraumatic stress disorder with depression symptoms. *Journal of Traumatic Stress*, 14, 445–452.

Southwick, S., Yehuda, R. & Giller, E. J. 1991. Characterization of depression in war-related posttraumatic stress disorder. *American Journal of Psychiatry*, 148, 179–183.

Statistics Canada 2002. Canadian Community Health Survey Cycle 1.2 – Mental Health and Well-being (Canadian Forces Supplement). *In:* Canada, S. (ed.). Ottawa: Statistics Canada.

Stein, D., Ipser, J. & Seedat, S. 2006 Pharmacotherapy for post traumatic stress disorder (PTSD). *Cochrane Database of Systematic Reviews,* 25, CD002795.

Stein, M. B., Kline, N. A. & Matloff, J. L. 2002. Adjunctive Olanzapine for SSRI-Resistant Combat-Related PTSD: A Double-Blind, Placebo-Controlled Study. *American Journal of Psychiatry,* 159, 1777-1779.

Stein, M. B., Walker, J. R., Hazen, A. L. & Forde, D. R. 1997. Full and partial posttraumatic stress disorder: Findings from a community survey. *American Journal of Psychiatry,* 154, 1114-1119.

Switzer, G. E., Dew, M. A., Thompson, K., Goycoolea, J. M., Derricott, T. & Mullins, S. D. 1999. Posttraumatic stress disorder and service utilization among urban mental health center clients. *Journal of Traumatic Stress,* 12, 25-39.

Tuerk, P. W., Yodera, M., Grubaugha, A., Myricka, H., Hamnera, M. & Aciernoa, R. 2011. Prolonged exposure therapy for combat-related posttraumatic stress disorder: An examination of treatment effectiveness for veterans of the wars in Afghanistan and Iraq. *Journal of Anxiety Disorders,* 25, 397-403.

Turnbull, G. & McFarland, A. 1996. Acute Treatments. *In:* Van Der Kolk BA, McFarland A & L, W. (eds.) *Traumatic Stress.* New York: The Guilford Press.

Vaiva, G., Ducrocq, F., Jezequel, K., Averland, B., Lestavel, P., Brunet, A. & Marmar, C. 2003. Immediate treatment with propranolol decreases posttraumatic stress disorder two months after trauma. *Biological Psychiatry,* 54, 947–949.

Van Der Kolk, B., McFarland, A. & Van Der Hart, O. 1996a. A General Approach to Treatment of Posttraumatic Stress Disorder. *In:* Van Der Kolk, B., McFarland, A. & Weisaeth, L. (eds.) *Traumatic Stress.* New York: The Guilford press.

Van Der Kolk, B., McFarland, A. & Weisaeth, L. 1996b. A Pharmacological Treatment of Post-Traumatic Stress Disorder. *In:* Davidson Jonathan & Bessel, V. D. K. (eds.) *Traumatic Stress: The Effects of Overwhelming Experience and Mind, Body and Society.* The Guilford Press

Weathers, F. W., Litz, B. T., Herman, D. S., Huska, J. A. & Keane, T. M. The PTSD checklist: Reliability, validity, & diagnostic utility. annual meeting of the International Society for Traumatic Stress Studies, October 1993 San Antonio, Texas. International Society for Traumatic Stress Studies.

Weiden, P. J. & Rao, N. 2005. Teaching Medication Compliance to Psychiatric Residents: Placing an Orphan Topic Into a Training Curriculum. *Acad Psychiatry,* 29, 203-210.

Weiss, D. S., Marmar, C. R., Schlenger, W. E., Fairbank, J. A., Jordan, B. K., Hough, R. L. & Kulka, R. A. 1992. The prevalence of lifetime and partial post-traumatic stressdisorder in Vietnam theater veterans. *Journal of Traumatic Stress,* 5, 365-376.

Yehuda, R. 1999. *Risk Factors for Posttraumatic Stress Disorder,* , Washington DC, American Psychiatric Press Inc.

Yehuda, R., McFarlane, A. & Shalev, A. 1998. Predicting the development of posttraumatic stress disorder from the acute response to a traumatic event. *Biological Psychiatry,* 44, 1305-1313.

Zlotnick, C., Franklin, C. & Zimmerman, M. 2002. Does "subthreshold" posttraumatic stress disorder have any clinical relevance? *Comprehensive Psychiatry,* 43, 413-419.

Part 2

Review of Etiological Factors

4

Acquisition of Active Avoidance Behavior as a Precursor to Changes in General Arousal in an Animal Model of PTSD

Thomas M. Ricart, Richard J. Servatius and Kevin D. Beck
Veteran Affairs New Jersey Health Care System,
University of Medicine & Dentistry of New Jersey – New Jersey Medical School,
USA

1. Introduction

1.1 Increased defensive reactions as a sign of PTSD

Post-traumatic stress disorder (PTSD) is a multi-symptom condition that includes three primary psychological features: reexperiencing, avoidance and emotional numbing, and hyperarousal (American Psychiatric Association, 2000). Historically, reexperiencing and hyperarousal have been the most studied features, from a neurobiological perspective, using various animal models. In these animal models, changes in defensive reflexive behaviors serve as the assessment measures for these symptoms; thus, both startle reactivity and freezing are now commonly used measures. Freezing behavior is advantageous because of its easy implementation; either the naked eye or an automated motor-tracking system can determine the duration and/or frequency of freezing behavior. In addition, freezing can occur in response to a specific fear-eliciting stimulus or to a fear-experienced context (Doyle & Yule, 1959; Bouton & Bolles, 1980; Fanselow, 1980). Because of these stimulus-response properties, freezing is the response commonly used to assess the experiencing of memories of conditioned stimuli that previously caused a heightened state of fear. At times, the acoustic startle response is used as an assessment of stimulus-elicited fear reactions (Davis, 1986; Hitchcock & Davis, 1987). Under this guise, similar stimuli used in conditioned freezing are experienced by the animal prior to a quick onset, relatively loud, acoustic stimulus. The result is a startle response that is enhanced over baseline levels, which is termed fear-potentiated startle.

However, in the case of PTSD, arousal is not necessarily tied to a memory or triggered by an explicit learned association. There are several examples of patients with PTSD exhibiting exaggerated startle responses in the absence of a known trigger (Butler et al., 1990; Orr et al., 1995; Yehuda et al., 1998; Orr et al., 2002). In fact, human longitudinal studies have found changes in startle reactivity occur over a period of time following the associated trauma (Shalev et al., 1998). Although there are possible confounding variables with any repeated test, such as developing an aversion to the startle testing context, there is a difference with PTSD patients as they fail to habituate to the startle test over months (Shalev et al., 1998). Increases in startle magnitudes can be elicited in rats in order to model this feature of PTSD by exposing them to inescapable shock. Interestingly, like some of the symptoms of PTSD, this procedure increases startle magnitude in male rats several days following stressor exposure (Servatius et al., 1994; Servatius et al. 1995; Beck et al., 2002; Manion et al., 2007; Manion et al., 2010); in other words, there is a delayed sensitization of the startle response (Beck et al., 2002). As shown in Figure 1, enhanced startle reactivity may then be observed up to several days later (Servatius et al., 1995; Beck et al., 2002; Manion et al., 2007; Manion et al., 2010). The delayed expression of this enhanced defensive response could be due to competing processes that are similarly elicited by inescapable shock. For instance, some female rats exhibit a suppression of the startle response following stressor exposure that is clearly linked to the immune response elicited by shock exposure (Beck & Servatius, 2005; Beck & Servatius, 2006), but this suppression of startle reactivity is only evident under certain ovarian hormone conditions (Beck et al., 2008). Although males have not been shown to exhibit a reduction in startle reactivity, there may be similar acute physiological reactions that negate an immediate increase in startle reactivity. Hence, these inescapable stress models of enhanced startle reactivity are largely based on the concept that PTSD-like hyperarousal is a product of a single, uncontrollable, traumatic event (or period of repeated trauma over consecutive days).

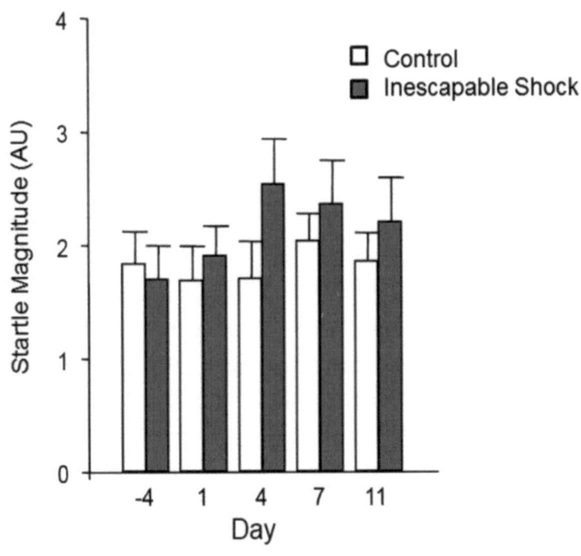

Fig. 1. Following 3 days of repeated tailshock exposure (40, 3 s, 2 mA shocks over 2 h), startle magnitudes are elevated in male Sprague Dawley rats 4 days thereafter [Stress x Day $F(4, 776) = 7.6$, $p < .001$]. The delayed expression of this increase has been used as a model of the emergent aspects of PTSD symptom expression. Startle magnitude is represented in arbitrary units (AU), as the amplitude is corrected for by body weight.

These models of hyperarousal in PTSD fail to account for potential innate differences in reactivity. The diathesis model of anxiety disorders indicates that anxiety symptomology are a combination of innate characteristics that are influenced by life events. Recent prospective research on PTSD symptomology has indicated elevated arousal may be predictive of anxiety symptoms (Guthrie & Bryant, 2005; Pole et al., 2009). Startle reactivity, as one measure, is elevated in children that upon follow-up were diagnosed with clinical anxiety (Merikangas et al., 1999). Thus, pathologic arousal exhibited in PTSD may reflect an innate elevation in arousal, an increase in arousal due to experience, or a combination of innate characteristics that are exacerbated by experience.

In developing our model of anxiety vulnerability, we sought a rodent that reliably exhibited a set of behaviors that are analogous to an identified human vulnerability condition. Behavioral inhibition (or inhibited temperament) is extreme withdrawal in face of novel social and nonsocial situations (Kagan et al., 1987; Kagan et al., 1989; Hirshfeld et al., 1992) and is identified as a risk factor for future symptoms of anxiety disorders (Hirshfeld-Becker et al., 2008; Jovanovic et al., 2010). The Wistar-Kyoto (WKY) rat demonstrates aspects of behavioral inhibition in terms of withdrawal in tests of social interaction (Pardon et al., 2002) and lack of exploration in the open field test (Pare, 1994; Servatius et al., 2008). Hence, we tested WKY rats in a multi-intensity startle procedure that allows for assessment of startle sensitivity (percentage of startles elicited at various acoustic intensities) and startle responsivity (magnitude of responses to the highest intensity). Given startle is a defensive-reflex, increased responsivity can be viewed as an increase in general arousal (i.e. more energy used to respond to threat), whereas increased sensitivity can be viewed as an enhancement of vigilance (i.e. greater signal-detection of threats). With no prior manipulations, WKY rats exhibited similar startle sensitivity measures as Sprague Dawley (SD) rats, but exhibited higher startle responsivity (see Figure 2). These temperament and reactivity characteristics are reasonable analogs of the predisposing factors for developing symptoms of anxiety disorders; therefore, we adopted the WKY rat as our anxiety vulnerability model (Servatius et al., 2008; Beck et al., 2010; Ricart et al., 2011a; Ricart et al., 2011b; Jiao et al., 2011; Beck et al., 2011).

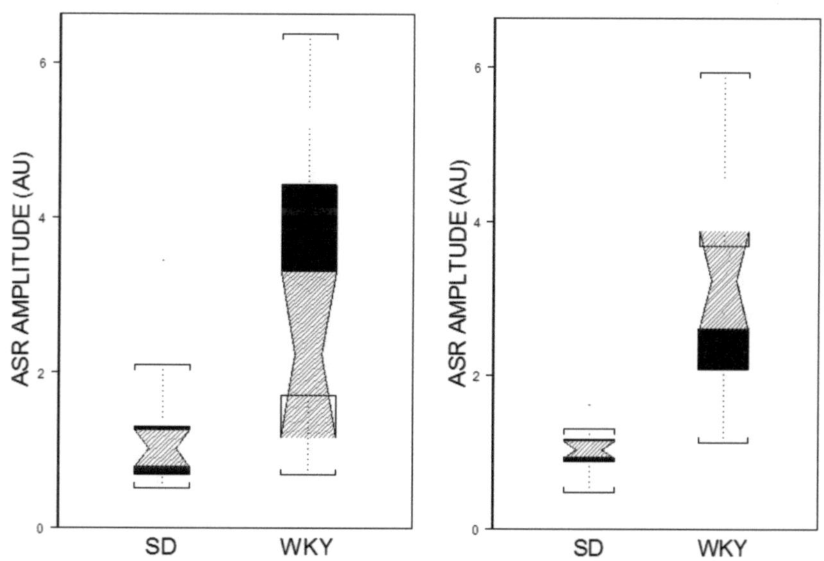

Fig. 2. Box plots of initial startle responsivity and latency of male SD and WKY rats (left) and female SD and WKY rats (right). Notched regions depict median ± 95% confidence interval. Boxed region depicts interquartile range (middle fifty). Whiskers depict range of data 1.5*interquartile range. Outliers depicted by '*'. WKY rats exhibit innate differences in ASRs, with greater amplitudes.

2. Avoidance as an alternative symptom to model

2.1 Behavioral avoidance

Avoidance, though not as well studied as hyperarousal or reexperiencing, is a common symptom to all anxiety disorders, including PTSD. Recent research indicates that the presentation of increased avoidant behavior has been found to track the general worsening of PTSD symptoms (Karamustafalioglu et al., 2006; O'Donnell et al., 2007) suggesting that acquisition of avoidance may have an etiological role in the development of PTSD. Many animal models of avoidance use tasks that are based on a tendency to exhibit passive avoidance. Tasks such as the elevated plus maze, measure how often rodents explore elevated open-arms (no walls) versus arms with high walled sides (closed arms). Rodents have a fundamental aversion to well-lit open spaces; therefore, prior exposure to stressors and anxiogenic drugs will reduce the limited tendency to explore the open arms (Pellow et al., 1985); conversely, anxiolytics increase exploration into the open arms. However, avoidance symptoms displayed in PTSD get progressively worse over time, and models such as the plus maze are not conducive to exhibit such a progression. In fact, avoidance and avoidant behaviors distinguish between those that develop PTSD and those who recover from trauma (Karamustafalioglu et al., 2006; Foa et al., 2006; O'Donnell et al., 2007). Thus the adoption of behavioral, cognitive, or emotional avoidance represents the detrimental process, which distinguishes those that successfully cope from those that develop pathological anxiety. Further, many avoidance behaviors in humans are active, that is, behaviors are not merely the absence of activity. Active avoidance is often more debilitating as the increased time and resources utilized for avoidance limit the capability of the individual to perform other tasks. Therefore, a task that allows for the observance of a methodical increase in active avoidant behaviors would be more akin to what is described in PTSD.

The process of learning active avoidance behaviors in rodents involves three steps. One, the rodent has to learn how to instrumentally dissociate itself from a noxious stimulus. This either may involve removing itself from the noxious stimulus (as in shuttle escape) or manipulating the environment such that the presentation of the noxious stimulus stops (as in lever-press escape or wheel-turn). Two, the rodent needs to recognize that the presence of certain stimuli in the environment precede the presentation of the noxious stimulus. These "warning signals" need to be distinct from the environment, with auditory stimuli serving as better signals than visual stimuli (Gilbert, 1971). This association may occur either during or following the acquisition of escape behavior. Three, the rodent needs to use the warning signal as a cue to emit an instrumental response prior to the actual onset of the noxious stimulus (i.e. avoidance behavior). The emitting of the response to the predictive warning signal removes the warning signal and the associated threat. In some regard, this process is more complicated for the rodent because it usually has to learn an escape response before it will reliably learn an avoidance response; although for some the transition from escape to avoidance behavior may involve less training sessions than for others. Humans can obviously learn to avoid people, animals, places, and situations without necessarily having to learn to escape from them first, but because this learning process is a bit more methodical in the rodents, we can understand how each stage of the process occurs and how it contributes to pathological avoidance.

Pathological avoidance is when the animal (human or non-human) responds to warning signals nearly 100% of the time. Although intuitively this may seem to be a responsible strategy for the animal, it does not allow for the individual to be sensitive to contingency changes (i.e. when the warning signal no longer reliably predicts the noxious stimulus). At that point, the avoidance responses are being emitted to remove a possible (not probable) threat. Therefore, the animal may be expending energy, by moving to the lever and subsequently depressing it, trial after trial to avoid a threat that the warning signal no longer reliably predicts. For individuals with severe anxiety disorders, this strategy of avoiding possible threats can become very disruptive if 1) the individuals expend more energy to avoid situations than what would be required to actually deal with them and 2) the perceived warning signals become generalized, which narrows their ability to interact with the world. Therefore, identifying an animal model that will acquire an exceptionally high asymptotic level of avoidant behavior, and subsequently exhibits the predictable slow extinction of the response, can provide us with a valuable system for identifying the vulnerability factors that predict such avoidant behaviors as well as the neural mechanisms that bias their behavioral strategies to such an extreme.

There are various forms of active avoidance that can be modeled in rats, but the desire to track the development of increased avoidant behavior over time led us to adopt distinct lever-press avoidance as our active avoidance procedure. Lever-press avoidance has been utilized for decades to study learning, but it also has a history as a prominent model of anxiety (Pearl, 1963; D'Amato & Fazzaro, 1966; Hurwitz & Dillow, 1968; Gilbert, 1971; Dillow et al., 1972; Berger & Brush, 1975). Derived initially from the 2-factor theory of threat/fear motivation and learned avoidance (Mowrer, 1939a; Mowrer, 1939b; Mowrer & Lamoreaux, 1942; Mowrer & Lamoreaux, 1946), the general premise of this approach is that a learned fear of signals is sufficient to support avoidant behavior without requiring a continued re-exposure to the actual noxious stimulus or event. Others have provided alternative interpretations of the development of active avoidance learning. Herrnstein, Hineline, and Sidman all focused on the reduction in shock density over time and a second internal factor (e.g fear or anxiety) need not be required in order to explain the acquisition of avoidance behavior (Sidman, 1962; Herrnstein & Hineline, 1966; Hineline & Herrnstein, 1970). This is an important consideration, for without the theoretical need for an internal state, there is no reason to assume a general state of arousal should be evident in the absence of shock exposure. In short, once asymptotic performance is attained, because of the adaptation of the instrumental response to minimize shock frequency, general arousal should be reduced compared to early acquisition (when shocks are more frequent). Still, others have suggested that there may be another component to this acquired behavior - the attainment of perceived safety (Dinsmoor, 1977; Dinsmoor, 2001). This is an interesting proposition because it also does not require any rumination upon the animal's part to "know" the shock is coming. In this approach, the animal learns to exhibit the behavior because it leads to the attainment of perceived safety, which could be in the form of an explicit stimulus only present during periods of non-threat or simply as the absence of the warning signal. At the foundation of this theoretical discussion is a fundamental difference in the view of how animals perceive learning: a molecular (trial by trial, stimulus by stimulus) or molar (general state) analysis (Hineline, 2001; Bersh, 2001). One could argue that lingering changes in general arousal outside of the avoidance learning context may reflect overall changes in the animals that would be proposed by molar analysis theory.

2.2 Avoidance susceptibility as a model of anxiety vulnerability

As mentioned above, it is well documented that approximately 10% of those people who experience a significant trauma develop PTSD; therefore, there has been recent interest in identifying vulnerability factors that cause some proportion of the public to be susceptible to developing PTSD symptoms. From a learning-diathesis approach, vulnerability for developing anxiety disorders comes from differences in acquiring associations. People self-ascribed as being behaviorally inhibited, as well as the rat model of behavioral inhibition (the WKY rat), exhibit quicker acquisition of classically conditioned responses (Ricart et al., 2011b; Myers et al., 2011; Beck et al., 2011). In addition, females also exhibit enhanced susceptibility to acquire associations. This is reflected in faster acquisition of predictive relationships (classical conditioning) (Spence & Spence, 1966; Wood & Shors, 1998; Shors et al., 1998; Holloway et al., 2011) and behavioral reactions to stimuli (instrumental learning) (Van Oyen et al., 1981; Heinsbroek et al., 1983; Heinsbroek et al., 1987; Saavedra et al., 1990; Dreher et al., 2007; Dalla et al., 2008; Lynch, 2008), the 2 primary components of avoidance learning (Mowrer & Lamoreaux, 1946). Based on these characteristics, it is not surprising that both female sex and behaviorally inhibited temperament are associated with a greater susceptibility to acquire active avoidance behaviors (Beck et al., 2010; Beck et al., 2011). As shown in Figure 3, male SD rats are slowest to acquire a lever-press avoidance response, compared to their same-strain female counterparts and WKY rats of both sexes. Interestingly, the relationships between the 4 groups change during extinction with male WKY rats extinguishing slower than both female groups and male SD rats. Females of both strains in this study both acquired and extinguished at the same rate.

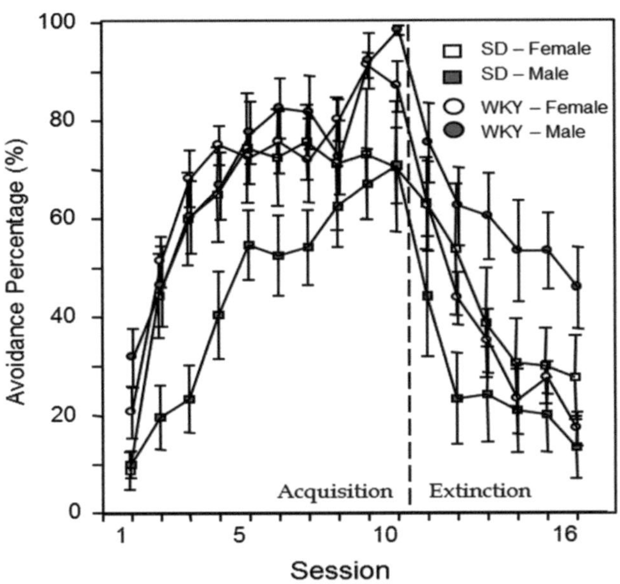

Fig. 3. Avoidance susceptibility can be observed by comparing rates active avoidance responses are acquired. In this example, WKY rats acquire a lever-press avoidance behavior quicker than male SD rats, main effect Strain $F(1, 36) = 9.0$, $p < .005$. Female SD rats also acquired the behavior quicker than male SD rats, Sex x Session $F(9, 324) = 2.6$, $p < .01$. Following session 10, the shock was removed in order to assess extinction of the avoidance response. In general, WKY rats extinguish the response slower than SD rats, but female SD rats were slower than male SD rats extinguishing the behavior, whereas male WKY rats were slower to extinguish the response compared to female WKY rats, Strain x Sex $F(1, 36) = 6.0$, $p < .02$ and Sex x Session $F(5, 179) = 2.2$, $p < .05$.

3. Behavioral avoidance as a precursor for increased arousal

For some time, the acquisition and performance of avoidance behavior was used as a tool to increase arousal in studies of physiological responsiveness in monkeys (e.g. stress-induced hypertension) because it was obvious to the investigators that control over the stimuli did not necessarily lead to a reduction in arousal (Forsyth, 1968; Forsyth, 1969; Natelson et al., 1976; Natelson et al., 1977). Since that time, avoidance learning fell out of favor as such a tool and was generally replaced by inescapable stressor paradigms. Therefore, with our rodent model of acquisition and extinction of active avoidance behavior, we questioned whether the process of acquiring avoidant behavior would influence general arousal outside of the avoidance-training context, which was not the case for some of the monkey studies (Forsyth, 1968; Forsyth, 1969).

Having established strain and sex differences in the acquisition and extinction of active avoidance, as well as differences in innate reactivity between strains, the question became whether the process of acquiring avoidant behavior would influence general arousal outside of the avoidance-training context. There are three possible periods of time startle reactivity may show changes as a function of acquiring lever-press avoidance and each would have associated with it a different theory of how the learning procedure was affecting general sensory reactivity. First, based on the above inescapable shock model, one could hypothesize that startle reactivity should be increased within days of the first few training trials, following the sessions the rats experience the most shock. Second, if the development of avoidant behavior follows the trajectory of developing anxiety, then one could hypothesize that startle reactivity should increase over acquisition. Yet, there is also a third option. That is, startle reactivity could increase if the association between the signals and the consequence becomes less certain. In this third possibility, startle reactivity could be increased if there is a change in the relationship between the signals that represent threat and the consequences following acquisition (such as conducting extinction trials). Another consideration is that only certain animals may be affected in a way that increases their general arousal. Strain differences in both acquiring the avoidant behavior and resistance to extinguish it may be a sign of anxiety vulnerability that could also be reflected in a change in general arousal (reflected as a persistent change in startle reactivity).

3.1 Acquisition of active avoidance and changes in startle reactivity

There are several examples of shock-induced changes in various behavioral indexes of anxiety-like reactions outside of the shock-exposure context (Servatius et al., 1994; Servatius et al., 1995; Beck et al., 2002; Cordero et al., 2003; Beck & Servatius, 2005; Manion et al., 2007; Daviu et al., 2010; Manion et al., 2010), which may lead one to assume that stressors which cause pain have a particularly significant role in causing context-independent changes in general arousal. However, in the case of acquiring behavior that is conducive to active avoidance of shock, the acute role of shock exposure in early acquisition can be contrasted with the expression of stimulus control during asymptotic performance levels. This is an important distinction in any avoidance-based model of PTSD, since the clinical condition does not necessarily involve an acute increase in arousal (as defined by startle reactivity), but the development of avoidance does parallel the general worsening of symptoms (O'Donnell et al., 2007). The implication of this correlation is that other symptoms, such as hyperarousal, may come as a result of increasing stimulus control, as active avoidance coping strategies strengthen.

Male SD and WKY rats were trained to acquire a lever-press, active avoidance behavior (or served as non-trained homecage controls), and were tested weekly to assess changes in startle sensitivity and responsivity. In concert with prior studies, male WKY rats demonstrated greater baseline startle magnitudes and equivalent startle sensitivity. Subsequently, acquisition of avoidance appeared to be equivalent between the strains. In addition, WKY rats reached greater asymptotic avoidance performance, as seen in previous studies (see Figure 4).

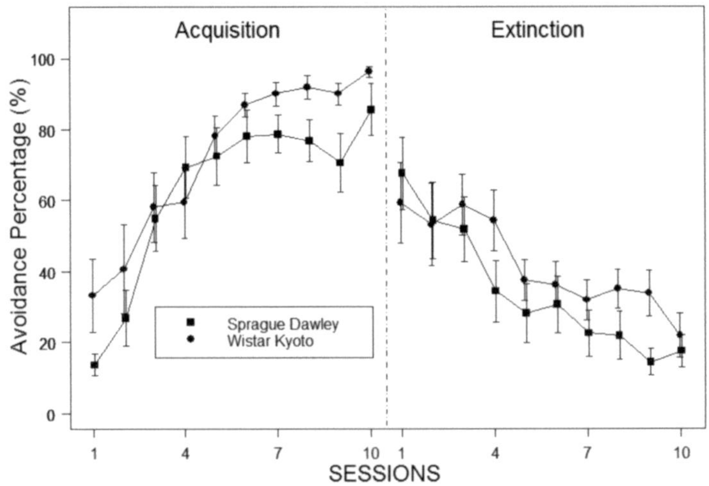

Fig. 4. Lever-press behavior to avoid intermittent footshock was reliably acquired in both strains of male rats (main effect Session, $F(9, 135) = 25.0$, $p < .001$ and Session x Trial interaction, $F(171, 2394) = 1.4$, $p < .001$), but WKY rats exhibited more avoidance responses per session in the later sessions of the acquisition phase. During extinction, differences between strains were not apparent across sessions, but they were significant within sessions (Strain x Trial interaction, $F(19, 266) = 3.3$, $p < .001$.

Increases in startle sensitivity and responsivity are expected early in acquisition if shock exposure causes changes in vigilance and arousal, but, if learning and performing an avoidance behavior causes anxiety, startle measures should be elevated during the later weeks of acquisition. Thus, as shown in Figure 5, there was a strain-independent elevation in startle sensitivity displayed by those being trained in avoidance behavior. This difference is evident from the beginning of acquisition, when animals are receiving the most of shocks, and dissipates by the end of acquisition. Conversely, startle responsivity, as demonstrated by relative increases in startle magnitudes above baseline, is largely unchanged early in acquisition, when enhanced startle sensitivity is greatest. Startle responsivity in avoidance-trained animals increased toward the end of acquisition, during the refinement of avoidance behavior when few (if any) shocks are being received. Therefore, as startle sensitivity differences dissipate (possibly a sign of normalizing vigilance), enhancements in responsivity appear (a possible sign of increased arousal).

Historically, female rats are known to generally acquire discrete lever-press avoidance, as well as other active avoidance behaviors, better than their male counterparts (Beatty & Beatty, 1970; Scouten et al., 1975; Van Oyen et al., 1981; Heinsbroek et al., 1983; Beck et al.,

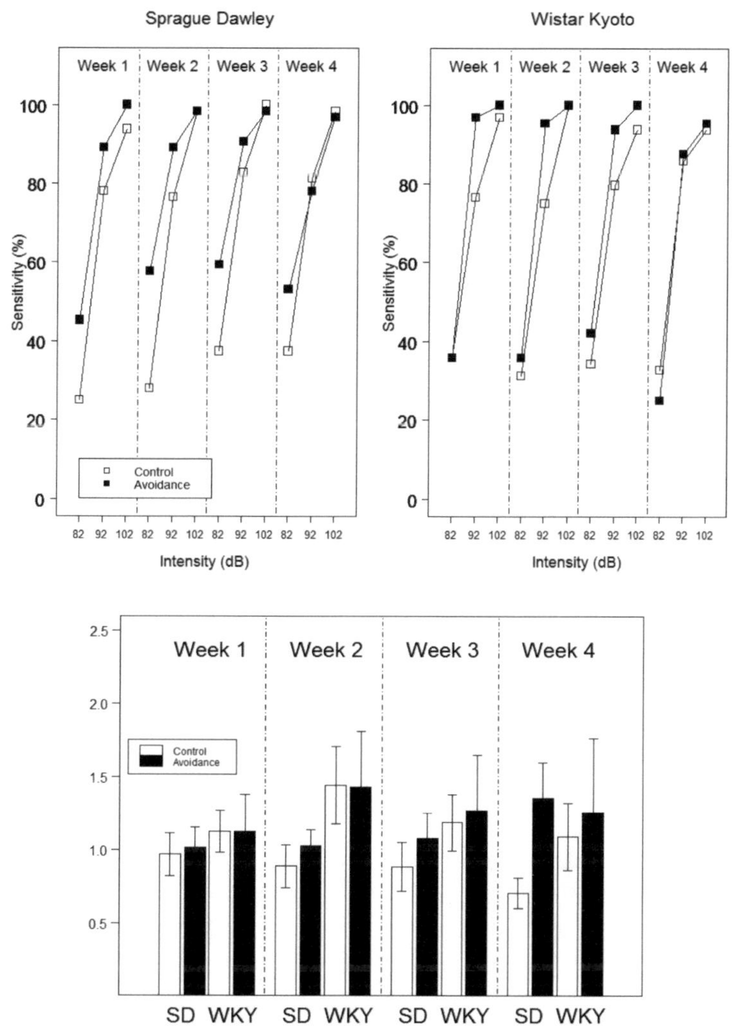

Fig. 5. Exposure to avoidance learning was associated with an increase in startle sensitivity in both strains of male rats (main effect Avoidance, $F(1, 28) = 4.2$, $p < .05$. Otherwise, significant differences across the three stimulus intensities used (main effect Intensity $F(2, 56) = 595.5$, $p < .001$) were coupled with a marginally significant difference in the percentage of startles elicited across the 4 weeks of acquisition (Avoidance x Week interaction $F(3, 84) = 2.4$, $p < .07$. By the fourth week of avoidance acquisition, differences in startle sensitivity between the avoidance-trained group and the homecage control group were greatly reduced. Although not statistically significant, signs of increased startle magnitude began to become apparent in SD rats by the last week of acquisition.

2010; Beck et al., 2011), suggesting that they may exhibit greater changes in general arousal, as indexed by startle reactivity measures, with the experience of shock and/or the acquisition of the avoidant behavior. Using the same procedures as described for the male rats, female rats demonstrated similar patterns of initial startle reactivity: greater startle magnitudes in WKY rats but equivalent startle sensitivity across strain. During avoidance training, female rats exhibited rapid acquisition of the lever-press avoidance response (see Figure 6). However, the effect avoidance had on arousal was different than the effects observed in male rats of these strains. As shown in Figure 7, startle sensitivity was transiently elevated in SD rats training in avoidance learning, but WKY rats exhibited much less elevations in sensitivity early in acquisition. In stark contrast, at the end of acquisition, WKY rats trained in avoidance were showing a reduction in startle sensitivity. Thus, while female SD rats demonstrate an enhancement in sensitivity similar to male rats, WKY rats demonstrate a unique decrease in reactivity compared to their untrained controls in the later phase of acquisition. Also, unlike the male rats, no consistent changes in startle responsivity were observed in SD or WKY females as was seen in male rats.

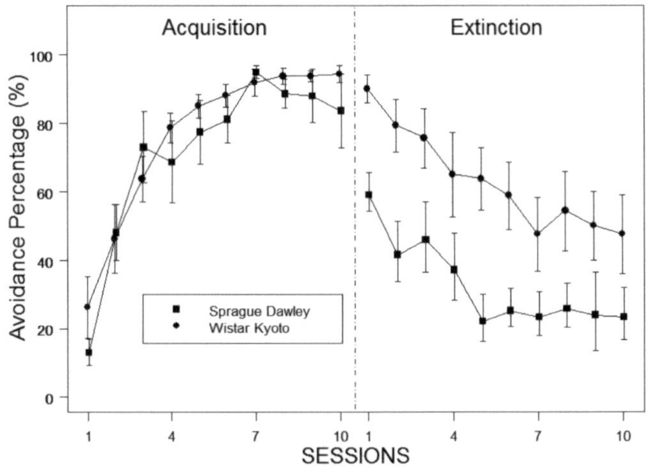

Fig. 6. Lever-press behavior to avoid intermittent footshock was reliably acquired in both strains of female rats (main effect Session, $F(9, 135) = 33.4$, $p < .001$ and Session x Trial interaction, $F(171, 2394) = 1.5$, $p < .001$). During extinction, differences between strains became evident within the first session and continued throughout the extinction phase of the experiment (main effects, Strain, $F(1, 14) = 7.4$, $p < .05$, Session, $F(9, 135) = 13.7$, $p < .001$, and Trial, $F(19, 266) = 10.9$, $p < .001$).

These data suggest there are different aspects to an avoidance-induced change in vigilance and/or general arousal (as reflected by an increase in startle sensitivity and responsivity, respectively). Increases in startle sensitivity generally occurred proximal to experiences with periodic foot-shock during the early phase of acquisition (when rats are slowly transitioning from a majority of escape responses to an increasing number of avoidance responses). This pattern appears consistent, albeit to varying degrees, across both sexes of each strain, and suggests experience with a painful stimulus is increasing vigilance in those animals. However, strain and sex differences become evident as avoidance responses occur in a greater majority. Of note are the changes in startle responsivity in male SD rats and startle sensitivity in female WKY rats. These groups exhibited divergent changes, with male SD rats exhibiting enhanced startle responsivity and female WKY rats exhibiting decreased startle sensitivity. The enhancement in startle responsivity, as a possible index of general arousal, observed in male SD rats is akin to the described usage of avoidance decades ago, where daily sessions of non-cued avoidance (i.e. Sidman avoidance) over several months

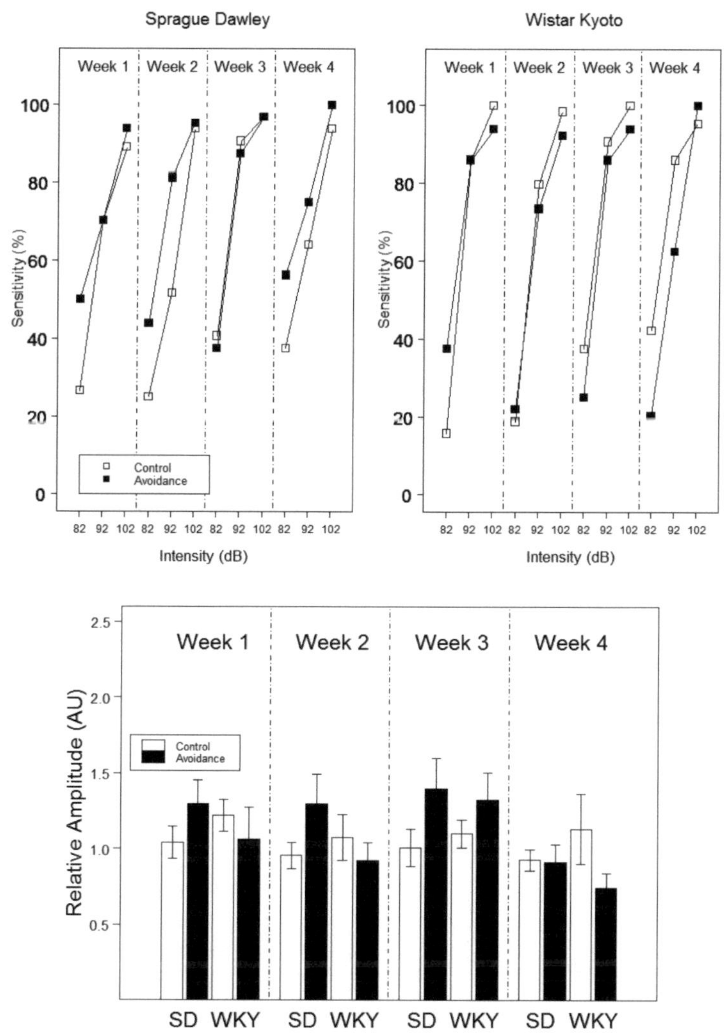

Fig. 7. Exposure to avoidance learning was associated with strain and intensity-dependent changes in startle sensitivity. Early acquisition showed elevations in the startles elicited at the lowest intensity. For SD females this continued into the second week and expanded to the middle intensity. Differences within the SD strain declined over the later two weeks, but, at the same time, a decrease in startle sensitivity became apparent in the female WKY rats. These impressions were confirmed by significant Strain x Avoidance, $F (1, 28) = 5.2$, $p < .03$, Strain x Intensity, $F (2, 56) = 11.3$, $p < .001$, and Avoidance x Week x Intensity, $F (6, 168) = 3.0$, $p < .007$, interactions. Startle magnitudes did not significantly differ between groups across the four weeks of acquisition.

caused reported increases in agitation towards the researchers and chronic increases in mean systolic and diastolic blood pressure in well-avoiding monkeys (Forsyth, 1969). In contrast, the reduction in startle sensitivity, in female WKY rats, is suggestive of an avoidance-induced reduction in general vigilance. Might this suggest female WKY rats adapt better to a stressful environment? If so, would these changes in vigilance and arousal be maintained in the absence of the actual threat? On the other hand, is the reduction in startle responsivity evidence for a difference in the underlying associations made during the acquisition of the avoidance behavior? Continued assessment of startle sensitivity and responsivity during the extinction phase should provide further evidence for or against these interpretations.

3.2 Extinction of active avoidance, removal of avoidance context and persistent changes in startle reactivity

As shown in Figure 4, unlike previous studies, there was minimal difference in the extinction rates of male SD and WKY rats; still the WKY rats extinguished slightly slower. Nonetheless, the resultant effects on the indexes of startle were rather clear. As with the end of the acquisition phase, startle sensitivity did not appreciably change during the extinction phase; however, differences in startle responsivity grew in appearance (see Figure 8). Differences first observed at the end of acquisition in the male SD rats continued to be present during the first two weeks of extinction. Interestingly, by the end of the extinction phase, avoidance-trained WKY rats were also exhibiting greater startle responsivity than their non-trained counterparts. This suggests the arousal displayed by male SD rats is contingent upon emitting a certain level of avoidance behavior, even in the absence of shock. In contrast, the male WKY rats show differences as they extinguish the avoidant behavior. This may reflect an increase in arousal due to the slow abandonment of the avoidant behavior. Following this logic, the male WKY rats could have perceived that their behavior does not control the presence or absence of the shock anymore; the result is an increase in general arousal during extinction.

The female rats exhibited a substantial difference in their rates of extinction (see Figure 6), with female WKY rats extinguishing much slower than female SD rats. This may be attributable to differences in how the females of these strains extinguish the response, since both groups exhibited very similar acquisition rates and attained a similar asymptotic level of responding. However, under similar avoidance learning conditions we have not observed such a difference (see Figure 3). The one difference in procedure from our previous experiments is the addition of the weekly startle tests. This may be an example of the vigilance/arousal test influencing performance on subsequent avoidance acquisition/extinction session days. The results of the startle test showed that intra-strain differences in the startle sensitivity continued from the end of avoidance acquisition. As shown in Figure 9, female WKY rats showed reduced responding, which eventually normalized through the extinction period. Female SD rats still showed signs of increased startle sensitivity early in the extinction phase. These data suggest that elevated vigilance continues to be expressed in female SD rats for a period of time even in the absence of shock. Conversely, the decreased vigilance in female WKY rats, attributed to the reduction in shock exposure, continues, as the shocks remain absent in extinction, then eventually normalize.

Following extinction training, all rats were allowed to remain in their home cages for a number of weeks, with startle measures taken every week, as during acquisition and extinction. These sessions allowed for the measurement of long-term changes in arousal and vigilance following avoidance training. In male rats, the elevations in startle responsivity returned to baseline. Thus changes in arousal that were evident during avoidance training did not persist following cessation of training. The same was seen in female rats, where the

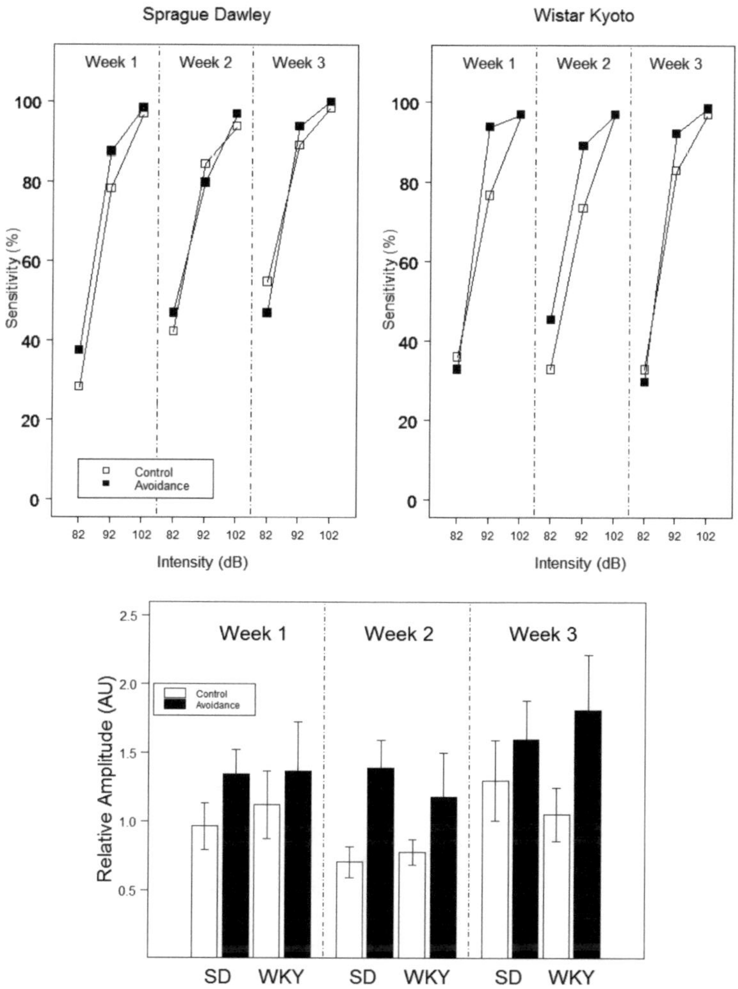

Fig. 8. During the extinction phase, avoidance-trained male SD and WKY rats did not show significantly different startle sensitivity measures compared to their same-strain controls. However, differences in startle magnitude continued into the extinction phase for those male SD rats that had been trained in avoidance behavior. In addition, in the midst of extinction sessions, startle magnitudes of male WKY rats were elevated compared to their homecage control counterparts (main effects of Avoidance, $F (1, 28) = 4.1$, $p < .05$ and Week, $F (2, 56) = 10.5$, $p < .001$).

decreases in startle sensitivity seen in WKY female rats did not persist. Innate differences in startle behavior, however, remained. WKY rats of both sexes demonstrated higher startle responsivity compared to SD rats.

Since these between-group differences in startle responsivity are not observed during the subsequent month where they remained in the home-cage, these results suggest there is a connection between displaying some level of active-avoidance behavior, on a regular basis, and persistent changes in vigilance or general arousal. For males, differences were observed

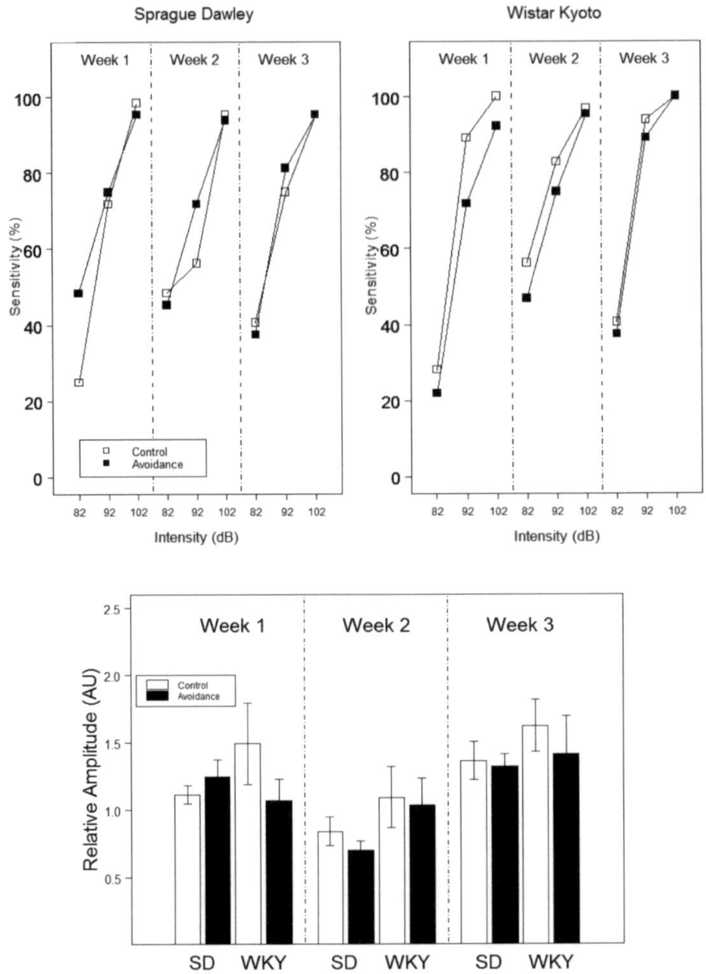

Fig. 9. Exposure to avoidance learning was associated with strain and intensity-dependent changes in startle sensitivity in female rats that persisted into the extinction phase. For SD rats startle sensitivity was still elevated during the first week of extinction. In a similar fashion, female WKY rats this continued to show a reduction in startle sensitivity into the initial week of extinction. These impressions were confirmed by significant Strain x Avoidance, $F(1, 28) = 4.6$, $p < .04$, Strain x Intensity, $F(2, 56) = 6.5$, $p < .002$, and Week x Intensity, $F(4, 112) = 12.6$, $p < .001$, interactions. There were no significant differences during the extinction phase measures of startle responsivity attributable to prior avoidance learning.

during extinction (in both strains) suggesting that the behavior itself (not actual response to shocks) may be sufficient to induce a state of general arousal. Herein lies the possible connection to the growth of general PTSD symptoms with the trajectory of avoidance symptoms (O'Donnell et al., 2007). Similarly, the persistent elevation in vigilance observed in female SD rats can also fit the description of PTSD. It may be that avoidance learning causes changes in the brain systems underlying fundamental defensive behaviors. For example, since the males learned to manipulate the environment to avoid noxious stimuli, maybe defensive reactions are enhanced; the animal is more "at-the-ready". Increases in vigilance may cause the female SD rats to perceive their environment in a more apprehensive manner. Interestingly, inescapable shock causes a transient reduction in SD startle responsivity without affecting startle sensitivity (Beck et al., 2002; Beck & Servatius, 2005). With this in mind, it was surprising to see reductions in startle sensitivity in the female WKY rats, although we should acknowledge it was long after the type of shock experienced during inescapable shock that reduces female SD startle responsivity. We could speculate that this lack of sensitivity to the startle pulses in the WKY females is the counterpoint to "increased vigilance", being post-stress "numbness", but we do not have any other data to substantiate such a claim. Additional testing across other modalities may give us a better idea of the scope of this reduction in reactivity that appears to be correlated with high levels of active avoidance behavior.

These data provide an interesting example of how startle reactivity can be enhanced by prior exposure to an escapable and avoidable stressor. Moreover, as was observed following inescapable stress (Servatius et al., 1995; Beck et al., 2002; Manion et al., 2007; Manion et al., 2010), the presentation of enhanced startle reactivity in male rats did not occur proximal to any period of significant shock exposure. This finding is important for 2 reasons. First, it shows that inescapable and uncontrollable stress is not necessary to increase startle reactivity, and yet, the appearance of the startle enhancement is still delayed following predictable and controllable shocks. Second, these features are suggestive that a mechanism not specifically triggered by the shock is causing: 1) general arousal to increase over time in male rats; 2) vigilance to remain elevated in female SD rats; and 3) vigilance to be reduced in female WKY rats (possibly a transient numbing effect). It is important to also note the differences in these patterns across subject groups can be translated into different symptoms associated with PTSD: arousal, vigilance, and numbness. These group differences may help us understand why different individuals present with certain symptoms yet, in total, still constitute a PTSD diagnosis.

3.3 Active avoidance and general physiology

Reactivity to stressors may be further characterized by their effect on the general physiology of the rats. Growth, as measured by changes in bodyweight from the beginning of the experiment, provides a measure to investigate the effects of avoidance on general physiology, and to see if correlations exist between these changes and startle reactivity. In male SD rats, growth was suppressed during acquisition, but recovered by the termination of extinction training (see Figure 10). In WKY rats, the exact opposite effect was observed. Differences in bodyweight slowly emerged across training, with decreased growth in avoidance trained rats becoming evident in extinction and further developing following the end of training. Thus while male SD rats demonstrate a decrease in growth proximal to shock administration, WKY rats demonstrate a continual decrement in growth that developed after the removal of the shock. In females, SD rats demonstrated a similar pattern to male WKY rats, with differences in growth emerging over training. In female WKY rats, no differences in growth are seen at any point during training. Thus it appears that changes in bodyweight, often thought to reflect physiological reactions to stress, do not mimic all of the changes observed in startle sensitivity and responsivity across all groups; although, changes in startle responsivity in male rats do somewhat follow periods of lower body weight (especially in WKY rats).

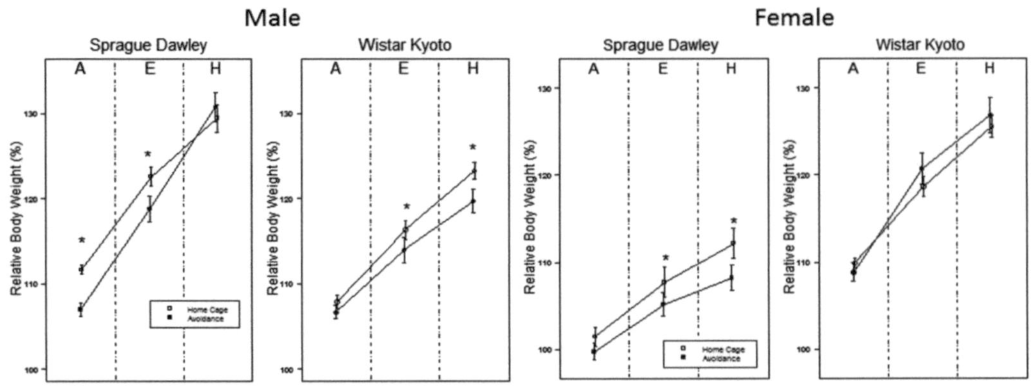

Fig. 10. Relative body weight of male and female SD and WKY rats. Data are averaged into phases: A-Acquisition, E-Extinction, and H-Home Cage. Relative body weights were determined by dividing average weight during a phase by the weight during the pre-training startle session. Male SD rats gain more relative weight than male WKY rats. Male SD rats that underwent avoidance demonstrated suppressed growth during acquisition and extinction. Male WKY rats demonstrated a diverging growth pattern; rats that underwent avoidance learning gained less weight than their home cage controls, an effect that did not emerge until extinction and persisted into the home cage phase. Female WKY rats gained more relative weight than female SD rats. Female SD rats that underwent avoidance demonstrated suppressed growth during extinction and the home cage phase relative to home cage controls. These differences in growth were not observed in female WKY rats. *$p < .05$

3.4 Active avoidance, shock exposure and changes in startle reactivity

The confounding variables of shock exposure controllability and exposure to shock limited our ability to make firm conclusions regarding what caused the enhanced startle responses to occur in avoidance-trained rats. Hence, we designed a follow-up study that substituted an additional control group for our baseline comparison of avoidance-trained rats. This control group was placed in the training boxes at the same time the others were being trained to avoid the shock. The rats in this new condition were each paired to a rat in the avoidance-training condition such that when an avoidance rat was shocked, so was the paired control (yoked condition). Thus, the yoked rats in this experiment heard, saw, and felt the same stimuli as their avoidance learning paired counterparts, but the lever in their chambers was disabled. For these yoked controls, they may learn the predictive relationship between the stimuli and the shocks, but they will not learn or experience any perceived control over the occurrence of shock.

As shown in Figure 11, the rats of both strains exhibited a clear acquisition of a lever-press avoidance response. WKY rats exhibited a much higher level of asymptotic performance than did SD rats. Furthermore, the subsequent extinction of the response was less apparent in WKY rats compared to SD rats. The analysis of startle sensitivity found that both strains of rats trained in avoidance learning increased their sensitivity to the acoustic startle pulses the day following the first training session, as did the yoked controls from baseline. This confirms that the short-term change in startle sensitivity is a product of shock exposure, not the acquisition process involved in learning to escape or avoid the shocks. Moreover, the

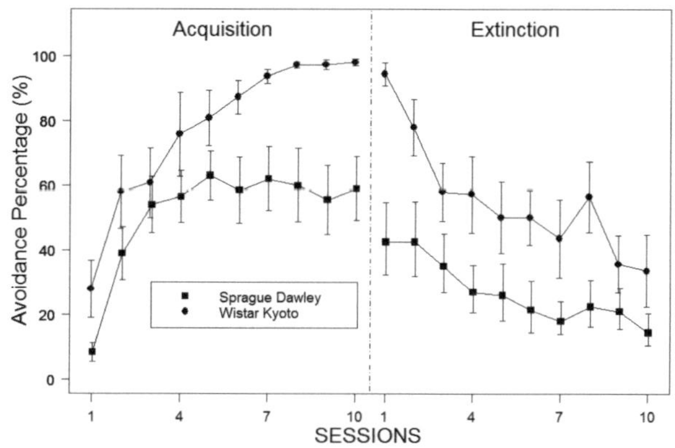

Fig. 11. Avoidance performance of SD and WKY rats. Strain designations are in the figure legend. WKY rats acquire the avoidance response faster and reach higher asymptotic performance than SD rats (main effect, Strain, $F(1, 15) = 9.2$, $p < .01$ and Session x Trial interaction, $F(171, 2565) = 1.3$, $p < .001$. WKY rats also extinguished the response less than SD rats (main effect, Strain, $F(1, 15) = 7.0$, $p < .01$, Strain x Trial, $F(19, 285) = 2.1$, $p < .01$ and Session x Trial, $F(171, 2565) = 1.4$, $p < .01$ interactions).

effect is not strain dependent and dissipates as the rats are exposed to fewer shocks. With respect to startle magnitude, WKY rats exhibited higher startle magnitudes than SD rats (see Figure 12). In SD rats, no differences in relative startle magnitudes were observed between avoidance and yoked rats, but in WKY rats, elevations in startle responsivity developed late

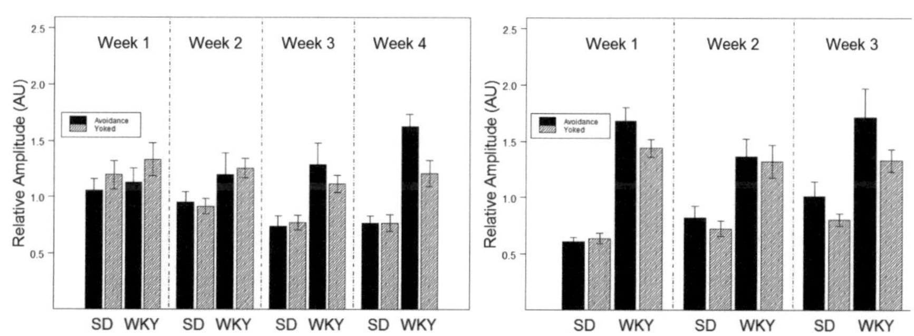

Fig. 12. Differences in startle magnitude emerged between the avoidance and yoked-shock group during the later phase of acquisition in male WKY rats. This impression was confirmed by a significant Strain x Week interaction, $F(3, 90) = 6.1$, $p < .001$ and a marginally significant Avoidance x Week interaction, $F(3, 90) = 2.3$, $p < .07$. Unlike the previous experiment, male SD rats did not attain a level of avoidance performance proximal to that of the WKY rats, and failed to develop enhanced startle responsivity over training (left). The differences in startle magnitude between avoidance and yoke-control male WKY rats continued during the extinction phase of the experiment (right). This impression was confirmed by a significant Strain x Week interaction, $F(2, 60) = 7.5$, $p < .001$ and a marginally significant Avoidance x Week interaction, $F(2,60) = 2.5$, $p < .08$. Two of the three weekly startle tests found higher startle magnitudes between those trained in avoidance behavior and their yoked controls in the WKY strain.

in acquisition in both avoidance and yoked rats. These differences persisted into extinction and following the termination of training. Because each yoked rat experienced the same stimuli as an avoidance-trained rat, they should have learned the predictability of the stimuli. However, what is not learned is that they have any controllability. The poorer learning by the avoidance-trained SD rats in the yoked-avoidance experiment would conform to this idea. Because they did not acquire the response to the same level as those in the earlier experiment, this may have led them to also fail to show any difference in startle responsivity from their yoked controls as training progressed. Observed differences in avoidance-trained WKY rats suggest they too may have similar increased arousal with the adoption and use of active avoidance behavior.

4. Conclusions

Active avoidance requires attentiveness, memory, and, arguably, anxiety (Mowrer, 1940; Miller, 1948; Solomon & Wynne, 1954; Hoffman & Fleshler, 1962; Dinsmoor, 1977). Others have utilized active avoidance procedures as a tool to study chronic stress effects on physiology (Forsyth, 1968; Forsyth, 1969; Natelson et al., 1976; Natelson et al., 1977), but an explicit connection between the behavioral symptoms of PTSD needed to be tested. In both experiments, an initial increase in startle sensitivity occurred in response to the initial experiences with shock. This supports our contention that an acute increase in startle sensitivity occurs as a function of shock exposure - not startle responsivity. In other words, the level of vigilance is changed for a period of time following shock exposures. Conversely, startle responsivity, general arousal, increases as the avoidance response is refined and can even persist into extinction sessions. This supports our and others' past work that found inescapable shock only increases startle reactivity after some period of time has passed since the acute exposure to the shock (in the range of days)(Manion et al., 2007; Servatius et al., 1995; Beck et al., 2002), unless the subject under study is female (Beck & Servatius, 2005; Beck et al., 2008). Thus, we also replicated that sex difference by failing to find any increases in startle responsivity in female rats following shock exposure (Beck & Servatius, 2005; Beck et al., 2008).

One could surmise that the strain and sex differences in startle sensitivity and responsivity during later acquisition and through extinction occur as a function of differences in learning controllability over their general environment. Controllability over stressful situations cause different long-term effects, but these effects are commonly thought to be regarded as a means to buffer the deleterious effects of stress (Baratta et al., 2007). This work questions that general assumption, and, in fact, suggests for some individuals diagnosed with PTSD, hyperarousal could be the result of being overly controlling and not "testing the waters" to see if perceived warning signals still predict aversive events. This appears to be the male pattern, which may vary in expression across strain, but is most consistent in the WKY rats. For the females, some have proposed emotional numbing in PTSD patients is actually a result of hyperarousal to negative valenced stimuli being contrasted to a lack of arousal to positive stimuli (Litz et al., 1997; Litz & Gray, 2002). Our data do not support that relationship, but the reduction in startle sensitivity in female WKY rats appears to parallel documented lower startle reactions in female PTSD patients with comorbid depressive symptoms (Medina et al., 2001). WKY rats have been touted as a model for depression and maybe this sex-specific response reflects some aspect of those characteristics (Pare & Redei, 1993). Overall, these divergent changes in arousal and vigilance are likely a bi-product of strain and sex differences in learning processes that are involved in forming predictive associations under conditions where some level of stress is involved (Wood & Shors, 1998; Ricart et al., 2011a; Ricart et al., 2011b; Beck et al., 2011). When the requirement to cope is brought to the fore, these inherent differences in learning processes can be seen both in their rate to acquire an avoidance coping response (i.e. gaining control) and any subsequent resistance to cease that response (maintaining control).

The problems associated with anxiety disorders, such as PTSD, are multifaceted and variable. In part, this is because different individuals perceive and cope with stressors differently. Active-avoidance behavior, using a lever-press, is rarely uniform within a group of animals, and, as evidenced from the data from our lab; the resulting effects on startle responses can be variable in when they emerge over time. Yet, the variability caused by having a subject-controlled manipulation of stressor exposure is important for understanding the disorder. Controllability may selectively influence certain individuals in a manner that causes increases in general arousal whereas others are not so affected. Moreover, when we consider vulnerability factors, such as demonstrated by the WKY rats (behavioral inhibition and higher baseline startle responses), individual differences in coping with stressor and different rates of acquisition of avoidant strategies should occur – as in the human condition. Gaining an understanding of the relationship between different symptoms, as demonstrated here between avoidance and arousal, will provide us with the knowledge to broaden our expectations for how different populations may develop the symptoms of PTSD. As shown here, our data suggest that the acute experience of pain is not sufficient to immediately increase startle responsivity, and it may not be a good marker for tracking the development of PTSD. Our data suggests that the avoidance process is already well-acquired when this other symptom becomes evident. As has been suggested from the clinical literature, increased expression of avoidance may be a very good marker for tracking the development of the disorder (Karamustafalioglu et al., 2006; O'Donnell et al., 2007). Further, breaking the adoption and utilization of avoidance strategies may lead to a reduction in general arousal (at least in males); therefore, some non-pharmacological therapeutic approaches (e.g. cognitive-behavioral therapy) may have beneficial effects on these two core features of PTSD. Additional research is required to better understand and track the developmental course of symptom expression in different subpopulations (e.g. women) such that our animal model systems can be better tailored to reflect the cascade of changes occurring in those people, especially for those at risk for developing the symptoms of PTSD.

5. Acknowledgements

The presented work was supported by grants from the Department of Veterans Affairs Office of Biomedical Laboratory Research & Development (1I01BX000218) and the Department of Defense Psychological Health/Traumatic Brain Injury Research Program of the Office of the Congressionally Directed Medical Research Programs (W81XWH-08-2-0657) to KDB. The opinions expressed are those of the authors and do not reflect the official position of the U.S. Army, U.S. Department of Defense, or U.S. Department of Veterans Affairs.

6. References

American Psychiatric Association (2000). Diagnostic and Statistical Manual of Mental Disorders. (4th-TR ed.) Washington, DC: American Psychiatric Association.

Baratta, M. V., Christianson, J. P., Gomez, D. M., Zarza, C. M., Amat, J., Masini, C. V. et al. (2007). Controllable versus uncontrollable stressors bi-directionally modulate conditioned but not innate fear. Neuroscience, 146, 1495-1503.

Beatty, W. W. & Beatty, P. A. (1970). Hormonal determinants of sex differences in avoidance behavior and reactivity to electric shock in the rat. J.Comp Physiol Psychol., 73, 446-455.

Beck, K. D., Brennan, F. X., & Servatius, R. J. (2002). Effects of stress on nonassociative learning processes in male and female rats. Integr.Physiol Behav.Sci., 37, 128-139.

Beck, K. D., Jiao, X., Cominski, T. P., & Servatius, R. J. (2008). Estrus cycle stage modifies the presentation of stress-induced startle suppression in female Sprague-Dawley rats. Physiol Behav., 93, 1019-1023.

Beck, K. D., Jiao, X., Pang, K. C., & Servatius, R. J. (2010). Vulnerability factors in anxiety determined through differences in active-avoidance behavior. Prog. Neuropsychopharmacol. Biol.Psychiatry, 34, 852-860.

Beck, K. D., Jiao, X., Ricart, T. M., Myers, C. E., Minor, T. R., Pang, K. C. et al. (2011). Vulnerability factors in anxiety: Strain and sex differences in the use of signals associated with non-threat during the acquisition and extinction of active-avoidance behavior. Prog.Neuropsychopharmacol.Biol.Psychiatry, 35, 1659-1670.

Beck, K. D. & Servatius, R. J. (2005). Stress-induced reductions of sensory reactivity in female rats depend on ovarian hormones and the application of a painful stressor. Horm.Behav., 47, 532-539.

Beck, K. D. & Servatius, R. J. (2006). Interleukin-1beta as a mechanism for stress-induced startle suppression in females. Ann.N.Y.Acad.Sci., 1071, 534-537.

Berger, D. F. & Brush, F. R. (1975). Rapid acquisition of discrete-trial lever-press avoidance: Effects of signal-shock interval. Journal of the Experimental Analysis of Behavior, 24, 227-239.

Bersh, P. J. (2001). The molarity of molecular theory and the molecularity of molar theory. J.Exp.Anal.Behav., 75, 348-350.

Bouton, M. E. & Bolles, R. C. (1980). Conditioned fear assessed by freezing and by the suppression of three different baselines. Animal Learning & Behavior, 8, 429-434.

Butler, R. W., Braff, D. L., Rausch, J. L., Jenkins, M. A., Sprock, J., & Geyer, M. A. (1990). Physiological evidence of exaggerated startle response in a subgroup of Vietnam veterans with combat-related PTSD. Am.J.Psychiatry, 147, 1308-1312.

Cordero, M. I., Venero, C., Kruyt, N. D., & Sandi, C. (2003). Prior exposure to a single stress session facilitates subsequent contextual fear conditioning in rats. Evidence for a role of corticosterone. Horm.Behav., 44, 338-345.

D'Amato, M. R. & Fazzaro, J. (1966). Discriminated lever-press avoidance learning as a function of type and intensity of shock. Journal of Comparative and Physiological Psychology, 61, 313-315.

Dalla, C., Edgecomb, C., Whetstone, A. S., & Shors, T. J. (2008). Females do not express learned helplessness like males do. Neuropsychopharmacology, 33, 1559-1569.

Davis, M. (1986). Pharmacological and anatomical analysis of fear conditioning using the fear-potentiated startle paradigm. Behav.Neurosci., 100, 814-824.

Daviu, N., Fuentes, S., Nadal, R., & Armario, A. (2010). A single footshock causes long-lasting hypoactivity in unknown environments that is dependent on the development of contextual fear conditioning. Neurobiol.Learn.Mem., 94, 183-190.

Dillow, P. V., Myerson, J., Slaughter, L., & Hurwitz, H. M. (1972). Safety signals and the aquisition and extinction of lever-press discriminated avoidance in rats. British Journal of Psychology, 63, 583-591.

Dinsmoor, J. A. (1977). Escape, avoidance, punishment: where do we stand? J.Exp.Anal.Behav., 28, 83-95.

Dinsmoor, J. A. (2001). Stimuli inevitably generated by behavior that avoids electric shock are inherently reinforcing. J.Exp.Anal.Behav., 75, 311-333.

Doyle, G. & Yule, E. P. (1959). Grooming activities and freezing behaviour in relation to emotionality in albino rats. Animal Behaviour, 7, 18-12.

Dreher, J. C., Schmidt, P. J., Kohn, P., Furman, D., Rubinow, D., & Berman, K. F. (2007). Menstrual cycle phase modulates reward-related neural function in women. Proc.Natl.Acad.Sci.U.S.A, 104, 2465-2470.

Fanselow, M. S. (1980). Conditioned and unconditional components of post-shock freezing. Pavlov.J.Biol.Sci., 15, 177-182.

Foa, E. B., Stein, D. J., & McFarlane, A. C. (2006). Symptomatology and psychopathology of mental health problems after disaster. J.Clin.Psychiatry, 67 Suppl 2, 15-25.

Forsyth, R. P. (1968). Blood pressure and avoidance conditioning. A study of 15-day trials in the Rhesus monkey. Psychosom.Med., 30, 125-135.

Forsyth, R. P. (1969). Blood pressure responses to long-term avoidance schedules in the restrained rhesus monkey. Psychosom.Med., 31, 300-309.

Gilbert, R. M. (1971). Signal functions in discriminated avoidance behavior. J.Exp.Anal.Behav., 15, 97-108.

Guthrie, R. M. & Bryant, R. A. (2005). Auditory startle response in firefighters before and after trauma exposure. Am.J.Psychiatry, 162, 283-290.

Heinsbroek, R. P., Van Oyen, H. G., van de Poll, N. E., & Boer, G. J. (1983). Failure of dexamethasone to influence sex differences in acquisition of discriminated lever press avoidance. Pharmacol.Biochem.Behav., 19, 599-604.

Heinsbroek, R. P., van, H. F., Zantvoord, F., & van de Poll, N. E. (1987). Sex differences in response rates during random ratio acquisition: effects of gonadectomy. Physiol Behav., 39, 269-272.

Herrnstein, R. J. & Hineline, P. N. (1966). Negative reinforcement as shock-frequency reduction. J.Exp.Anal.Behav., 9, 421-430.

Hineline, P. N. (2001). Beyond the molar-molecular distinction: we need multiscaled analyses. J.Exp.Anal.Behav., 75, 342-347.

Hineline, P. N. & Herrnstein, R. J. (1970). Timing in free-operant and discrete-trial avoidance. J.Exp.Anal.Behav., 13, 113-126.

Hirshfeld, D. R., Rosenbaum, J. F., Biederman, J., Bolduc, E. A., Faraone, S. V., Snidman, N. et al. (1992). Stable behavioral inhibition and its association with anxiety disorder. J.Am.Acad.Child Adolesc.Psychiatry., 31, 103-111.

Hirshfeld-Becker, D. R., Micco, J., Henin, A., Bloomfield, A., Biederman, J., & Rosenbaum, J. (2008). Behavioral inhibition. Depress.Anxiety., 25, 357-367.

Hitchcock, J. M. & Davis, M. (1987). Fear-potentiated startle using an auditory conditioned stimulus: effect of lesions of the amygdala. Physiol Behav., 39, 403-408.

Hoffman, H. S. & Fleshler, M. (1962). The course of emotionality in the development of avoidance. J.Exp.Psychol., 64, 288-294.

Holloway, J. L., Beck, K. D., & Servatius, R. J. (2011). Facilitated acquisition of the classically conditioned eyeblink response in females is augmented in those taking oral contraceptives. Behav.Brain Res., 216, 301-307.

Hurwitz, H. M. & Dillow, P. V. (1968). The effects of the warning signal on response characteristics in avoidance learning. Psychological Record, 18, 351-360.

Jiao, X., Pang, K. C., Beck, K. D., Minor, T. R., & Servatius, R. J. (2011). Avoidance perseveration during extinction training in Wistar-Kyoto rats: An interaction of innate vulnerability and stressor intensity. Behav.Brain Res., 221, 98-107.

Jovanovic, T., Norrholm, S. D., Blanding, N. Q., Davis, M., Duncan, E., Bradley, B. et al. (2010). Impaired fear inhibition is a biomarker of PTSD but not depression. Depress.Anxiety., 27, 244-251.

Kagan, J., Reznick, J. S., & Gibbons, J. (1989). Inhibited and uninhibited types of children. Child Dev., 60, 838-845.

Kagan, J., Reznick, J. S., & Snidman, N. (1987). The physiology and psychology of behavioral inhibition in children. Child Dev., 58, 1459-1473.

Karamustafalioglu, O. K., Zohar, J., Guveli, M., Gal, G., Bakim, B., Fostick, L. et al. (2006). Natural course of posttraumatic stress disorder: a 20-month prospective study of Turkish earthquake survivors. J.Clin.Psychiatry, 67, 882-889.

Litz, B. T. & Gray, M. J. (2002). Emotional numbing in posttraumatic stress disorder: current and future research directions. Aust.N.Z.J.Psychiatry, 36, 198-204.

Litz, B. T., Schlenger, W. E., Weathers, F. W., Caddell, J. M., Fairbank, J. A., & LaVange, L. M. (1997). Predictors of emotional numbing in posttraumatic stress disorder. J.Trauma Stress., 10, 607-618.

Lynch, W. J. (2008). Acquisition and maintenance of cocaine self-administration in adolescent rats: effects of sex and gonadal hormones. Psychopharmacology (Berl), 197, 237-246.

Manion, S. T., Figueiredo, T. H., roniadou-Anderjaska, V., & Braga, M. F. (2010). Electroconvulsive shocks exacerbate the heightened acoustic startle response in stressed rats. Behav.Neurosci., 124, 170-174.

Manion, S. T., Gamble, E. H., & Li, H. (2007). Prazosin administered prior to inescapable stressor blocks subsequent exaggeration of acoustic startle response in rats. Pharmacol.Biochem.Behav., 86, 559-565.

Medina, A. M., Mejia, V. Y., Schell, A. M., Dawson, M. E., & Margolin, G. (2001). Startle reactivity and PTSD symptoms in a community sample of women. Psychiatry Res., 101, 157-169.

Merikangas, K. R., Avenevoli, S., Dierker, L., & Grillon, C. (1999). Vulnerability factors among children at risk for anxiety disorders. Biol.Psychiatry, 46, 1523-1535.

Miller, N. E. (1948). Studies of fear as an acquirable drive: I. Fear as motivation and fear-reduction as reinforcement in the learning of new responses. Journal of Experimental Psychology, 38, 89-101.

Mowrer, O. H. (1939a). A stimulus-response analysis of anxiety and its role as a reinforcing agent. Psychological Review, 46, 553-566.

Mowrer, O. H. (1939b). Anxiety and learning. Psychological Bulletin, 36, 517-518.

Mowrer, O. H. (1940). Anxiety-reduction and learning. Journal of Experimental Psychology, 27, 497-516.

Mowrer, O. H. & Lamoreaux, R. R. (1942). Avoidance conditioning and signal duration--a study of secondary motivation and reward. Psychological Monographs, 54, No. 5, 34.

Mowrer, O. H. & Lamoreaux, R. R. (1946). Fear as an intervening variable in avoidance conditioning. Journal of Comparative Psychology, 39, 29-50.

Myers, C. E., VanMeenen, K. M., McAuley, J. D., Beck, K. D., Pang, K. C. H., & Servatius, R. J. (2011). Behaviorally-inhibited temperament is associated with severity of PTSD symptoms and faster eyeblink conditioning in veterans. Stress, accepted manuscript.

Natelson, B. H., Kotchen, T. A., Stokes, P. E., & Wooten, G. F. (1977). Relationship between avoidance-induced arousal and plasma DBH, glucose and renin activity. Physiol Behav., 18, 671-677.

Natelson, B. H., Krasnegor, N., & Holaday, J. W. (1976). Relations between behavioral arousal and plasma cortisol levels in monkeys performing repeated free-operant avoidance sessions. J.Comp Physiol Psychol., 90, 958-969.

O'Donnell, M. L., Elliott, P., Lau, W., & Creamer, M. (2007). PTSD symptom trajectories: from early to chronic response. Behav.Res.Ther., 45, 601-606.

Orr, S. P., Lasko, N. B., Shalev, A. Y., & Pitman, R. K. (1995). Physiologic responses to loud tones in Vietnam veterans with posttraumatic stress disorder. J.Abnorm.Psychol., 104, 75-82.

Orr, S. P., Metzger, L. J., & Pitman, R. K. (2002). Psychophysiology of post-traumatic stress disorder. Psychiatric Clinics of North America, 25, 271-293.

Pardon, M. C., Gould, G. G., Garcia, A., Phillips, L., Cook, M. C., Miller, S. A. et al. (2002). Stress reactivity of the brain noradrenergic system in three rat strains differing in their neuroendocrine and behavioral responses to stress: implications for susceptibility to stress-related neuropsychiatric disorders. Neuroscience, 115, 229-242.

Pare, W. P. (1994). Open field, learned helplessness, conditioned defensive burying, and forced-swim tests in WKY rats. Physiol Behav., 55, 433-439.

Pare, W. P. & Redei, E. (1993). Sex differences and stress response of WKY rats. Physiol Behav., 54, 1179-1185.

Pearl, J. (1963). Intertrial interval and acquisition of a lever press avoidance response. Journal of Comparative and Physiological Psychology, 56, 710-712.

Pellow, S., Chopin, P., File, S. E., & Briley, M. (1985). Validation of open:closed arm entries in an elevated plus-maze as a measure of anxiety in the rat. J.Neurosci.Methods, 14, 149-167.

Pole, N., Neylan, T. C., Otte, C., Henn-Hasse, C., Metzler, T. J., & Marmar, C. R. (2009). Prospective prediction of posttraumatic stress disorder symptoms using fear potentiated auditory startle responses. Biol.Psychiatry, 65, 235-240.

Ricart, T. M., De Niear, M. A., Jiao, X., Pang, K. C., Beck, K. D., & Servatius, R. J. (2011a). Deficient proactive interference of eyeblink conditioning in Wistar-Kyoto rats. Behav.Brain Res., 216, 59-65.

Ricart, T. M., Jiao, X., Pang, K. C., Beck, K. D., & Servatius, R. J. (2011b). Classical and instrumental conditioning of eyeblink responses in Wistar-Kyoto and Sprague-Dawley rats. Behav.Brain Res., 216, 414-418.

Saavedra, M. A., Abarca, N., Arancibia, P., & Salinas, V. (1990). Sex differences in aversive and appetitive conditioning in two strains of rats. Physiol Behav., 47, 107-112.

Scouten, C. W., Groteleuschen, L. K., & Beatty, W. W. (1975). Androgens and the organization of sex differences in active avoidance behavior in the rat. J.Comp Physiol Psychol., 88, 264-270.

Servatius, R. J., Jiao, X., Beck, K. D., Pang, K. C., & Minor, T. R. (2008). Rapid avoidance acquisition in Wistar-Kyoto rats. Behav.Brain Res., 192, 191-197.

Servatius, R. J., Ottenweller, J. E., Bergen, M. T., Soldan, S., & Natelson, B. H. (1994). Persistent stress-induced sensitization of adrenocortical and startle responses. Physiol Behav., 56, 945-954.

Servatius, R. J., Ottenweller, J. E., & Natelson, B. H. (1995). Delayed startle sensitization distinguishes rats exposed to one or three stress sessions: further evidence toward an animal model of PTSD. Biol.Psychiatry, 38, 539-546.

Shalev, A. Y., Freedman, S., Peri, T., Brandes, D., Sahar, T., Orr, S. P. et al. (1998). Prospective study of posttraumatic stress disorder and depression following trauma. Am.J.Psychiatry, 155, 630-637.

Shors, T. J., Lewczyk, C., Pacynski, M., Mathew, P. R., & Pickett, J. (1998). Stages of estrous mediate the stress-induced impairment of associative learning in the female rat. Neuroreport, 9, 419-423.

Sidman, M. (1962). Classical avoidance without a warning stimulus. J.Exp.Anal.Behav., 5, 97-104.

Solomon, R. L. & Wynee, L. C. (1954). Traumatic avoidance learning: the principles of anxiety conservation and partial irreversibility. Psychol.Rev., 61, 353-385.

Spence, K. W. & Spence, J. T. (1966). Sex and anxiety differences in eyelid conditioning. Psychological Bulletin, 65, 137-142.

Van Oyen, H. G., Walg, H., & van de Poll, N. E. (1981). Discriminated lever press avoidance conditioning in male and female rats. Physiol Behav., 26, 313-317.

Wood, G. E. & Shors, T. J. (1998). Stress facilitates classical conditioning in males, but impairs classical conditioning in females through activational effects of ovarian hormones. Proc.Natl.Acad.Sci.U.S.A, 95, 4066-4071.

Yehuda, R., McFarlane, A. C., & Shalev, A. Y. (1998). Predicting the development of posttraumatic stress disorder from the acute response to a traumatic event. Biol.Psychiatry, 44, 1305-1313.

5

Peritraumatic Distress in Accident Survivors: An Indicator for Posttraumatic Stress, Depressive and Anxiety Symptoms, and Posttraumatic Growth

Daisuke Nishi[1,2], Masato Usuki[1,2,4] and Yutaka Matsuoka[1,3]
[1]National Disaster Medical Center,
[2]Japan Science and Technology Agency,
[3]National Center for Neurology and Psychiatry,
[4]Kyushu University,
Japan

1. Introduction

In 1997, the Global Burden of Disease Study (Murray, 1997) predicted that by 2020 motor vehicle accident would be the third biggest contributor to worldwide burden of disease. With more than 50 million people reported in 2007 to be injured each year in road traffic accidents worldwide (Derriks & Mark, 2007), motor vehicle accidents are indeed contributing highly to burden of disease. Moreover, such accidents are regarded as one of the leading causes of posttraumatic stress disorder in today's world. As advances in injury care systems have increased the number of seriously injured people who are able to survive their injuries (MacKenzie et al., 2006), this has drawn increasing attention to psychiatric morbidity after injury among such survivors.
Recent studies have shown that accident-related posttraumatic stress disorder is fairly common. The prevalence of posttraumatic stress disorder determined by structured clinical interviews with injured patients consecutively admitted to the intensive care unit or emergency department ranges from 5–30% at 0–3 months after injury to 2–23% at 4–12 months after it (Bryant et al., 2010; Hamanaka et al., 2006; Hepp et al., 2008; Matsuoka et al., 2008; Matsuoka, Nishi, Yonemoto, Nakajima et al., 2010; O'Donnell et al., 2004; Schnyder, Moergeli, Klaghofer et al., 2001; Schnyder et al., 2008; Shalev et al., 1998). Recent large epidemiological studies using questionnaires have reported a 17–23% point prevalence of clinically significant posttraumatic stress disorder symptoms at 4–12 months after injury (Zatzick et al., 2007; Mayou et al., 2001). It is well known that this disorder can be associated with higher psychiatric comorbidity, attempted suicide, and physical illnesses such as asthma, hypertension, and peptic ulcer (Davidson et al., 1991), as well as carry high healthcare costs (O'Donnell et al., 2005; Walker et al., 2003). It remains, therefore, a serious public health problem that needs to be addressed (Kessler et al., 1995; Kessler et al., 2005.)

1.1 Depression and other anxiety disorders after motor vehicle accidents

Major depression is also highly prevalent in individuals injured in a motor vehicle accident. The prevalence of major depression as determined by structured clinical interviews ranges from 10–19% at 0–3 months after the accident (O'Donnell et al., 2004; Matsuoka et al., 2008; Shalev et al., 1998) to 10–14% at 4–12 months after it (O'Donnell et al., 2004; Shalev et al., 1998). Although many symptoms overlap between posttraumatic stress disorder and major depression, the high comorbidity cannot be explained solely by this (Franklin & Zimmerman, 2001). Exposure to traumatic events has been shown to be linked not only to posttraumatic stress disorder, but also to depression (Duncan et al., 1996; Kilpatrick et al., 1987), and a recent study suggested that traumatic experiences during young adulthood and middle age are strong predictors of anxiety and depression among older adults (Dulin & Passmore, 2010). The treatment of psychiatric morbidity after injury is thus a matter of some urgency, especially for high-risk individuals. However, as it is difficult for emergency department staff to screen patients early after the event using a conventional questionnaire-based tool, given the large number of motor vehicle accident survivors they handle (Nishi et al., 2006), it is desirable to find indicators for posttraumatic stress disorder which can be easily assessed in order to provide preventive strategies as early as possible.

1.2 The importance of assessing peritraumatic distress

Among the indicators for posttraumatic stress disorder, peritraumatic distress is a good candidate for screening individuals at high risk of developing the disorder. Peritraumatic stress can enhance trauma-related memory and sensitize the neurobiological systems (Charney et al., 1993), which links to the development of posttraumatic stress disorder. Many clinical studies and a meta-analysis have shown that perceived threat to life is a predictor of posttraumatic stress disorder (Holbrook et al., 2001; Matsuoka et al., 2008; Ozer et al., 2003; Schnyder, Moergeli, Trentz et al., 2001) and psychiatric morbidity (Matsuoka et al., 2008; Schnyder, Moergeli, Trentz et al., 2001). Peritraumatic distress is also linked with posttraumatic growth, which Tedeschi & Calhoun (2004) define as the positive psychological change experienced as a result of the struggle with highly challenging life circumstances. They state that only psychologically 'seismic' events shake the assumptive world, which leads to posttraumatic growth. Accordingly, peritraumatic distress can be an indicator for posttraumatic growth. A better understanding of peritraumatic distress would be significant for both prevention, especially in emergency settings, and treatment of posttraumatic stress disorder.

The aim of this chapter is to elucidate the predictive usefulness of peritraumatic distress and to examine the future directions for prevention with a focus on the use of the Peritraumatic Distress Inventory, an assessment tool for peritraumatic distress.

2. Method

2.1 Participants

Participants were selected from the Tachikawa Cohort of Motor Vehicle Accidents study conducted at the National Disaster Medical Center in Tokyo, Japan (Matsuoka et al., 2009). The inclusion criteria in the present study were as follows: 1) motor vehicle accident-related severe physical injury causing a life-threatening or critical condition; 2) age between 18 and 69 years; and 3) native Japanese speaking ability. The exclusion criteria were the following:

1) diffuse axonal injury, brain contusion, and subdural and subarachnoidal bleeding detected by either computed tomography or magnetic resonance imaging or both (with the exception of concussion), because the presence of traumatic brain injury creates considerable difficulties when assessing psychological responses to injury; 2) cognitive impairment, defined as a score of <24 on the Mini Mental State Examination; 3) currently suffering from schizophrenia, bipolar disorder, drug dependence or abuse, or epilepsy before the accident; 4) marked serious symptoms such as suicidal ideation, self-harm behavior, dissociation, or a severe physical condition preventing the patient from tolerating the interview; and 5) living or working at a location more than 40 km from the National Disaster Medical Center.

The above-mentioned study was conducted between 30 May 2004 to 8 January 2008, and the present study is part of that larger study. Patients with motor vehicle accident-related physical injury were consecutively admitted to the intensive care unit of the National Disaster Medical Center between 18 August 2005 and 8 January 2008. Of the 221 patients who met the inclusion criteria, 189 agreed to participate in the study. Fifty-nine patients were excluded because their peritraumatic distress could not be assessed due to memory loss. Ultimately, 130 patients participated in this study.

2.2 Procedures

The study protocol was approved by the Institutional Review Board and Ethics Committee of the National Disaster Medical Center. After providing a complete description of the study to the subjects, written informed consent was obtained from them. The median number of days between the motor vehicle accident and the initial assessment was 2 days (range, 0–23 days). The initial assessment was conducted after cognitive function was assessed by a trained research nurse or psychiatrist using the Mini Mental State Examination.

In a structured interview, data was collected on general socio-demographics, the motor vehicle accident in detail, injury severity score (Baker & O'Neill, 1976), Glasgow Coma Scale score (Teasdale & Jennett, 1974), status during the accident (e.g., vehicle driver), vital signs first recorded on admission to the emergency room, lifestyle, and family history of psychopathology. Also, the Peritraumatic Distress Inventory was conducted at initial assessment. Follow-up assessments were performed at 1 month (median, 37 days, range, 24-76 days) and 18 months (median, 561.5 days, range, 442-700 days) after the accident. The Impact of Event Scale-Revised and the Hospital Anxiety and Depression Scale were conducted at 1 month post accident, and the Posttraumatic Growth Inventory was conducted at 18 month post accident. The participants were asked to visit the National Disaster Medical Center or to return the completed self-report questionnaires in a stamp-addressed envelope. After each assessment, participants were given a gift voucher for their participation (1,000 JPY [12 USD]).

2.3 Measures

2.3.1 The Peritraumatic Distress Inventory

The Peritraumatic Distress Inventory is a 13-item self-report questionnaire which assesses not only any threat to life experienced but various emotional responses experienced during and immediately after a critical incident (Brunet et al., 2001). Responses are provided on a 5-point Likert scale ranging from 0 to 4 (0, not at all to 4, extremely true). It typically takes only several minutes to complete all of the items, meaning the Inventory can be used immediately after a motor vehicle accident.

The original Peritraumatic Distress Inventory has been demonstrated to be internally consistent, stable over time, and with good to excellent correlations between item and total scores (Brunet et al., 2001). Moreover, it was found to be valid against posttraumatic symptoms and peritraumatic dissociation as assessed by the Impact of Event Scale-Revised and the civilian version of the Mississippi Scale for Combat-Related Posttraumatic Stress Disorder.

With the original authors' permission, we translated the original English Peritraumatic Distress Inventory into Japanese. We followed the standard procedure of back-translation. Namely, the first author (DN) translated the English version into Japanese. This preliminary Japanese version was then backtranslated into English by an independent translator. The backtranslated version was examined by the original authors. Then we corrected the Japanese translation accordingly. This process was repeated until both sets of authors agreed that the original and backtranslated versions matched closely. Subsequently, we verified the internal consistency, test-retest reliability, concurrent validity with measures of peritraumatic dissociation and posttraumatic symptoms, and divergent validity of the Japanese version of the Peritraumatic Distress Inventory (Nishi et al., 2009).

2.3.2 The Impact of Event Scale-Revised

The posttraumatic stress symptoms as assessed using the Impact of Event Scale- Revised at follow-up were considered to be the outcome. The Impact of Event Scale-Revised is a 22-item self-report questionnaire used to determine the level of symptomatic responses to a specific traumatic stressor (motor vehicle accident in the present study) in the past week (Asukai et al., 2002; Wolfe & Kimerling, 1997). The degree of distress for each item is rated on a 5-point scale (0, not at all to 4, extremely; range, 0-88).

2.3.3 The Hospital Anxiety and Depression Scale

Depressive and anxiety symptoms as assessed using the Hospital Anxiety and Depression Scale were also considered as the outcome. The Scale is comprised of a 7-item anxiety subscale and a 7-item depression subscale that assess general psychological distress for the preceding week (Kugaya et al., 1998; Zigmond & Snaith, 1983). Each item is rated on a scale of 0–3, with high scores denoting greater psychological distress (range, 0-42).

2.3.4 The Posttraumatic Growth Inventory

The Posttraumatic Growth Inventory, which assesses posttraumatic growth, measures the degree of change experienced in the aftermath of a traumatic event. The 21-item Inventory evaluates five factors: relating to others, new possibilities, personal strength, spiritual change, and appreciation of life. The degree of posttraumatic growth for each item is rated on a 6-point scale (range, 0-105) (Taku et al., 2007; Tedeschi & Calhoun, 1996).

2.4 Statistical analysis

Univariate regression analysis was used to examine the relationships of total score and individual item scores on the Peritraumatic Distress Inventory with posttraumatic symptoms and depressive and anxiety symptoms. In a model for determining the predictive value of the Peritraumatic Distress Inventory, multivariate regression analysis was used to examine the relationships of the Peritraumatic Distress Inventory with posttraumatic stress symptoms and depressive and anxiety symptoms adjusted for 7 other covariates based on the following theoretical considerations.

For the covariates, age at motor vehicle accident, being female, history of psychiatric illness, family history of psychopathology, and lower education level are well-established pretraumatic risk factors across trauma type (Brewin et al., 2000; Ozer et al., 2003). As for educational level, we used graduation from junior high school as a reference (0), and assigned 1 to graduation from high school, 2 to graduation from junior or technical college, and 3 to graduation from university or higher educational institutions according to the Japanese educational system. Heart rate on admission was selected because some reports in the literature on motor vehicle accident showed its association with posttraumatic stress disorder (Bryant et al., 2000; Shalev et al., 1998; Zatzick et al., 2005). Injury Severity Score divided into 10–point increments was assigned as the objective accident-related variable. Injury Severity Score is a scoring system that provides a total score for patients with multiple injuries, and it correlates with measures of severity such as mortality and hospital stay (Baker & O'Neill, 1976).

Univariate regression analysis was also conducted to examine the relationships of total score on the Peritraumatic Distress Inventory with total score and individual subscale scores on the Posttraumatic Growth Inventory. Any association between the dependent variable and the independent variable was expressed as a regression coefficient (beta weight) and quantified by the 95% confidence interval (95% CI).

All statistical analyses used two-tailed tests. Statistical significance was established at a P value < 0.05. All data analyses were performed using SPSS statistical software version 19.0J for Windows (SPSS, Tokyo, Japan).

3. Results

Of the 130 patients participating, 79 (60.8%) attended the 1-month follow-up assessment and 51 (39.2%) attended the 18-month one. The patients who dropped out of the study did not differ significantly from those who participated in terms of the variables selected for investigation in this study, including total Peritraumatic Distress Inventory score.

Of the 79 participants at first follow-up, 16 (20.3%) were women and median age was 37.0 years (mean, 39.7; range 18-69), and 7 (8.9%) reported a past history of psychiatric illness. Median ISS was 6.0 (range 1-41) and median Peritraumatic Distress Inventory score was 15.0 (range 0-40).

The relationships of total score and individual item scores on the Peritraumatic Distress Inventory with posttraumatic stress symptoms and depressive and anxiety symptoms are shown in Table 1. The Peritraumatic Distress Inventory was an independent predictor for posttraumatic stress symptoms and depressive and anxiety symptoms after adjusting for potential confounders.

PDI item	IES-R		HADS	
	Beta (95% CI)	P	Beta (95% CI)	P
Univariate regression analysis				
1. I felt helpless to do more	4.00 (2.05, 5.94)	<0.01	1.80 (0.86, 2.74)	<0.01
2. I felt sadness and grief	3.05 (1.03, 5.06)	<0.01	0.94 (-0.06, 1.93)	0.06
3. I felt frustrated or angry I could not do more	2.99 (1.11, 4.87)	<0.01	0.74(-0.20, 1.68)	0.12

PDI item	IES-R		HADS	
4. I felt afraid for my safety	3.02 (1.04, 5.00)	<0.01	1.21 (0.24, 2.17)	0.02
5. I felt guilt that more was not done	1.43 (-0.97, 3.83)	0.24	0.23 (-0.92, 1.38)	0.69
6. I felt ashamed of my reactions	0.76 (-2.47, 3.99)	0.81	0.53 (-1.01, 2.07)	0.50
7. I felt worried about others	1.40 (-0.66, 3.46)	0.18	0.20 (-0.80, 1.19)	0.70
8. I was about to lose control	2.31 (-0.82, 5.45)	0.15	0.07 (-1.45, 1.59)	0.93
9. I had difficulty controlling my bowel and bladder	1.39 (-3.89, 6.67)	0.60	0.84 (-1.68, 3.36)	0.51
10. I was horrified	3.41 (1.46, 5.37)	<0.01	0.99 (0.01, 1.97)	0.047
11. I had physical reactions like pounding heart	4.04 (2.14, 5.93)	<0.01	1.64 (0.71, 2.58)	<0.01
12. I felt I might pass out	0.90 (-1.12, 2.92)	0.38	0.51 (-0.45, 1.48)	0.29
13. I felt I might die	2.63 (0.74, 4.52)	<0.01	1.00 (0.08, 1.92)	0.03
PDI total score	0.61 (0.34, 0.89)	<0.01	0.21 (0.07, 0.35)	<0.01
Multivariate regression analysis*				
PDI total score	0.49 (0.18, 0.80)	<0.01	0.15 (0.00, 0.30)	0.046

*In the multivariate analysis, the predictive value of the PDI was adjusted for 7 covariates; age at MVA, being a female, history of psychiatric illness, family history of psychopathology, education level, heart rate at admission and Injury Severity Score.

P, p value; CI, confidential interval; HADS, Hospital Anxiety and Depression Scale; IES-R, the Impact of Event Scale-Revised; PDI, the Peritraumatic Distress Inventory

Table 1. The predictive value of the Peritraumatic Distress Inventory for Impact of Event Scale-Revised and Hospital Anxiety and Depression Scale at follow-up (N=79)

The relationships between total score on the Peritraumatic Distress Inventory and total score and individual subscale scores on the Posttraumatic Growth Inventory are shown in Table 2.

	PDI total score (independent variable)			
	Beta	95%CI	P value	R^2
PTGI subscales (dependent variables)				
Relating to others	0.25	0.02 – 0.48	0.03	0.09
New possibilities	0.14	-0.02 – 0.30	0.08	0.06
Personal strength	0.06	-0.07 – 0.18	0.36	0.02
Spiritual change	0.07	0.01 – 0.11	0.03	0.10
Appreciation of life	0.20	0.10 – 0.30	<0.01	0.26
PTGI total score	0.72	0.12 – 1.31	0.02	0.11

CI, confidential interval; PDI, Peritraumatic Distress Inventory; PTGI, Posttraumatic Growth Inventory; R^2, multiple correlation coefficient, the index of goodness fitness in the model

Table 2. The predictive value of the PDI for PTGI at follow-up (N=51)

4. Discussion

4.1 Summary in the present study

This study showed that the Peritraumatic Distress Inventory could predict posttraumatic stress and depressive and anxiety symptoms at 1 month after motor vehicle accident and posttraumatic growth at 18 months after the accident. The predictive value of the Peritraumatic Distress Inventory for the Impact of Event Scale-Revised and the Hospital Anxiety and Depression Scale remained after adjusting for covariates in a multivariate regression analysis.

4.2 An indicator for posttraumatic stress symptoms

As mentioned in the Introduction, it is no surprise that the Peritraumatic Distress Inventory predicted posttraumatic symptoms in the present study. Although some previous prospective studies have failed to show that this Inventory is an independent predictor of posttraumatic stress disorder, these studies used the Inventory from 2 weeks (Kuhn et al., 2006) to several months (Birmes et al., 2005; Simeon et al., 2005) following a traumatic event. The time of assessment in the present study was within several days following the traumatic event in most participants, in order to minimize the effects of inaccurate memory over time. It is likely that the Peritraumatic Distress Inventory has a better predictive value when used early after a traumatic event, making it well suited for use in emergency departments.

4.3 An indicator for depressive and anxiety symptoms

The Peritraumatic Distress Inventory also predicted depressive and anxiety symptoms in the present study, although the predictive value for these symptoms was lower than that for posttraumatic symptoms. A previous study reported that posttraumatic stress disorder symptoms were a reliable predictor for depressive symptoms (Erickson et al., 2001). The Impact of Event Scale-Revised is one of the tools used most frequently for measuring posttraumatic stress symptoms; however, the it was intended to assess posttraumatic stress disorder symptoms over the previous 7 days, whereas the Peritraumatic Distress Inventory can be used immediately after motor vehicle accident. Given our findings, the Peritraumatic Distress Inventory seems to be a useful indicator not only for posttraumatic stress disorder but also major depression or other anxiety disorders.

4.4 Two Peritraumatic Distress Inventory items showed high predictive values

Items 1 and 11 of the Peritraumatic Distress Inventory showed higher predictive values for both posttraumatic stress and depressive and anxiety symptoms than other items. Item 1 inquires about helplessness. The author and colleagues previously discussed that non-drivers (passengers, bicyclists, or pedestrians) might be susceptible to subsequent posttraumatic stress disorder and other psychiatric morbidity (Matsuoka et al., 2008). Loss of control in a motor vehicle accident is suggested to be an important risk factor. Regarding item 11, some studies showed that high heart rate shortly after a motor vehicle accident is a predictor for later posttraumatic stress disorder (Bryant et al., 2000; Shalev et al., 1998; Zatzick et al., 2005), although other studies reported that heart rate was not an independent predictor (Buckley et al., 2004; Kraemer et al., 2008) and a review indicated that it cannot be accurately used to identify individuals who are at high risk for later posttraumatic stress disorder (Bryant, 2006). To ask if survivors felt any physical reactions might be a better alternative to predict subsequent psychiatric morbidity.

4.5 An indicator for posttraumatic growth

The present results also suggested that the Peritraumatic Distress Inventory could predict posttraumatic growth, especially the 3 aspects of appreciation of life, spiritual change, and relating to others at 18 months after the accident. Multivariate regression analysis was not used to examine the predictive value of the Peritraumatic Distress Inventory for the Posttraumatic Growth Inventory because predictors for posttraumatic growth are not well established and our sample size was modest; however, the result was consistent with that of previous studies. According to Janoff-Balman, these 3 subscales can best be understood as existential reevaluation (Janoff-Bulman, 2004), and a previous study showed that they had a positive association with posttraumatic stress disorder (Taku et al., 2007). The author and colleagues also showed that appreciation of life and spiritual change were positively correlated with posttraumatic stress disorder symptoms, which can be regarded as signifying coping effort in the face of enduring distress, rather than an outcome of coping success (Nishi, Matsuoka and Kim, 2010). The predictive value of the Peritraumatic Distress Inventory for appreciation of life was quite high in the present study, so managing peritraumatic distress may need specific coping efforts. This would point to the importance of clinicians and researchers identifying and being attentive to the survivor's own meanings and interpretations.

4.6 The potential use of the Peritraumatic Distress Inventory in emergency departments

The author and colleagues previously showed that a cut-off score of 23 on the Peritraumatic Distress Inventory maximized the balance between sensitivity (77%) and specificity (82%) (Nishi, Matsuoka, Yonemoto et al., 2010). Further investigation is required to determine its adequate usage bearing in mind its low positive predictive value (53%). However, the early identification of motor vehicle accident survivors who appear not to be at risk of developing posttraumatic stress disorder is one potential use of the Peritraumatic Distress Inventory because of its high negative predictive value (93%). Given the typical limits on the psychiatric resources available, the Peritraumatic Distress Inventory would likely be a useful indicator for posttraumatic stress disorder and psychiatric morbidity in emergency departments.

4.7 Limitations

This study has some limitations. Firstly, the sample size was modest. Secondly, the attrition rate was relatively high, although the patients who dropped out were not significantly different from those who participated in the follow-up assessments in terms of the Peritraumatic Distress Inventory and other covariates. In an earlier publication, we revealed that the factors of being male, unconscious during MVA, low cooperativeness, and less severe injuries were significant predictors of dropout (Nishi et al., 2008). Participants with less severe injuries did not need to come to the National Disaster Medical Center for treatment after discharge which might have affected the attrition rate. Also, those with low cooperativeness might have been reluctant to continue participating in the study.

5. Future directions

5.1 Consolidation of fear memory

Fear memory, which is the important component of peritraumatic distress, has attracted considerable attention especially preclinically. An excellent review by Ressler & Mayberg (2007) has demonstrated that memories do not immediately become permanent at the time of initial experience but exist in a labile state for at least a period of hours and possibly days, during which time they become consolidated into more permanent memory. During this consolidation, molecular, synaptic, neurotransmitter, and system-level changes occur consecutively (McGaugh, 2000). The neural circuitry implicated in fear memory likely involves complex interactions between the hippocampus, the amygdala, and the medial prefrontal cortex (Nemeroff et al., 2006). Because the hippocampus processes and temporarily stores new memory before transferring labile memory to the cortex for permanent storage (Feng et al., 2001), it may be possible to modulate the consolidation of new fear memories while they are being formed (Pitman & Delahanty, 2005) .

5.2 Role of hippocampal neurogenesis in memory consolidation

A previous study showed that exercise on a running wheel, which promotes neurogenesis, increased the rate of loss of hippocampus-dependent contextual fear memory (Kitamura et al., 2009). The study suggested that the level of hippocampal neurogenesis could be modulated and was associated with a causal relationship between adult neurogenesis and the hippocampus-dependent period of fear memory. It is theoretically possible, therefore, that promoting adult neurogenesis early in the transition period might facilitate the clearance of fear memory from the hippocampus. Modulating memory consolidation would mean that posttraumatic stress disorder could be prevented in the aftermath of a traumatic event.

5.3 Omega-3 fatty acids and hippocampal neurogenesis

Fear consolidation can be blocked by an antagonist of noradrenergic activation, and the effectiveness of beta blockers for secondary prevention of posttraumatic stress disorder has been studied in clinical trials (Pitman et al., 2002; Vaiva et al., 2003). However, as traumatized individuals are not psychiatric patients, daily life-based intervention for the prevention of posttraumatic stress disorder is preferable.

Based on the animal research conducted to date, omega-3 fatty acids are the most promising candidate for dietary intervention in the aftermath of a traumatic event to facilitate adult hippocampal neurogenesis (Beltz, 2007; Calderon & Kim, 2004; Kawakita et al., 2006; Wu et al., 2004, 2008). The possible effects of omega-3 fatty acids on brain structures have also been observed clinically: a significant correlation was found between omega-3 fatty acid consumption and gray matter volume of the amygdala, hippocampus, and anterior cingulated gyrus in healthy adults (Conklin et al., 2007). Conversely, a selective deficit of docosahexaenoic acid was reported in the postmortem frontal cortex of patients with depressive disorder (McNamara et al., 2007). Following discussion of these results in the literature, Matsuoka proposed that promoting hippocampal neurogenesis by omega-3 fatty acid supplementation after trauma could reduce subsequent posttraumatic stress disorder symptoms (Matsuoka, 2011).

Support for the ability of omega-3 fatty acids to minimize subsequent posttraumatic stress disorder symptoms comes from one published but preliminary open trial (Matsuoka, Nishi, Yonemoto, Hamazaki et al., 2010). The author and colleagues recruited 15 consecutive patients admitted to the intensive care unit oft a Japanese general hospital immediately following accidental injury (mostly motor vehicle accidents). Patients received omega-3 fatty acid capsules containing 1,470 mg docosahexaenoic acid and 147 mg eicosapentaenoic acid daily for 12 weeks. The primary efficacy variable was score on the Clinician-Administered Posttraumatic Stress Disorder Scale (CAPS). Omega-3 fatty acid supplementation was well tolerated and resulted in a significantly increased docosahexaenoic acid concentration in erythrocytes. Compared with the hypothetical mean calculated in our previous cohort study (Matsuoka et al., 2009), omega-3 fatty acid supplementation resulted in a significantly reduced mean total score on the Clinician-Administered PTSD Scale (11 vs. 25, $p = 0.03$). This pilot study provided promising support for our hypothesis that omega-3 fatty acid supplementation started shortly after accidental injury may be efficacious in attenuating the symptoms of posttraumatic stress disorder. However, because of the open-label design and the lack of controls, no definitive conclusion could be drawn from the trial and we must wait for the results of an adequately powered randomized controlled trial (ClinicalTrials.gov Identifier: NCT00671099).

6. Conclusion

Peritraumatic distress can be assessed quickly and efficiently by using the Peritraumatic Distress Inventory and is an indicator for posttraumatic stress, depressive and anxiety symptoms, and posttraumatic growth in motor vehicle accident survivors. The Peritraumatic Distress Inventory can be used in the emergency department for early identification of motor vehicle accident survivors who are unlikely to develop posttraumatic stress disorder, and the combination of screening with the Peritraumatic Distress Inventory and supplementation with omega-3 fatty acids might be an effective preventive strategy for posttraumatic stress disorder in motor vehicle accident survivors.

7. References

Asukai, N., Kato, H., Kawamura, N., Kim, Y., Yamamoto, K., Kishimoto, J., Miyake, Y., & Nishizono-Maher, A. (2002). Reliability and validity of the Japanese-language version of the impact of event scale-revised (IES-R-J): four studies of different traumatic events. *J Nerv Ment Dis*, Vol. 190, No. 3, pp. (175-182), 1539-736X

Baker, SP., & O'Neill, B. (1976). The injury severity score: an update. *J Trauma*, Vol. 16, No. 11, pp. (882-825), 1529-8809

Beltz, BS., Tlusty, MF., Benton, JL., & Sandeman, DC. (2007). Omega-3 fatty acids upregulate adult neurogenesis. *Neurosci Lett*, Vol. 415, No. 2, pp. (154-158), 1872-7972

Birmes, PJ., Brunet, A., Coppin-Calmes, D., Arbus, C., Coppin, D., Charlet, JP., Vinnemann, N., Juchet, H., Lauque, D., & Schmitt, L. (2005). Symptoms of peritraumatic and acute traumatic stress among victims of an industrial disaster. *Psychiatr Serv*, Vol. 56, No. 1, pp. (93-95), 1557-9700

Brewin, CR., Andrews, B., & Valentine, JD. (2000). Meta-analysis of risk factors for posttraumatic stress disorder in trauma-exposed adults. *J Consult Clin Psychol*, Vol. 68, No. 5, pp. (748-766), 1939-2117

Brunet, A., Weiss, DS., Metzler, TJ., Best, SR., Neylan, TC., Rogers, C., Fagan, J., & Marmar, C R. (2001). The Peritraumatic Distress Inventory: a proposed measure of PTSD criterion A2. *Am J Psychiatry*, Vol. 158, No. 9, pp. (1480-1485), 1535-7228

Bryant, RA., Harvey, AG., Guthrie, RM., & Moulds, ML. (2000). A prospective study of psychophysiological arousal, acute stress disorder, and posttraumatic stress disorder. *J Abnorm Psychol*, Vol. 109, No. 2, pp. (341-344), 1939-1846

Bryant, RA. (2006). Longitudinal psychophysiological studies of heart rate: mediating effects and implications for treatment. *Ann N Y Acad Sci*, Vol. 1071, pp. (19-26), 1749-6632

Bryant, RA., O'Donnell, ML., Creamer, M., McFarlane, AC., Clark, CR., & Silove, D. (2010). The psychiatric sequelae of traumatic injury. *Am J Psychiatry*, Vol. 167, No. 3, pp. (312-320), 1535-7228

Buckley, B., Nugent, N., Sledjeski, E., Raimonde, AJ., Spoonster, E., Bogart, LM., & Delahanty, DL. (2004). Evaluation of initial posttrauma cardiovascular levels in association with acute PTSD symptoms following a serious motor vehicle accident. *J Trauma Stress*, Vol. 17, No. 4, pp. 317-324, 1573-6598

Calderon, F., & Kim, HY. (2004). Docosahexaenoic acid promotes neurite growth in hippocampal neurons. *J Neurochem*, Vol. 90, No. 4, pp. (979-988), 1471-4159

Charney, DS., Deutch, AY., Krystal, JH., Southwick, SM., & Davis, M. (1993). Psychobiologic mechanisms of posttraumatic stress disorder. *Arch Gen Psychiatry*, Vol. 50, No. 4, 295-305, 1538-3636

Conklin, SM., Gianaros, PJ., Brown, SM., Yao, JK., Hariri, AR., Manuck, SB., & Muldoon, MF. (2007). Long-chain omega-3 fatty acid intake is associated positively with corticolimbic gray matter volume in healthy adults. *Neurosci Lett*, Vol. 421, No. 3, pp. (209-212), 1872-7972

Davidson, JR., Hughes, D., Blazer, DG., & George, LK. (1991). Post-traumatic stress disorder in the community: an epidemiological study. *Psychol Med*, Vol. 21, No. 3, pp. (713-21), 1469-8978

Derriks, HM., & Mark, PM. (2007). *IRTAD special report: underreporting of road traffic casualties*. Ministry of Transport, Public Works and Water Management, The Hague

Dulin, PL., & Passmore, T. (2010) Avoidance of potentially traumatic stimuli mediates the relationship between accumulated lifetime trauma and late-life depression and anxiety. *J Trauma Stress*, Vol. 23, No. 2, pp. (296-299), 1573-6598

Duncan, RD., Saunders, BE., Kilpatrick, DG., Hanson, RF., & Resnick, HS. (1996). Childhood physical assault as a risk factor for PTSD, depression, and substance abuse: findings from a national survey. *Am J Orthopsychiatry*, Vol. 66, No. 3, pp. (437-448), 1939-0025

Erickson, DJ., Wolfe, J., King, DW., King, LA., & Sharkansky, EJ. (2001). Posttraumatic stress disorder and depression symptomatology in a sample of Gulf War veterans: a prospective analysis. *J Consult Clin Psychol*, Vol. 69, No. 1, pp. (41-49), 1939-2117

Feng, R., Rampon, C., Tang, YP., Shrom, D., Jin, J., Kyin, M., Sopher, B., Miller, MW., Ware, CB., Martin, GM., Kim, SH., Langdon, RB., Sisodia, SS., & Tsien, JZ. (2001). Deficient neurogenesis in forebrain-specific presenilin-1 knockout mice is associated with reduced clearance of hippocampal memory traces. *Neuron*, Vol. 32, No. 5, pp. (911-926), 1097-4199

Franklin, CL., & Zimmerman, M. (2001). Posttraumatic stress disorder and major depressive disorder: investigating the role of overlapping symptoms in diagnostic comorbidity. *J Nerv Ment Dis*, Vol. 189, No. 8, pp. (548-551), 1539-736X

Hamanaka, S., Asukai, N., Kamijo, Y., Hatta, K., Kishimoto, J., & Miyaoka, H. (2006). Acute stress disorder and posttraumatic stress disorder symptoms among patients severely injured in motor vehicle accidents in Japan. *Gen Hosp Psychiatry*, Vol. 28, No. 3, pp. (234-241), 1873-7714

Hepp, U., Moergeli, H., Buchi, S., Bruchhaus-Steinert, H., Kraemer, B., Sensky, T., & Schnyder, U. (2008). Post-traumatic stress disorder in serious accidental injury: 3-year follow-up study. *Br J Psychiatry*, Vol. 192, No. 5, pp. (376-383), 1472-1465

Holbrook, TL., Hoyt, DB., Stein, MB., & Sieber, WJ. (2001). Perceived threat to life predicts posttraumatic stress disorder after major trauma: risk factors and functional outcome. *J Trauma*, Vol. 51, No. 2, pp. (287-293), 1529-8809

Janoff-Bulman, R. (2004). Posttraumatic growth: three explanatory models. *Psychological Inquiry*, Vol. 15, pp. (30-34), 1532-7965

Kawakita, E., Hashimoto, M., & Shido, O. (2006) Docosahexaenoic acid promotes neurogenesis in vitro and in vivo. *Neuroscience*, Vol.139, No. 3, pp. (991-997), 1873-7544

Kessler, RC., Chiu, WT., Demler, O., Merikangas, KR., & Walters, EE. (2005). Prevalence, severity, and comorbidity of 12-month DSM-IV disorders in the National Comorbidity Survey Replication. *Arch Gen Psychiatry*, Vol.62, No. 6, pp. (617-627), 1538-3636

Kessler, RC., Sonnega, A., Bromet, E., Hughes, M., & Nelson, CB. (1995). Posttraumatic stress disorder in the National Comorbidity Survey. *Arch Gen Psychiatry*, Vol. 52, No. 12, pp. (1048-1060), 1538-3636

Kilpatrick, DG., Saunders, BE., Veronen, LJ., Best, CL., & Von, JM. (1987). Lifetime prevalence, reporting to police, and psychological impact. *Crime and Delinquency*, Vol. 33, pp. (479-489), 0011-1287

Kitamura, T., Saitoh, Y., Takashima, N., Murayama, A., Niibori, Y., Ageta, H., Sekiguchi, M., Sugiyama, H., & Inokuchi, K. (2009). Adult neurogenesis modulates the hippocampus-dependent period of associative fear memory. *Cell*, Vol. 139, No. 4, pp. (814-827), 1097-4172

Kraemer, B., Moergeli, H., Roth, H., Hepp, U., & Schnyder, U. (2008). Contribution of initial heart rate to the prediction of posttraumatic stress symptom level in accident victims. *J Psychiatr Res*, Vol. 42, No. 2, pp. (158-162), 1879-1379

Kugaya, A., Akechi, T., Okuyama, T., Okamura, H., & Uchitomi, Y. (1998). Screening for psychological distress in Japanese cancer patients. *Jpn J Clin Oncol*, Vol. 28, No. 5, pp. (333-338), 1465-3621

Kuhn, E., Blanchard, EB., Fuse, T., Hickling, EJ., & Broderick, J. (2006). Heart rate of motor vehicle accident survivors in the emergency department, peritraumatic psychological reactions, ASD, and PTSD severity: a 6-month prospective study. *J Trauma Stress*, Vol. 19, No. 5, pp. (735-740), 1573-6598

MacKenzie, EJ., Rivara, FP., Jurkovich, GJ., Nathens, AB., Frey, KP., Egleston, BL., Salkever, DS., & Scharfstein, DO. (2006). A national evaluation of the effect of trauma-center care on mortality. *N Engl J Med*, Vol. 354, No. 4, pp. (366-378), 1533-4406

Matsuoka, Y. (2011). Clearance of fear memory from the hippocampus through neurogenesis by omega-3 fatty acids: a novel preventive strategy for posttraumatic stress disorder? *Biopsychosoc Med*, Vol. 5, No. 1, p. (3), 1751-0759

Matsuoka, Y., Nishi, D., Nakajima, S., Kim, Y., Homma, M., & Otomo, Y. (2008). Incidence and prediction of psychiatric morbidity after a motor vehicle accident in Japan: the Tachikawa Cohort of Motor Vehicle Accident Study. *Crit Care Med*, Vol. 36, No. 1, pp. (74-80), 1530-0293

Matsuoka, Y., Nishi, D., Nakajima, S., Yonemoto, N., Hashimoto, K., Noguchi, H., Homma, M., Otomo, Y., & Kim, Y. (2009). The Tachikawa cohort of motor vehicle accident study investigating psychological distress: design, methods and cohort profiles. *Soc Psychiatry Psychiatr Epidemiol*, Vol. 44, No. 4, pp. (333-340), 1433-9285

Matsuoka, Y., Nishi, D., Yonemoto, N., Hamazaki, K., Hashimoto, K., & Hamazaki, T. (2010). Omega-3 fatty acids for secondary prevention of posttraumatic stress disorder after accidental injury: an open-label pilot study. *Journal of Clinical Psychopharmacology*, Vol. 30, No. 2, pp. (217-219), 1533-712X

Matsuoka, Y., Nishi, D., Yonemoto, N., Nakajima, S., & Kim, Y. (2010). Towards an explanation of inconsistent rates of posttraumatic stress disorder across different countries: infant mortality rate as a marker of social circumstances and basic population health. *Psychother Psychosom*, Vol. 79, No. 1, pp. (56-57), 1423-0348

Mayou, R., Bryant, B., & Ehlers, A. (2001). Prediction of psychological outcomes one year after a motor vehicle accident. *Am J Psychiatry*, Vol. 158, No. 8, pp. (1231-1238), 1535-7228

McGaugh, JL. (2000). Memory-a century of consolidation. *Science*, Vol. 287, No. 5451, pp. (248-51), 1095-9203

McNamara, RK., Hahn, CG., Jandacek, R., Rider, T., Tso, P., Stanford, KE., & Richtand, NM. (2007). Selective deficits in the omega-3 fatty acid docosahexaenoic acid in the postmortem orbitofrontal cortex of patients with major depressive disorder. *Biol Psychiatry*, Vol. 62, No. 1, pp. (17-24), 1873-2402

Murray. CJ., & Lopez, AD. (1997)Alternative projections of mortality and disability by cause 1990-2020: Global Burden of Disease study. Lancet, Vol. 349, No. 9064, pp. (1498-1504), 0099-5355

Nemeroff, CB., Bremner, JD., Foa, EB., Mayberg, HS., North, CS., & Stein, MB. (2006). Posttraumatic stress disorder: a state-of-the-science review. *J Psychiatr Res*, Vol. 40, No. 1, pp. (1-21), 1879-1379

Nishi. D., Matsuoka, Y., Kawase, E., Nakajima, S., & Kim, Y. (2006). Mental health service requirements in a Japanese medical centre emergency department. *Emerg Med J*, Vol. 23, No. 6, pp. (468-469), 1472-0213

Nishi, D., Matsuoka, Y., Nakajima, S., Noguchi, H., Kim, Y., Kanba, S., & Schnyder, U. (2008). Are patients after severe injury who drop out of a longitudinal study at high risk of mental disorder? *Compr Psychiatry*, Vol. 49, No. 4, pp. (393-398), 1532-8384

Nishi, D., Matsuoka, Y., Noguchi, H., Sakuma, K., Yonemoto, N., Yanagita, T., Homma, M., Kanba, S., & Kim, Y. (2009). Reliability and validity of the Japanese version of the Peritraumatic Distress Inventory. *Gen Hosp Psychiatry*, Vol. 31, No. 1, pp. (75-79), 1873-7714

Nishi, D., Matsuoka, Y., & Kim, Y. (2010). Posttraumatic growth, posttraumatic stress disorder and resilience of motor vehicle accident survivors. *Biopsychosoc Med*, Vol. 4, No. 1, p. (7), 1751-0759

Nishi, D., Matsuoka, Y., Yonemoto, N., Noguchi, H., Kim, Y., & Kanba, S. (2010). Peritraumatic Distress Inventory as a predictor of post-traumatic stress disorder after a severe motor vehicle accident. *Psychiatry Clin Neurosci*, Vol. 64, No. 2, pp. (149-156), 1440-1819

O'Donnell, ML., Creamer, M., Pattison, P., & Atkin, C. (2004). Psychiatric morbidity following injury. *Am J Psychiatry*, Vol. 161, No. 3, pp. (507-514), 1535-7228

O'Donnell, ML., Creamer, M., Elliott. P., & Atkin, C. (2005). Health costs following motor vehicle accidents: the role of posttraumatic stress disorder. *J Trauma Stress*, Vol. 18, No. 5, pp. (557-561), 1573-6598

Ozer, EJ., Best, SR., Lipsey, TL., & Weiss, DS. (2003). Predictors of posttraumatic stress disorder and symptoms in adults: a meta-analysis. *Psychol Bull*, Vol. 129, No. 1, pp. (52-73), 1939-1455

Pitman, RK., Sanders, KM., Zusman, RM., Healy, AR., Cheema, F., Lasko, NB., Cahill, L., & Orr, SP. (2002). Pilot study of secondary prevention of posttraumatic stress disorder with propranolol. *Biol Psychiatry*, Vol. 51, No. 2, pp. (189-192). 1873-2402

Pitman, RK., & Delahanty, DL. (2005). Conceptually driven pharmacologic approaches to acute trauma. *CNS Spectr*, Vol. 10, No. 2, pp. (99-106), 1092-8529

Ressler, KJ., & Mayberg, HS. (2007). Targeting abnormal neural circuits in mood and anxiety disorders: from the laboratory to the clinic. *Nat Neurosci*, Vol. 10, No. 9, pp. (1116-1124), 1546-1726

Schnyder, U., Moergeli, H., Klaghofer, R., & Buddeberg, C. (2001). Incidence and prediction of posttraumatic stress disorder symptoms in severely injured accident victims. *Am J Psychiatry*, Vol. 158, No. 4, pp. (594-599), 1535-7228

Schnyder, U., Moergeli, H., Trentz, O., Klaghofer, R., & Buddeberg, C. (2001). Prediction of psychiatric morbidity in severely injured accident victims at one-year follow-up. *Am J Respir Crit Care Med*, Vol. 164, No. 4, pp. (653-656), 1535-4970

Schnyder, U., Wittmann, L., Friedrich-Perez, J., Hepp, U., & Moergeli, H. (2008). Posttraumatic stress disorder following accidental injury: rule or exception in Switzerland? *Psychother Psychosom*, Vol. 77, No. 2, pp. (111-118), 1423-0348

Shalev, AY., Freedman, S., Peri, T., Brandes, D., Sahar, T., Orr, SP., & Pitman, RK. (1998). Prospective study of posttraumatic stress disorder and depression following trauma. *Am J Psychiatry*, Vol. 155, No. 5, pp. (630-637), 1535-7228

Simeon, D., Greenberg, J., Nelson, D., Schmeidler, J., & Hollander, E. (2005). Dissociation and posttraumatic stress 1 year after the World Trade Center disaster: follow-up of a longitudinal survey. *J Clin Psychiatry*, Vol. 66, No. 2, pp. (231-237), 1555-2101

Taku, K., Calhoun, LG., Tedeschi, RG., Gil-Rivas, V., Kilmer, RP., & Cann, A. (2007). Examining posttraumatic growth among Japanese university students. *Anxiety Stress Coping*, Vol. 20, No. 4, pp. (353-367), 1477-2205

Teasdale, G., & Jennett, B. (1974). Assessment of coma and impaired consciousness. A practical scale. *Lancet*, Vol. 2, No. 7872, pp. (81-84), 0099-5355

Tedeschi, RG., & Calhoun, LG. (1996). The Posttraumatic Growth Inventory: measuring the positive legacy of trauma. *J Trauma Stress*, Vol. 9, No. 3, pp. (455-471), 1573-6598

Tedeschi, RG., & Calhoun, LG. (2004). Posttraumatic growth: conceptual foundations and empirical evidence. *Psychological Inquiry*, Vol. 15, pp. (1-18), 1532-7965

Vaiva, G., Ducrocq, F., Jezequel, K., Averland, B., Lestavel, P., Brunet, A., & Marmar, CR. (2003). Immediate treatment with propranolol decreases posttraumatic stress disorder two months after trauma. *Biol Psychiatry*, Vol. 54, No. 9, pp. (947-949), 1873-2402

Walker, EA., Katon, W., Russo, J., Ciechanowski, P., Newman, E., & Wagner, AW. (2003). Health care costs associated with posttraumatic stress disorder symptoms in women. *Arch Gen Psychiatry*, Vol. 60, No. 4, pp. (369-374), 1538-3636

Wolfe, J., & Kimerling, R. (1997). Gender Issues in the Assessment of PTSD, In *Assessing Psychological Trauma and PTSD*, Wilson, JP., & Keane, T M., pp. (399-411), Guilford Press, 1572301627, New York

Wu, A., Ying, Z., & Gomez-Pinilla, F. (2004). Dietary omega-3 fatty acids normalize BDNF levels, reduce oxidative damage, and counteract learning disability after traumatic brain injury in rats. *J Neurotrauma*, Vol. 21, No. 10, pp. (1457-1467), 1557-9042

Wu, A., Ying, Z., & Gomez-Pinilla, F. (2008). Docosahexaenoic acid dietary supplementation enhances the effects of exercise on synaptic plasticity and cognition. *Neuroscience*, Vol. 155, No. 3, pp. (751-759), 1873-7544

Zatzick, DF., Rivara, FP., Nathens, AB., Jurkovich, GJ., Wang, J., Fan, MY., Russo, J., Salkever, DS., & Mackenzie, EJ. (2007). A nationwide US study of post-traumatic stress after hospitalization for physical injury. *Psychol Med*, Vol. 37, No. 10, pp. (1469-1480), 1469-8978

Zatzick, DF., Russo, J., Pitman, RK., Rivara, F., Jurkovich, G., & Roy-Byrne, P. (2005). Reevaluating the association between emergency department heart rate and the development of posttraumatic stress disorder; a public health approach. *Biol Psychiatry*, Vol. 57, No. 1, pp. (91-95), 1873-2402

Zigmond, AS., & Snaith, RP. (1983). The hospital anxiety and depression scale. *Acta Psychiatr Scand*, Vol. 67, No.6, pp. (361-370), 1600-0447

6

Sex Differences in PTSD

Dorte Christiansen[1] and Ask Elklit[2]
[1]Aarhus University, Institute of Psychology,
[2]National Center for Psychotraumatology,
University of Southern Denmark,
Denmark

1. Introduction

Research into the psychological sequalae of trauma originally started out by focusing on two sex-specific trauma populations: male war veterans with "soldier's heart", "shellshock", "battle fatigue", or "war neurosis" and female victims of sexual assault or domestic violence with "rape trauma syndrome" or "battered woman syndrome". It was noted how the flashbacks and nightmares reported by rape survivors were similar to the symptoms reported by war veterans, and several researchers and clinicians started pointing out that these trauma specific syndromes might be more similar than different (Ray, 2008; Van der Kolk, 2007). Finally, it was the large number of male Vietnam veterans and the activities of feminist and student organisations, which led to the inclusion of the first PTSD diagnosis into the American DSM-III in 1980 (American Psychiatric Association, 1980). With the introduction of the PTSD diagnosis, the idea that the male war neurosis and the female rape trauma syndrome were ultimately manifestations of the same disorder was widely accepted. As a result, most research on PTSD has been based on the idea that males and females are traumatised in similar ways, and studies on sex differences in PTSD have primarily focused on examining and explaining sex differences in the prevalence and severity of PTSD, whereas studies on sex differences in the manifestation of PTSD are almost completely absent from this otherwise expanding area of research.
Most literature on sex differences in PTSD uses the terms sex and gender interchangeably. Traditionally, however, the term sex refers to the biological distinction between males and females, whereas gender refers to the much more complex cultural understanding of masculine and feminine gender roles as they are viewed in the context of not only sex, but also culture, subculture, age, race, class, and sexual orientation. Even though many studies on PTSD claim to examine gender differences, most studies have in fact studied sex differences and only few have looked into the effect of masculinity or femininity on PTSD. Although this chapter will focus primarily on sex, we acknowledge that gender is likely to affect the development and maintenance of PTSD in males and females as well. The contribution of sex versus gender based explanations for sex differences in PTSD will be discussed throughout the chapter, although the topic merits a more thorough discussion than is possible here.

2. Sex differences in PTSD prevalence

Males are exposed to more potentially traumatic events (PTE's) than females (Tolin & Foa, 2008), particularly in adolescence and young adulthood (Norris et al., 2007). In the National Comorbidity Survey, 60.7% of males and 51.2% of females reported at least one PTE, and significantly more males than females reported exposure to more than two trauma types (Kessler et al., 1995). Despite the finding that males are more likely to experience a PTE and experience more types of PTE's than females, the female-male ratio in the prevalence of PTSD is approximately 2:1 (Tolin & Foa, 2008) with females reporting higher levels of both re-experiencing, avoidance, and arousal (Ditlevsen & Elklit, 2010). The lifetime prevalence of PTSD is 10.4% for females and 5.0% for males and the conditional risk across trauma types is 20.4% for females and 8.2% for males (Kessler et al., 1995). Sex differences in the prevalence of PTSD become evident early in life, peak in early adulthood, and become weakened with increased age (Ditlevsen & Elklit, 2010; Norris et al., 2002). Across studies, the increased prevalence of PTSD in females compared to males appears to be particularly evident for lifetime PTSD (Tolin & Foa, 2008), indicating that PTSD tends to be of longer duration in females than in males, and in the Detroit Area Survey of Trauma the median time from onset of PTSD to remission was four years for females compared to one year for males (Breslau et al., 1998).

Despite findings that sex differences in PTSD have been found across cultures and thus appear to be culturally persistent, variations regarding how pronounced such sex differences are have been reported. Norris et al. (2001) compared sex differences in a Mexican, an African-American, and an Anglo-American sample and found that sex differences in the prevalence of PTSD were amplified in the Mexican sample and attenuated in the African-American sample, with the Anglo-American sample falling in between (Norris et al., 2001). It has been suggested that sex differences in PTSD are particularly evident in communities that emphasise traditional gender roles (Norris et al., 2007). This suggests that social gender and biological sex are both important in making up such differences. Interestingly, the cross-cultural variations on sex differences appear to be more pronounced for intrusion and avoidance than for arousal symptoms, which are thought to be rooted in biological processes. It is thus possible that sex differences in arousal are primarily related to biological sex, whereas sex differences in avoidance and intrusion are also affected by social gender.

3. Possible explanations for sex differences in the prevalence of PTSD

3.1 Sex differences in exposure to potentially traumatic events

The finding that more females than males develop PTSD has been reported independently of study type, population studied, culture, type of assessment, and other methodological variables (Tolin & Foa, 2008). Thus, the increased prevalence of PTSD in females compared to males does not appear to be simply a product of measurement error or methodological bias. Chung and Breslau (2008) conducted a latent class analysis and found no evidence of differential symptom reporting in males compared to females. This led them to the conclusion that the increased symptomatology reported by females is likely to reflect a substantive difference, rather than a sex-related reporting bias. Instead, it has been suggested that as a result of the different gender roles, which males and females hold in society, men and women are exposed to different stressors on a day-to-day basis (Barnyard & Graham-Bermann, 1993; Ptacek et al., 1992). This structural theory is likely to influence not only the types of traumatic events males and females are exposed to, but also how they generally respond to such events. Thus, one possible explanation for the sex difference in the prevalence and severity of PTSD is that males and females differ in the types of trauma they experience. A well-conducted meta-analysis by Tolin and Foa (2008) found that across studies, more males than females are exposed to accidents, non-sexual assaults, combat or war, disasters, illness, unspecified injuries, and witnessing the death or injury of others. In contrast, more females experience sexual assault and childhood sexual abuse (CSA). However, it should be kept in mind that these findings are based simply on whether or not the subjects report having been exposed to the different types of PTE's. Males and females may not only be subject to different reporting bias, but may also differ in the number of times they have been exposed to each event, and such differences are unlikely to be identified in this type of meta-analysis. Thus, even though females are less likely overall to be subjected to non-sexual assaults, overall, it is possible that females who are exposed tend to be assaulted repeatedly, such as is often the case in domestic violence. It remains to be seen, whether such differences in multiple exposures to the same type of PTE add to the sex difference in PTSD prevalence. In addition to an increased risk of sexual assault, females also appear to be more exposed to betrayal trauma, in that more females than males report having been exposed to interpersonal trauma, especially assault by a perpetrator close to the victim (Goldberg & Freyd, 2006).

Certain types of trauma (e.g. rape, CSA, combat) have been found to be more toxic (i.e. more likely to lead to PTSD) than others (e.g. accidents, bereavement; Kessler et al., 1995). It is therefore possible that the increased risk of sexual trauma in childhood, adolescence, and adulthood in females may account for differences in PTSD prevalence. However, studies have found that sex differences in PTSD prevalence persist even after trauma type is controlled, indicating that the high prevalence of PTSD in females is not simply a result of increased exposure to sexual trauma (Kessler et al., 1995; Tolin & Foa, 2008). However, even though studies have shown that sex differences in PTSD prevalence exist across trauma types, sexual assault prior to the index trauma is rarely controlled for and may still contribute to the increased PTSD prevalence in females following new traumas. Furthermore, even within the same type of trauma, males and females may differ in the characteristics as well as their interpretation of the event (Tolin & Foa, 2008). For example, a woman who is robbed in an isolated spot may fear that the robber will also rape her and thus have a stronger physiological reaction than a man in the same situation, who may be less likely to interpret the event as anything more than a robbery, although males as well as females may interpret the event as highly threatening and even fear for their lives.

Interestingly, studies based on military and police samples have generally failed to find an increased risk of PTSD in females compared to males (Lilly et al., 2009). Although male and female military veterans generally differ in the types of events they have been exposed to, it is also possible that the lack of reported sex difference is related to one or more variables on which police and military females differ from female civilians. Furthermore, the meta-analysis by Tolin and Foa, (2008) found that a significant sex difference in PTSD rates has not been established following adult and childhood sexual assault and abuse. This failure to discover significant sex differences in PTSD following sexual assault and abuse may be accounted for by the relatively low number of both-sex studies focusing on these two trauma types. Furthermore, males are more often assaulted by numerous perpetrators and sustain more physical injuries, both of which could add to the prevalence of PTSD in males compared to females (Tolin & Foa, 2008). However, another possibility is that there may be a ceiling effect, whereby high levels of PTSD following particularly toxic traumas (e.g. sexual assault, combat) in males as well as females will overrule any specific female vulnerability to traumatic stress (Gavranidou & Rosner, 2003). Although it is too soon to rule out this possibility, one study published after Tolin and Foa's meta-analysis found that adult female victims of childhood abuse and neglect were significantly more likely to meet criteria for a PTSD diagnosis than males (Koenen & Widom, 2009), suggesting that sex differences may exist in these types of traumas. Even more importantly, this sex difference was reduced but not eliminated when later rape and exposure to multiple traumas were controlled for. Similarly, a Danish study found that sex remained a significant predictor of lifetime PTSD symptoms in students after rape was controlled for (O'Connor & Elklit, 2008), although the way in which rape was assessed might have underestimated the true degree of sexual victimisation in the sample.

To sum up, although the extent of sex differences in PTSD following highly toxic trauma types needs to be studied more thoroughly, the increased risk of PTSD in females has been established across a wide variety of trauma types. Although the increased exposure of females to sexual assault and CSA may add to the sex difference in the prevalence of PTSD, differences in exposure to traumatic events do not appear to fully account for these differences. It thus appears that the structural theory is not sufficient to account for the observed sex differences in PTSD. Instead, the mediation hypothesis suggests that females are more vulnerable to PTSD, because they exhibit higher levels of certain risk factors associated with PTSD. From a gender based perspective on coping with traumatic stress, the socialisation theory holds that the way men and women are brought up and continue to be socialised in a context of gender role expectations affect how they react in the face of trauma (Ptacek et al., 1992; Rosario et al., 1988). As a result, men and women differ in the kind of events they interpret as threatening, and consequently their preferred coping strategies as well as the physiological reactions are likely to differ (Simmons & Granvold, 2005). Thus, according to both the socialisation theory and the mediation hypothesis, sex differences in PTSD prevalence may be related to sex differences in associated risk factors in the time leading up to, during, and following the traumatic event. Next, we will focus on some risk factors, which are more pronounced in females than in males, and which according to the mediation hypothesis may help account for the increased PTSD prevalence in females. Risk factors, which are not hypothesised to add to the higher PTSD severity in females, compared to males (e.g. social support, prior trauma exposure) will not be discussed here.

3.2 Sex differences in risk factors related to the development of PTSD

3.2.1 Pre-traumatic risk factors

It has been suggested that the higher degree of negative affectivity in females may result in more reactive emotional and somatic responses in females compared to males (Zeidner, 2006). The overlapping constructs of negative affectivity and neuroticism have been defined as the propensity to experience a wide variety of somatic and emotional dysphoric states including depression, anxiety, anger, and somatic symptoms (Kirmayer et al., 1994). High levels of neuroticism have been shown to increase sensitivity to stressful life events (Kendler et al., 2004). Furthermore, neuroticism and negative affectivity have been shown to play a role in the development of PTSD (Ahern et al., 2004) as well as in other psychiatric disorders, including depression and anxiety disorders (Hettema et al., 2004; Kendler et al., 2004). The influence of negative affectivity on all three disorders may explain, why both anxiety and depression have been found to be related to PTSD (Kessler et al., 1995). Females have been reported to score higher than males on measures of neuroticism (Hettema et al., 2004; Lynn and Martin, 1997) and negative affectivity (Joiner & Blalock, 1995), and both anxiety and depression are more commonly found in females than in males (Kessler et al., 1994). It could therefore be hypothesised that the higher prevalence of PTSD in females is a result of their higher negative affectivity/neuroticism as well as pre-existing anxiety and depression. One study has reported that the higher risk of PTSD in females in the general population was mainly due to their exposure to more toxic trauma types in combination with a higher prevalence of pre-existing psychiatric disorders (Hapke et al., 2006). However, Spindler et al. (2010) found that even though sex was no longer significantly associated with PTSD status after trait anxiety was controlled for, there was still a trend towards significance, and Fullerton et al. (2001) found that neither prior PTSD, major depression, nor other anxiety disorders could account for the increased PTSD prevalence in females. Finally, Breslau et al. (1997) found that although prior depression and anxiety disorder did reduce the sex difference in PTSD prevalence, they did not eliminate it. Thus, at the present time there is not convincing evidence that sex differences in PTSD can be fully accounted for by negative affectivity/neuroticism and pre-existing symptomatology.

3.2.2 Peritraumatic risk factors

In relation to the risk factors related directly to the traumatic event, sex differences have been widely reported in primary appraisal, and it has been suggested that the higher risk for stress-related disorders in females may be due to such differences. The A2 criterion of the DSM-IV PTSD diagnosis states that in order for an event to be considered traumatic, the person must have experienced intense fear, horror, or helplessness (American Psychiatric Association, 2000). Females have generally been found to be more likely than males to report such feelings in response to a PTE (Irish et al., 2011; Norris et al., 2002). Another common peritraumatic experience is dissociation, which is defined as a disruption in the usually integrated functions of consciousness, memory, identity, and perception (American Psychiatric Association, 2000). Dissociative reactions during or following a traumatic event have been found to be important risk factors for PTSD (Ehring et al., 2006; Ozer et al., 2003). Although there do not appear to be major sex differences in the prevalence of dissociative reactions in the general population (Spitzer et al., 2003), several studies have reported higher levels of trauma-related dissociation in females compared to males (Bryant & Harvey, 2003; Irish et al., 2011). It is thus possible that sex differences in such peritraumatic rections may account for sex differences in PTSD prevalence.

It has been suggested that the professional training of police officers, which is in accordance with a traditionally masculine minimisation of emotional reactivity, can account for the previously mentioned lack of reported sex differences in PTSD prevalence in police samples (Pratchett et al., 2010), which appears to be caused by a lower degree of traumatisation in female police officers compared to female civilians. Lilly et al. (2009) compared female police officers to female civilians and found that despite a higher degree of traumatic exposure, female police officers reported lower levels of PTSD. This could be accounted for by lower levels of peritraumatic emotional distress in the police officers. However, although female police officers also reported lower levels of peritraumatic dissociation, this did not account for any additional difference in PTSD levels. In contrast, Irish et al. (2011) found that perceived life threat could not account for sex differences in PTSD severity after an MVA, but that peritraumatic dissociation served as a partial mediator of 6 week and 6 month PTSD severity. Finally, Spindler et al. (2010) found that sex was no longer significantly associated with PTSD status in a logistic regression analysis after perceived life threat was controlled for, whereas two other studies have found that neither peritraumatic dissociation or perceived threat (Ehlers et al., 1998) nor peritraumatic helplessness or horror (O'Connor & Elklit, 2008) could eliminate sex as a risk factor for PTSD. Thus, contradictory findings have been reported on whether sex differences in peritraumatic reactions can account for sex differences in PTSD.

3.2.3 Post-traumatic risk factors

When it comes to post-traumatic risk factors associated with PTSD, sex differences in the use of coping strategies have been repeatedly reported. Females tend to use more emotion-focused coping strategies, such as support seeking, than males (e.g. Tamres et al., 2002). Furthermore, although males do not use any coping style more often than do females, overall, they do appear to use relatively more problem-focused and avoidance coping than females (Tamres et al., 2002), and they are more likely than females to choose problem-focused coping as their initial coping strategy (Ptacek et al., 1992). Whereas a problem-focused coping style has shown to serve as a protective factor against PTSD, both avoidant and emotion-focused coping have been reported to be related to increased PTSD severity (Bödvarsdóttir & Elklit, 2004; Gil, 2005). It is thus possible that the use of emotion-focused coping may to some degree account for the increased prevalence of PTSD in females (Peirce et al., 2002). However, O'Connor & Elklit (2008) found that neither problem-focused nor avoidant coping was significantly associated with lifetime PTSD severity, and that sex remained a significant predictor of PTSD severity after emotion-focused coping was controlled for.

In addition to differences in coping strategies, females are more likely to blame themselves for the trauma and to see themselves and the world in a negative light following a PTE, and such post-traumatic cognitions have been found to predict PTSD (Cromer & Smyth, 2010). Thus, it could be expected that sex differences in post-traumatic cognitions may help account for the increased rate and severity of PTSD in females compared to males. However, one study found that self-blame and negative cognitions about the world and oneself failed to account for sex differences in students' PTSD symptoms (Cromer & Smyth, 2010). And another study reported that although sex was no longer significantly associated with PTSD symptoms after the same three post-traumatic cognitions and interactions with sex had been controlled for, negative cognitions about oneself was actually more strongly associated with PTSD symptoms in males than in females, suggesting that such post-traumatic cognitions in females are unlikely to account for sex differences in PTSD severity.

3.3 Evaluation of the mediation hypothesis

Together, the structural and socialisation theories offer a valid gender-based approach to sex differences in PTSD. However, gender influences are generally very difficult to discern from sex differences and can often be equally well explained by biologically based theories. For example, it was reported that a traditionally masculine response style could at least partly explain why female police officers reported lower levels of PTSD than female civilians, and it was suggested that such differences were due to the masculine socialisation of police officers (Lilly et al., 2009). However, it could be argued that the women who choose to become police officers already differ from women who do not on a number of variables relevant to the development of PTSD. Such differences (which may be accounted for either by prior socialisation processes or differences in hormone levels and other physiological factors) may account for the similar PTSD prevalence found in male and female police officers better than any police-specific socialisation processes.

In sum, sex differences in the risk factors associated with PTSD may account for at least part of the increased prevalence of PTSD in females, and several studies have shown that when included in hierarchical regression models, sex often becomes non-significant after other variables are controlled for. This has led many researchers to conclude that the role of sex in PTSD research is not as important as has previously been assumed (e.g. Ozer et al., 2003). However, other studies have found that sex remains a significant predictor of PTSD severity even after risk factors, which are more prevalent in females have been controlled for (e.g. Ehlers et al., 1998, O'Conner & Elklit, 2008). In order to fully test the mediation hypothesis, specific mediation studies need to be conducted, which make an effort to include all risk factors known to be more prevalent in females. However, research on sex differences in PTSD is still in its childhood, and viewing sex simply as a risk factor or as a control variable may be too simplistic. We believe that sex differences in PTSD go much deeper than simple mediation effects. In addition to sex differences in the prevalence of PTSD, sex differences may also exist in the physiological response to trauma and in how such reactions may shape symptom development. The latter part of this chapter will focus on these less studied sex differences in PTSD.

4. Sex differences in initial trauma response

It is generally accepted, that confrontation with a stressor results in immediate activation of the sympathetic nervous system (SNS) and release of the catecholamines epinephrine and norepinephrine, which encourage either fighting or fleeing behaviour. The activation of the SNS further stimulates the slower stress response of the hypothalamic-pituitary-adrenal (HPA) axis. This triggers the release of corticotropin releasing hormone (CRH), adrenocorticotropin hormone (ACTH), and glucocorticoids, particularly cortisol. However, the physiological stress response in males has been much more extensively studied than is the case for females (Peirce et al., 2002). Furthermore, both-sex studies are often based on small samples without sufficient power to detect sex differences. This is highly problematic because important sex differences have been reported in the HPA response to stress (Kirschbaum et al., 1999).

Arginine vasopressin (AVP) and oxytocin are two peptide hormones induced by the HPA axis. Even though AVP and oxytocin are structurally similar and differ from each other by only two amino acids (Klein & Corwin, 2002), the two hormones differ widely in the roles they play in response to stress. Whereas AVP stimulates the fight-or-flight response and HPA axis activation (Klein & Corwin, 2002), oxytocin appears to suppress the HPA response to stress and be regulated by the parasympathetic nervous system (PNS; Klein & Corwin, 2002; Neumann, 2008; Neumann et al., 2000). AVP levels are higher in males than in females and may be regulated by testosterone (Rasmusson et al., 2002). In contrast, oxytocin levels are higher in females, and the bio-behavioural effects of oxytocin are enhanced by oestrogen (Klein & Corwin, 2002). Furthermore, there is preliminary evidence that the two female sex hormones oestrogen and progesterone are associated with blunted vascular reactivity to acute stressors (Peirce et al., 2002). Thus, oxytocin appears to play a greater role in females compared to males, and as a result females may be more prone to react to acute stress with a suppression of the HPA axis and the SNS in response to stress. In contrast, higher levels of AVP in males are associated with increases in HPA and SNS activity in the face of threat, consistent with a fight-or-flight response. This apparent sex difference is supported by the finding that males appear to respond to stressful events with greater increases in cortisol than do females, which may reflect a tendency to HPA hyper-reactivity in males and hypo-reactivity in females (Kudielka & Kirschbaum, 2005).

Taylor et al. (2000) have suggested that whereas males respond to stress with the well-known fight-or-flight system regulated by the SNS, evolutionary demands have favoured an alternative tend-and-befriend system in females regulated by the PNS. This alternative response to stress is hypothesised to have developed because it has been more adaptive for females in times of threat to tend to offspring and seek protection among other members of the group. The tend-and-befriend system is hypothesised to be linked to oxytocin, which is implicated in a variety of social behaviours relevant to how females respond in the aftermath of trauma, including social support and coping (Taylor, 2006; Taylor et al., 2002). In support of the tend-and-befriend theory, it has been documented that males generally respond to stressful events with physiological hyperarousal and an increase in aggressive behaviours, whereas females are more likely to group together and seek social support – especially from other females (Taylor et al., 2000). Kivlighan and colleagues (2005) found that activation of the HPA axis evidenced by increases in cortisol levels in preparation for a rowing competition was associated with increased competitiveness and more pre-race mental preparation in males but with bonding and social affiliation with teammates in females. Furthermore, a study of 18-month-old children has reported sex differences in the response to frightening maternal behaviour (David & Lyons-Ruth, 2005). Whereas males responded by displaying pronounced avoidance, resistance, and conflict behaviour, females responded by approaching their mothers while simultaneously displaying behaviours such as hesitation, fearfulness, and freezing. In accordance with this dissociation-like behaviour in female toddlers, adult females use more dissociative mechanisms in the face of trauma compared to males (Bryant & Harvey, 2003; Irish et al., 2011). Dissociation appears to be linked to a suppression of activity in the sympathetic nervous system and the HPA axis (Olff et al., 2007).

Some of the differences in the male and female response to stress, which have been discussed here, appear to be best explained by biological sex differences. For instance, several of the hormones involved in HPA activation appear to be affected by the female menstrual cycle (Kudielka & Kirschbaum, 2005), and sex differences in the physiological response to stress appear to be present very early in life (David & Lyons-Ruth, 2005; Kudielka & Kirschbaum, 2005). However, depression research has indicated ways in which both environmental and genetic factors can lead to changes in HPA reactivity (Swaab et al., 2000), and studies indicate that sex differences in the epinephrine response to cognitive challenge is mediated by gender role (Davis & Emory, 1995). Further support for the influence of gender roles can be found in one study, which reported that confrontation with achievement challenges involving performance failures resulted in increased free cortisol levels in males but not in females (Kudielka & Kirschbaum, 2005). In contrast, social rejection challenges resulted in significant increases in females, but not in males, indicating that different stressors may lead to activation of the HPA axis in males and females. This suggests that gender roles may affect, which characteristics of an event are considered stressful by males and females and such differences affect the psychobiological stress response and may further affect the risk of developing PTSD. Thus, even though the fight-or-flight system exists in females, the tend-and-befriend system is assumed to dominate in times of danger. However, despite sex differences in the normal response to trauma, it is important to note that there are intra-sex differences. Geary and Flynn (2002) have pointed out that males do not always react to stress with fight-or-flight behaviour and that in some situations males respond to stress in ways similar to the tend-and-befriend response in females. With the tend-and-befriend and the fight-or-flight system operating in both sexes, intra-sex differences in the initial trauma response are likely to depend upon intra-sex variations in hormone levels as well as socialisation and gender role expectations.

Little is known about which factors contribute to the shift from adaptive responses during and immediately following the traumatic event to maladaptive responses, which may be related to PTSD in the long run (Paris et al., 2010). However, it is likely that sex differences in the initial stress response may persist beyond the traumatic event and affect the development and maintenance of PTSD. Perry and colleagues (1995) have suggested that neural responses to stress may become sensitised, resulting in the same neural responses being elicited by decreasingly intense stimuli. The tend-and-befriend versus fight-or-flight theory proposed by Taylor and colleagues may thus be highly relevant to include in a theory of sex differences in PTSD, as it is possible that sex differences in the initial response to trauma may lead males and females to follow different pathways to PTSD, as has previously been suggested (Christiansen & Elklit, 2008).

5. Sex-specific pathways to PTSD

Perry and colleagues (1995) suggested that two neuronal response patterns could be identified in traumatised children. The first behaviour pattern was assumed to be associated with a sensitisation of the catecholamine system (in accordance with increased activation of the SNS) resulting in hyperarousal in response to traumatic reminders. The other behaviour pattern was assumed to be associated with a decrease in blood pressure and heart rate (consistent with a down-regulation of the SNS). This latter behaviour pattern was initiated by an attempt to seek protection from the caretaker, which in the case of persistent threat was followed by a range of more and more dissociative symptoms, including compliance, freezing, and complete dissociation (Perry et al., 1995). In accordance with this theory, two separate pathways to PTSD have been identified in a study of child burn victims (Saxe et al., 2005). One pathway was mediated by dissociation, the other by anxiety/arousal. The two pathways were separated by different risk factors, suggesting that different bio-behavioural systems may be involved in the development and maintenance of PTSD. More specifically, the anxiety/arousal pathway is likely to be related to the fight-or-flight system controlled by the SNS and the HPA axis, whereas the dissociation pathway may be connected to the animal "freeze" response, which appears to be controlled by the PNS (Saxe et al., 2005). This study did not report any sex differences in the two pathways, but another study focusing on sexually abused children found support for the existence of an anxiety/arousal pathway and a dissociation pathway as well as a third avoidance pathway, which was more pronounced in boys than in girls (Kaplow et al., 2005).

We do not know of any studies examining the existence of different pathways to PTSD in adults. However, the argument that the anxiety/arousal pathway should be linked to an activation of the HPA axis and the SNS corresponds to the male stress response proposed above. Similarly, the idea that the dissociation pathway should be linked to the PNS is in accordance with the female stress response. There thus appears to be some overlap between the findings concerning different pathways to PTSD in children and sex differences in the initial stress response in adults. This suggests, that even though the two studies on pathways in children did not report any sex differences in the anxiety/arousal and dissociation pathways, males may be more likely to follow an anxiety/arousal pathway and females a dissociation pathway to PTSD. This is in accordance with the ideas proposed by Perry et al. (1995), who argued that activation and sensitisation of the dissociative response pattern was more common in females than in males, whereas the opposite appeared to be the case for the hyperarousal pattern. Furthermore, although the avoidance pathway identified in the study by Kaplow et al. (2005) was not associated with a specific physiological system, it could be argued that avoidance is a form of fleeing behaviour associated with the fight-or-flight system (Taylor, 2006), and as this pathway was more pronounced in boys than in girls, males may also be more likely to follow an avoidance pathway than females.

If males and females follow different pathways to PTSD, we would expect to find sex differences in the relationship between certain risk factors and PTSD. Whereas research has mainly focused on mediation effects between sex and PTSD, the moderation hypothesis states that there may be sex differences in the relationship between different risk factors and symptomatology (Stone et al., 2010), suggesting that some risk factors may predict PTSD in females but not in males, whereas others may better predict PTSD in males. Furthermore, if such sex-specific pathways are related to the fight-or-flight versus tend-and-befriend response to stress, we would hypothesise that particularly behaviours related to either the tend-and-befriend or the fight-or-flight response to stress would show sex differences in their ability to predict PTSD. For example, physiological processes involved in the tend-and-befriend response, such as activation of the PNS and the role of oxytocin, could make emotion-focused coping, particularly support seeking, social support, and possibly dissociation more closely related to PTSD in females than in males. In contrast, arousal, anxiety, and possibly problem-focused and avoidant coping strategies could be hypothesised to be more important in the development or prevention of PTSD in males compared to females.

6. Sex as a moderator

6.1 Social support

Empirical evidence suggests that high levels of oxytocin may enhance the effect of social support on stress (Taylor et al., 2006), suggesting that the tend-and-befriend response may not only affect how females respond to stress, but may also affect how they benefit from social support in the aftermath of trauma. Not only do females seek, receive, and provide more support in stressful situations than males (Andrews et al., 2003; Tamres et al., 2002), there are also some indications that the positive effects of social support as well as the negative effects of negative social attention on PTSD may be more pronounced in females than in males (Ahern et al., 2004; Andrews et al., 2003; Gavrilovic et al., 2003). However, other studies have not found this (Farhood et al., 1993), and the relationship between sex, social support, and PTSD appears to be complex (Christiansen & Elklit, 2008). In order to test the hypothesis that sex moderates the impact of social support on PTSD, we conducted a number of new analyses on previously published data based on two separate samples (see Christiansen & Elklit, 2008). One sample consisted of residents in an area, which had been almost completely destroyed by an industrial disaster (the explosion sample), the other consisted of high school students who had witnessed a stabbing incident, were a young girl was killed (the high school sample). In contrast to the original analysis, this time we used only the symptoms included in the DSM-IV to measure PTSD severity (i.e. the first 17 items of the Harvard Trauma Questionnaire; Mollica et al., 1992). As can be seen in Table 1, we found no significant sex differences in correlations between feeling let down by others and PTSD severity, and positive social support was unrelated to PTSD severity in both males and females in the explosion sample. However, we did find that positive social support was significantly more strongly correlated with PTSD severity in females compared to males in the high school sample. There thus appears to be some support for the hypothesis that sex moderates the relationship between social support and PTSD.

	Positive support [a]	Feeling let down [a]	Dissociation [a]	Anxiety	Em. cop.	Probl. cop.	Avoi. cop.
Males	-.18 / -.05	.50**/ .33**	.56** / .44**	.73**	.66**	-.02	.32**
Females	-.14 / -.35**	.36** / .45**	.62** / .68**	.57**	.75**	-.16	.27*
z	n.s. / 2.51*	n.s. / n.s.	n.s. / -2.96**	1.73 [b]	n.s.	n.s.	n.s.

Pearson coefficients between risk factors and PTSD severity (combined score on the 17 first HTQ items) are shown separately for males and females from the explosion study and the high school stabbing incident.

[a] r values from the explosion study / r values from the stabbing incident study * $p < .05$; ** $p < .01$;
[b] $p = .08$; n.s.: not significant

Em. cop.: emotion-focused coping; rat. cop.: rational coping; Avoi. cop.: avoidance coping.

Table 1. Correlations between risk factors and PTSD severity for males and females

The impact of sex on the relationship between social support and PTSD is further complicated by the possibility that different types of support may have differential impact on PTSD symptomatology in males and females. For example, it is possible that talking about the traumatic event may be preferable for females, whereas males may appreciate practical assistance more, because such types of support are most compatible with the preferred coping strategies used by each sex. Furthermore, males as well as females depend mostly on female providers of support, and whereas females often have multiple support sources, adult males often cite their spouses as their only confidantes (Shumaker & Hill, 1991). This may be the reason why being unmarried serves as a risk factor for mortality in males to a greater extent than is the case for females (Shumaker & Hill, 1991). Thus, whereas females may be more vulnerable to lack of support and negative reactions from others more generally, males may be more dependent on the support of their spouse and thus be relatively more vulnerable to PTSD in case of separation or spousal bereavement. In accordance with this idea, a study on elderly bereaved has reported a lack of sex differences in PTSD severity (Elklit & O'Connor, 2005), which is in contrast to the otherwise well-established increased prevalence of PTSD among females.

6.2 Anxiety/arousal

The apparent role of AVP in the acute male stress response points to the possibility that arousal may predict PTSD in males more strongly than in females, and that males may be more prone to follow the anxiety/arousal pathway to PTSD. As a result, we would expect that anxiety as well as arousal during or immediately following a traumatic event would be stronger predictors of PTSD in males than in females. One study has found that a pre-existing anxiety disorder predicted PTSD in males, but not in females (Bromet et al., 1998), and a study on survivors of an industrial accident found that anxiety levels measured three months after the incident were significantly related to PTSD one year later in males, but not in females (the explosion study; Christiansen & Elklit, 2008). Furthermore, as can be seen in Table 1, new analyses revealed that there was a non-significant trend towards anxiety being more highly correlated with PTSD severity in males compared to females ($r = .73$ vs. $r = .57$; $z = .73$; $p = .08$; unpublished analyses). Finally, a study of PTSD in MVA victims found that the interaction between sex and prior anxiety disorder was not significant (Fullerton et al., 2001).

Arousal in the acute aftermath of trauma has primarily been studied in MVA victims. Two studies (Bryant et al., 2000; Shalev et al., 1998) have reported that increased heart rate, but not blood pressure, measured in the emergency department following an MVA was associated with later PTSD. In contrast, Blanchard et al. (2002) found that increased heart rate was associated with lower levels of PTSD. Interestingly, both studies, which found heart rate to be positively associated with later PTSD, were based on predominantly male samples, whereas Blanchard and colleagues based their results on a predominantly female sample. Furthermore, the relationship between lower heart rate and later PTSD was significant only in females (Blanchard et al. 2002). It is possible, that the lack of significance in males was due to the low number of male participants, but another possibility is that decreased heart rate measured in the emergency room was associated with later PTSD in females but not males because of a suppressed HPA and SNS response and increased rates of dissociation in females. It is thus possible that the results reflect the existence of different pathways leading to PTSD in males and females. Finally, one study examining the influence of HPA hormones on PTSD separately in males and females found that increased levels of epinephrine, norepinephrine, and cortisol were related to PTSD status in males, but not females, one month after an MVA (Hawk et al., 2000). Thus, there are some indications that anxiety and particularly physiological arousal may be more strongly related to PTSD in males than in females.

6.3 Dissociation

To the best of our knowledge, no study to date has examined a possible role of oxytocin in dissociative responses. However, there are some arguments for a possible role of oxytocin in dissociation. First, dissociation appears to be associated with a down-regulation of the HPA-axis (Olff et al., 2007) as well as a reduced heart rate, consistent with a down-regulation of autonomic reactivity (Griffin et al., 1997). This is consistent with the effects of oxytocin on the SNS and HPA-axis. Secondly, in the above mentioned study by Bryant and colleagues (2000) on MVA victims, acute stress disorder (ASD) was associated with a lower heart rate compared to subclinical ASD without dissociation, indicating that dissociative reactions may suppress SNS arousal, just as oxytocin appears to do. Finally, David and Lyons-Ruth (2005) reported that when faced with frightening maternal behaviour, female toddlers reacted with affiliative behaviour in combination with inhibition of all other behaviour, consistent with the freeze response. These findings are in line with the ideas of Perry and colleagues (1995), that dissociation in children is associated with help-seeking behaviour.

Such results suggest that dissociative reactions in the female stress response may coexist with affiliative behaviour, in which the role of oxytocin has repeatedly been demonstrated (Taylor, 2006; Taylor et al., 2002).

Not only do females report higher levels of traumatic dissociation than do males, peritraumatic dissociation also appears to be a stronger risk factor for PTSD in females. An initial diagnosis of ASD, which is very much based on the presence of dissociation, has been reported to be a more accurate predictor of PTSD in females than in males, and it has been suggested that this can be explained by an increased prevalence and predictive power of persistent dissociation in females compared to males (Bryant & Harvey, 2003). Similarly, peritraumatic dissociation has been found to predict higher levels of PTSD in female but not male victims of both motor vehicle accidents (Fullerton et al., 2001) and an industrial disaster (the explosion sample; Christiansen & Elklit, 2008). However, in the latter study, new correlation analyses revealed that although dissociation appeared to be more strongly related to PTSD severity in females than in males (see Table 1), this difference was not significant. Furthermore, dissociation was reported to significantly predict PTSD severity in both males and females in a study of Palestinians (Punamäki et al., 2005) and in a study of adolescents who witnessed a stabbing incident (the high school sample; Christiansen & Elklit, 2008). However, in the latter study, further analyses shown in Table 1 revealed that although there were no significant sex differences after other variables had been controlled for, dissociation and PTSD severity were more strongly correlated in females than in males in a simple correlation analysis ($r = .68$ vs. $r = .44$; $z = -2.96$; $p = .003$; unpublished analyses). This suggests that there may be at least a partial moderation effect of gender on the relationship between dissociation and PTSD.

6.4 Coping

Banyard and Graham-Bermann (1993) have criticised that female coping, and particularly the efficacy of coping in females, have only been studied based on theories of what constitutes adaptive coping in males, and that this has often led to the conclusion that females are inferior copers compared to males. They further criticise coping studies of being almost exclusively conducted on male samples. Although the field of coping and stress research is no longer as deserving of this critique as it used to be, the possibility of sex differences in coping efficacy is still largely ignored in the coping literature and research. However, it is possible that different coping styles are effective for males and females in different situations. Both oxytocin and AVP have been linked to the long-term effects of stress on coping (Neumann et al., 2000), and these two hormones may thus be involved in sex differences in the use of different coping strategies.

The tend-and-befriend response is in its nature related to coping through support seeking, which tends to be categorised as an emotion-focused coping strategy, although seeking practical support is often categorised as a problem-focused strategy. In contrast, Taylor (2006) has linked the use of avoidance coping strategies to the fight-or-flight response. Furthermore, whereas avoidance coping may be viewed as a form of fleeing behaviour, it could be argued that problem-focused or rational coping is a form of fighting behaviour. As a result, we hypothesise that emotion-focused coping, particularly support seeking may be more adaptive in females, whereas problem-focused and avoidant coping may be more strongly associated with PTSD in males. In accordance with this hypothesis, studies, which have evaluated the relationship between coping and psychopathology separately for males and females, have generally found important sex differences (Hovanitz & Kozora, 1989). Although this has not been examined specifically for PTSD, there is some support that problem-focused coping may be more adaptive for males, whereas emotion-focused coping may be less maladaptive for females than has been found for males (Hovanitz & Kozora, 1989). Furthermore, emotion approach coping, which is a certain kind of emotion-focused coping involving active processing and expression of emotions, has been shown to decrease depression and increase life satisfaction in females, but to have the opposite effect on males (Folkman & Moskowitz, 2004). In contrast, problem-focused coping has been reported to be generally more adaptive for males (Tamres et al., 2002). Although coping was not included in the original analyses by Christiansen & Elklit (2008), new analyses based on data from the explosion study showed that emotion-focused and avoidant coping were associated with higher PTSD levels in both males and females, whereas rational coping was not associated with PTSD severity for either sex (see Table 1). *Z*-tests revealed no significant sex differences.

Thus, although there is some support for the hypothesis that different coping strategies may be adaptive for males and females in relation to symptomatology in general, this has not been documented in relation to PTSD. Furthermore, the studies mentioned above, which have found sex differences in the effectiveness of coping have generally focused on more specific coping strategies rather than broader coping styles, such as emotion- or problem-focused coping. This may suggest, that such coping styles include strategies, which are too heterogeneous for the detection of moderation effects. In particular, we find it problematic that emotion-focused coping is often operationalised to include both behaviours which would normally be considered positive, such as seeking social support, and behaviours which are more likely to represent symptomatology, such as rumination (Stanton, Danoff-Burg, Cameron, & Ellis, 1994). In fact, in the new analyses introduced above, in which we found emotion-focused coping to be significantly and positively related to PTSD severity in both males and females, the questionnaire used to assess coping (a revised version of the Coping Style Questionnaire; Elklit, 1996) did not include any items on support seeking behaviour. In order to shed more light on sex differences in the relationship between emotion-focused coping and PTSD, future studies should look more closely at which specific emotion-focused behaviours may be more adaptive for females than for males.

The question of whether sex moderates the effect of coping on symptomatology becomes even more complex when one considers that the efficacy of different coping styles may depend upon the characteristics of the situation in question. Araya et al. (2007) examined mental distress and quality of life in a sample of internally displaced Ethiopians and found that the relationship between coping and mental distress varied not only by sex but also by degree of trauma exposure. For males, emotion-focused coping was associated with higher levels of mental distress only when trauma exposure was high, whereas for females, emotion-focused coping was only significantly associated with higher levels of symptomatology when exposure was low (J. Chotai, personal communication; see also Araya et al., 2007). This suggests, that for women who have been repeatedly traumatised, emotion-focused coping may not be maladaptive, whereas that appears to be the case for men who have been exposed to multiple trauma. Although the study did not specifically examine PTSD symptoms, it is possible that a similar relationship between sex, degree of exposure, and emotion-focused coping exists for PTSD. Furthermore, because the dose-response model of trauma may refer to both repeated exposure and exposure to more toxic trauma types (Elklit & Christiansen, in press), it is possible that this relationship holds true not only for number of traumatic events but also for severity of traumatic events. Many studies on the relationship between coping and PTSD examine coping across situations in general, instead of specifically in relation to the traumatic event. As the preference, and especially the efficacy, of different coping strategies are likely to differ according to the characteristics of the situation at hand, studies which assess coping in relation to the specific traumatic event may be best suited to illuminate the complex relationship between coping and PTSD in males and females.

6.5 Evaluation of the moderation hypothesis

To sum up, there is some support for the moderation hypothesis, although results are inconsistent. Studies indicate that peritraumatic physiological arousal may be predictive of later PTSD in males but not in females. There is also some support that peritraumatic and persistent dissociation are more predictive of later PTSD in females. Bryant et al. (2000) found that later PTSD status was best identified through either increased heart rate or a diagnosis of ASD, indicating that two distinct pathways may lead to PTSD. There is thus some support for the hypothesis that males may primarily follow an arousal pathway and females a dissociation pathway to PTSD, although these findings are only preliminary and more research is needed. Some support was also found for a moderation effect on social support and PTSD. Although the association between social support and PTSD was stronger in females than in males, consistent with the tend-and-befriend theory and the greater role of oxytocin in females, it is unknown how such findings relate to the proposed sex-specific pathways to PTSD. One possibility is that the positive effects of social support and the negative effects of dissociation are somehow combined in the dissociation pathway to PTSD, possibly because dissociation occurs when tend-and-befriend behaviour proves ineffective, as suggested by Perry et al. (1995). However, another possibility is that lack of social support may represent an independent pathway to PTSD. Furthermore, support for the moderation hypothesis did not appear to be as strong for anxiety as was the case for arousal. It has been suggested that physiological arousal associated with the fight-or-flight system is related to fear but not anxiety (Grillon, 2008). It is therefore possible that anxiety and physiological arousal may not be combined in a single pathway to PTSD, but may instead represent similar but independent pathways. Finally, there was not much support for the moderation hypothesis in relation to coping. It is thus possible, that if the avoidance pathway can be found in adults, it may not be moderated by sex.

Figure 1 shows a preliminary model of how sex differences in the acute response to trauma may affect the development of PTSD. It is hypothesised that exposure to a traumatic stressor results in activation of the SNS in both sexes. However, once the SNS has activated the HPA axis, sex differences in the stress response begin to show. In males, the SNS response is amplified by AVP, which is secreted as part of the HPA response and fuelled by testosterone. This sustained activation of the SNS and increased activity in the HPA axis prepares the body for fighting or fleeing. If the SNS and the HPA axis are not downregulated and brought back to homeostasis, this fight-or-flight response may persist and make up one or more pathways to PTSD. Such pathways are hypothesised to be mediated by physiological arousal and anxiety and possibly by avoidance coping, and problem-focused coping. In contrast, HPA activation in females is associated with the secretion of oxytocin, the effects of which are amplified by oestrogen. Oxytocin is assumed to suppress SNS and HPA activity and instead be associated with activation of the PNS. This triggers the tend-and-befriend response. Although the role played by dissociation in the tend-and-befriend response is unknown, it appears that both are associated with a suppression of SNS activity, and dissociation seems to often co-occur with affiliative behaviour. Whether sensitisation of the tend-and-befriend response and dissociative reactions make up separate or combined pathways to PTSD is unknown. However, such pathways may be mediated by risk factors such as social support, dissociation, and possibly emotion-focused coping. Testosterone and oestrogen levels are not hypothesised to be affected by trauma exposure in the model. Rather, they should be viewed as ways in which sex may influence the initial stress response in males and females. Even though there are obvious differences in the distribution of sex hormones, both oestrogen and testosterone are present and may influence the stress response in both sexes. Thus, although the tend-and-befriend response is hypothesised to dominate in females and the fight-or-flight response is hypothesised to dominate in males, both response patterns and the associated pathways to PTSD may exist in both sexes.

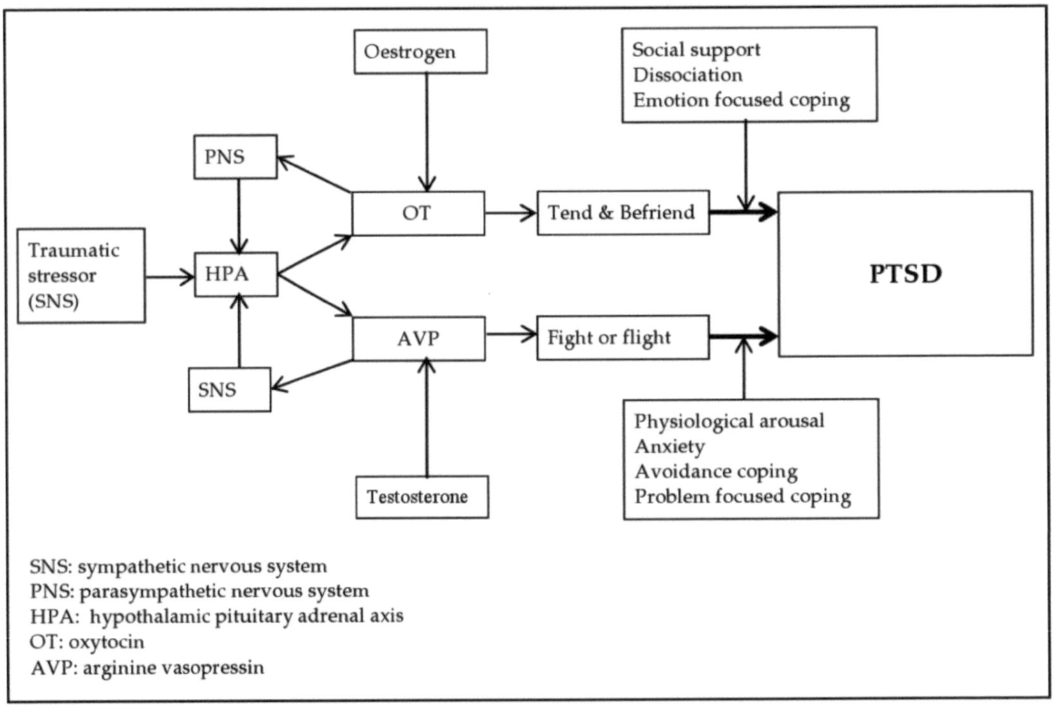

Fig. 1. A model of potential sex differences in initial stress response and pathways to PTSD

On a final note, the hypothesised involvement of oxytocin and social support in the development of PTSD in females should not be taken to suggest that the release of oxytocin and support seeking behaviour are maladaptive. In fact, higher oxytocin levels have been associated with faster HPA recovery in females after an acute stress laboratory challenge (Olff et al., 2007). Thus, it is most likely that it is a dysregulated oxytocin response which is related to PTSD in females, possibly because the downregulation of the SNS associated with the secretion of oxytocin may trigger dissociative reactions, which appear to be directly associated with the later development of PTSD (e.g. Ozer et al., 2003). Furthermore, because oxytocin appears to play a greater role in the female stress response as well as in other aspects of female behaviour, females are likely to be more vulnerable to dysregulation of the oxytocin system than males. Similarly, it is not the tend-and-befriend response in and of itself that leads to higher rates of PTSD in females. Rather, it has been suggested that PTSD develops when the tend-and-befriend response is compromised (Olff et al., 2007). This idea is consistent with studies suggesting that social support serves as a protective factor to a greater extent in females than what appears to be the case in males. Thus, it is possible that females rely on the tend-and-befriend response as protection against stressors, but when it is not possible to seek protection among others, as is often the case with interpersonal assault, the same physiological processes underlying the tend-and-befriend response may lead to dissociation as an alternative protection against the stressor. Thus, the females stress response may be more vulnerable to outside influences (e.g. a perpetrator who does not respond to the pleas of the victim or significant others who fail to understand the victim's distress) than the male stress response. This may be part of the reason why females appear to be more vulnerable to the negative sequalae of trauma.

Finally, it should be mentioned that moderation effects have been reported in relation to other risk factors not reported here. For example, depression, negative affectivity/ neuroticism, and younger age have sometimes been reported to be related to PTSD in females more than in males, whereas prior traumatic exposure has been reported to predict PTSD in males only (for an overview see Christiansen & Elklit, 2008). Sex differences in such risk factors do not appear to fit easily in the model introduced above. In fact, depression has been linked to increased SNS activity (Veith et al., 1994) as well as HPA hyper-reactivity (Pariante & Lightman, 2008). In the proposed model, PTSD in females is associated with a suppression of SNS activity and hypo-reactivity in the HPA axis, and it is thus unclear why depression should be particularly related to PTSD in females. More research on sex differences in the physiological correlates of PTSD as well as depression is needed in order to understand these apparently contradictory results. Furthermore, based on data from the NCS, Bromet, Sonnega, and Kessler (1998) reported that more risk factors predicted PTSD in female than male trauma survivors. Similarly, King and colleagues (1999) reported that the relative importance of pre-, peri-, and post-traumatic variables in predicting PTSD differed for male and female Vietnam veterans. These findings suggest not only that the relationship between different risk factors and the presence or severity of PTSD may be different for males and females, but also that we have identified more of the risk factors involved in the development of PTSD in females than in males. Therefore, in order to better understand and prevent the development of PTSD, more studies need to examine risk factors on PTSD separately in males and females.

Even though gender based theories may account for mediation effects of sex in PTSD, such as why females use more emotion-focused coping strategies, we do not believe that gender-specific socialisation and differences in the daily stressors that males and females are exposed to are likely to account for the moderation effects of sex in PTSD, which have been discussed here. Such moderation effects are more likely to arise from sex differences in the physiological response to stress and its sequalae. For example, isolation has been reported to serve as a stressor in female but not male rats. Similarly being with other rats has been found to reduce the adverse effects of chronic foot shock stress in female rats, while increasing such effects in males (Westenbroek et al, 2003). This suggests that sex differences in the beneficial effects of social support cannot fully be accounted for by gender differences. Instead, such results are consistent with an evolutionary pressure towards a tend-and-befriend response in females. However, whereas evolutionary-based sex differences in the initial stress response may best account for any consistent sex differences in different pathways to PTSD, the results reviewed here suggest that although there may be sex differences in the importance of the different risk factors, none of them appeared to be uniquely associated with either males or females. Thus, the different risk factors are likely to work together in very complex ways, and whether any particular individual follows a specific pathway to PTSD is likely to depend upon a number of different factors, including biology, contextual factors, and gender role. We believe that the role played by sex in the development of PTSD is one of mediated moderation, with sex influencing both the distribution of risk and protective factors and the relationship between such factors and the development and severity of PTSD.

7. Two disorders or one? Sex differences in the expression of traumatisation

When the PTSD diagnosis was introduced into the DSM-III (American Psychiatric Association, 1980), the different syndromes created to describe the symptoms reported by male war veterans and female survivors of rape and domestic violence were combined. However, if males and females follow different pathways to PTSD, which are mediated by different risk factors, it is possible that there are sex differences in the symptomatology of PTSD. It may therefore be time to reconsider, whether the post-traumatic symptomatology experienced by males and females is best captured by the present PTSD diagnosis, or whether we should go back to diagnosing trauma victims according to sex-specific syndromes. However, this question may prove very difficult to answer. Whereas numerous studies have examined sex differences in exposure to trauma and in the prevalence and severity of PTSD, sex differences in the expression of PTSD have been less extensively studied.

Preliminary support has been reported for a dissociative subtype of PTSD, which is associated with diminished physiological reactivity, and which may be more common in females than in males (Griffin et al., 1997; Olff et al., 2007). Furthermore, there is some support for the hypothesis that males and females with PTSD differ on measures of arousal and HPA disturbances. Although HPA activity is greater among healthy females than healthy males in response to acute stressors, studies have shown that male MVA victims with PTSD appear to have greater HPA arousal than females, although both sexes still display lower overall arousal levels than healthy controls (Paris et al., 2010). Based on these findings, it would be easy to assume, that males will score higher on symptoms of arousal than females. However, as mentioned earlier females experience more PTSD symptoms across all three symptom clusters than do males. The reason for this may be that arousal in PTSD research is often used to cover a wide range of increases in physiological activity. However, certain types of arousal may be associated with PTSD, whereas others may not. For example, PTSD in males has been reported to be related to elevated heart rate but not blood pressure (Bryant et al., 2000; Shalev et al., 1998), and it has been suggested that this is due predominantly to adrenergic rather than noradrenergic activation. Therefore, when considering the specific role of physiological arousal in male compared to female PTSD, it is important to note that the increased acute arousal, which may be associated with increased PTSD severity at later stages, may not be the same type of arousal as that included in the DSM-IV PTSD diagnosis (Hawk et al., 2000).

One study has found that males and females seeking treatment for PTSD related to interpersonal assault expressed comparable levels of PTSD severity, depression, guilt, and trait anger (Galovski et al., 2011). However, the same study found that males reported higher levels of state anger and females reported more health-related complaints. Unfortunately, symptoms of dissociation and physiological arousal were not included in this study. The findings suggest that there are minor differences in relation to features associated with PTSD in males and females, but that PTSD levels are comparable. In accordance with this view, it has been reported that males and females with PTSD show equal levels of functional impairment (Schonfeld et al., 1997). However, despite presenting with PTSD symptomatology very similar to that in females, the increased levels of state anger in males suffering from PTSD may result in specific problems. If males respond with anger when being reminded of the trauma, they may be more likely than females to push others away. PTSD constitutes a burden to the friends and family of the person suffering from PTSD and may lead to decreases in support availability over time (Jakupcak et al., 2006; Kaniasty & Norris, 2008). Although both males and females with PTSD may experience a decrease in social support as their symptoms persist, this phenomenon has been particularly reported for male combat veterans with PTSD (Kaniasty & Norris, 2008). In contrast, an increase in health-related complaints is less likely to result in the withdrawal of support and may in fact cause females to seek more support from both health professionals and informal support sources. Longitudinal studies are needed to illuminate how the relationship between social support and PTSD develops over time, and whether such developments differ between males and females. For example, it is possible that social support is a better predictor of PTSD symptomatology in females, but that PTSD leads to greater deterioration of social support in males.

In accordance with the relative similarity in symptomatology found in males and females with PTSD, studies of comorbidity generally find more similarities than differences. One study on treatment seeking patients with PTSD reported that males and females presented with fairly comparable clinical profiles overall, but that males with PTSD experienced more comorbid disorders than did females, particularly substance abuse and antisocial personality disorder (APD; Zlotnick et al., 2001). De Bellis & Keshavan (2003) have demonstrated that males show evidence of more adverse neurobiological consequences of childhood trauma than do females, and that PTSD related to childhood maltreatment appears to be more strongly associated with conduct disorder and criminal offences in males than in females. It is therefore possible that the increased prevalence of APD in males compared to females with PTSD may be particularly apparent in relation to childhood trauma. However, we have no knowledge of any studies examining this possibility. Research on sex differences in comorbidity is scarce as most comorbidity research has been conducted on male samples or has not looked at sex differences.

Although the studies mentioned above suggest that males and females display similar symptoms of post-traumatic stress, such findings have been based on participants already diagnosed with PTSD. It is possible that males tend to exhibit posttraumatic symptomatology other than the re-experiencing, avoidance, and arousal symptoms captured by the PTSD diagnosis, and that this can account for why there are more females than males with PTSD. As mentioned above, PTSD is more often associated with APD in males than in females. However, this sex difference in APD is also found in the general population (Cale & Lilienfeld, 2002). As APD has been repeatedly reported to be related to childhood trauma (Lobbestael & Arntz, 2010), it may be a common way for post-traumatic symptomatology to be expressed in males independently of the PTSD diagnosis. Although irritability and aggressive outbursts are not uncommon in people suffering from PTSD, particularly in males, the remaining symptoms are mostly internalising. It has been reported that females are more likely to suffer from internalising disorders such as anxiety disorders and depression, whereas males are more likely to suffer from externalising disorders, such as substance abuse or conduct disorder in the general population (Kessler et al., 1995) and following traumatic exposure (Pratchett et al., 2010). Similarly, although results have been mixed, increases in substance abuse have been registered primarily in male trauma survivors, whereas increases in somatisation, anxiety, and suicide attempts have been documented primarily in female trauma victims (Tolin & Foa, 2008). Furthermore, PTSD in females tends to be more chronic than in males, and females are more prone to develop complex PTSD. The concept of complex PTSD was specifically developed to capture posttraumatic symptoms other than the ones included in the PTSD diagnosis following prolonged and chronic trauma (Cloitre et al., 2002). Complex PTSD covers symptoms similar to those associated with borderline personality disorder (e.g. impulsivity, aggression, self-destructive behaviour, dissociation, and interpersonal problems) as well as feelings of alienation and trust issues (Cloitre et al., 2002).

In sum, there appear to be some sex differences on measures of physiological arousal, state anger, dissociation, somatisation, and personality disorders. Such differences are likely to be caused by a combination of gender differences in how males and females express symptomatology and sex differences in how trauma affects brain development. However, it appears that even though females are at a higher risk of developing PTSD than males, once males do develop PTSD their overall symptomatology does not appear to differ much from that found in females. This suggests that although males and females appear to differ in their initial trauma response and may follow different pathways to PTSD, there is at present no support for the idea that the symptomatology of male and female trauma victims is best captured by more sex-specific syndromes. However, more research focusing on sex differences once PTSD has developed is needed in order to draw any conclusions in relation to this question, especially as both males and females also display some sex-specific symptomatology (e.g. complex PTSD, antisocial personality disorder), which is not fully captured by the PTSD diagnosis.

8. Clinical implications of sex differences in PTSD

If males and females differ in their expression of PTSD symptomatology, it follows that they may respond differently to treatment. Even if there are no major differences in symptomatology, sex differences in the risk and protective factors associated with the development and maintenance of PTSD may still result in sex differences in treatment outcome. Finally, males and females may differ in how comfortable they feel in different treatment settings and with different treatment paradigms. It has been suggested that gender socialisation plays a role in the treatment of PTSD, and that males express less affect and are more cognitively oriented in therapy than females (Cason et al., 2002). It could be argued that such behaviour represents problem-focused coping and is consistent with activation of the fight-or-flight system.

Based on the idea that different pathways lead to PTSD in males and females, and that different risk factors may thus be important for the development of PTSD in the two sexes as illustrated in Figure 1, it might be expected that females will benefit more from therapy, which aims to reduce levels of dissociation and increase levels of social support. In contrast, a therapeutic approach, which aims to reduce physiological arousal and dampen anxiety may be more beneficial to males. In accordance with these ideas, a recent pilot study has indicated that propranolol, a beta-blocker known to reduce heart rate and blood pressure, may decrease PTSD severity in males but increase PTSD severity in females (Nugent et al., 2010). Whereas there is some support for the hypothesis that emotion-focused coping is more beneficial in females than in males, it is possible that males as well as females will benefit more from a therapeutic approach, which aims to strengthen the coping strategies, which they do not automatically use. The idea is, that once PTSD has developed, it appears that the preferred coping strategies have proved ineffective. Some support has been reported for this idea in grief counselling, as problem-focused counselling was found to be more effective in females compared to males, whereas an emotion-focused intervention form appeared to be more effective in males compared to females 11 months post-bereavement. However, such sex differences may not be transferable to less chronic populations or to PTSD treatment.

Finally, an important implication of the influence of multiple factors on HPA reactivity is the possibility that the disturbed HPA reactivity found in patients with PTSD can be treated through either pharmacological, psychotherapeutic, or social interventions. The efficacy of different treatment strategies may therefore not be dependent on what caused the HPA disturbances in the first place. Unfortunately, sex differences in treatment outcome are grossly understudied. Many PTSD treatment effect studies are based on sex-specific trauma samples, such as war veterans and sexual assault victims. Furthermore, of the few studies, which do include both-sex samples, few examine the impact of sex on treatment efficacy. As a result, little is known about whether the same treatments are equally effective (or non-effective) in males and females. Blain, Galovski, and Robinson (2010) reviewed the literature on sex differences in response to treatment and found only nine randomised controlled trials, which assessed sex differences in primary PTSD treatment outcome. Most of these studies were based on samples, which were too small to detect minor sex differences. Furthermore, 6/9 studies were based on samples exposed to a variety of trauma types, and not a single study examined sex differences in treatment outcome following interpersonal violence. Results from randomised as well as non-randomised clinical trials were mixed regarding the impact of sex on treatment outcome. The majority of studies reported no sex differences, but others found that females may respond better to trauma focused therapy than males, and that males may be more likely to drop out of treatment. However, more than anything this review highlights the need for more research in the area.

9. Future research on sex differences in PTSD

As stated earlier, we believe that the research on sex differences in PTSD is still in its childhood. Consistent sex differences still remain to be documented following certain types of trauma (e.g. sexual assault, combat exposure). Furthermore, the degree to which sex differences in risk factors associated with PTSD may account for the increased vulnerability of PTSD in females is still unknown and deserves more attention. However, future research should move beyond simply focusing on establishing and explaining sex differences in exposure and PTSD.

Sex differences have been reported in the initial response to threat, but the degree to which the tend-and-befriend response is dominant in females in response to specifically traumatic stressors remains to be documented across trauma types. One study of sex differences in the stress response of young children reported that sex differences were only significant under high levels of stress (David & Lyons-Ruth, 2005). This suggests that sex differences may exist in response to traumatic incidents and in the development of PTSD, which may not necessarily be detectable under lower levels of stress. Therefore, research on sex differences in PTSD and in the initial response to trauma should rely less on studies of non-traumatic stressors, such as the ones set up in laboratories, and instead focus more on the male and female response to actual trauma, both in the peritraumatic and post-traumatic phases.

While there is some support for the existence of different pathways to PTSD in children, such pathways remain to be confirmed in adults. Particularly, it is up to future research to establish how many pathways may exist, and whether males and females tend to follow different pathways to PTSD. Different pathways are likely to stem from different physiological and behavioural reactions in the acute trauma phase and to be mediated by different risk factors. Thus, the degree to which physiological systems, which are activated in the face of trauma, are sensitised in trauma victims with and without PTSD remains to be examined.

Although research has identified numerous risk factors associated with PTSD in both sexes, there are some indications that such risk factors may not predict PTSD equally well in males and females, and more studies need to focus on sex as a possible moderator of the impact of risk factors on PTSD development. The model proposed in the present chapter suggests that moderation effects may be particularly likely to be found for emotional support, physiological arousal, anxiety, dissociation, and coping. However, research should by no means be limited to these risk factors, as there may exist numerous pathways to PTSD, all of which may be moderated by sex. Haines, Beggs, and Hurlbert (2008) argued that even if sex only has a small moderating effect on the relationship between different risk factors and PTSD, such effects are still important to identify. Even small sex differences may point to different mechanisms involved in the development of PTSD in males and females, and such different mechanisms may call for different intervention strategies. Finally, as stated earlier, it appears that we have identified more of the risk factors involved in the development and maintenance of PTSD in females than in males. This suggests, that important risk factors related to the development of PTSD in males remain to be identified. It is our belief that by routinely including peritraumatic physiological arousal as a risk factor of PTSD, more variance in the symptomatology of males can be accounted for.

Furthermore, sex differences in symptomatology once PTSD has developed need to be examined. In particular, males and females with PTSD could be expected to differ on levels of dissociation and physiological arousal, and these two variables should be included in future studies on posttraumatic symptomatology in males and females. Research on sex differences in PTSD should include differences in symptomatology not covered by the PTSD diagnosis, in order to help us understand how the symptomatology of male and female trauma victims is best categorised, assessed, and treated.

As mentioned in the beginning of this chapter, most studies claiming to study gender have in fact studied sex. The extent to which gender roles, masculinity and femininity, sexuality, and other gender related concepts play a role in the development and maintenance of PTSD remains to be studied. Such variables may be relevant in relation to PTSD prevalence, initial stress response, different pathways to PTSD, symptomatology, and treatment efficacy. Any moderation effects which sex may have on the relationship between other risk factors and PTSD are likely to be affected by gender.

Finally, although sex is generally thought of as a relatively simple concept, different levels of the sex hormones (including oxytocin and AVP) included in the model presented in this chapter vary within each sex as well as between males and females. Thus, one last question for future research to answer, is how much sex differences in PTSD are affected by intra-sex variations in levels of testosterone, oestrogen, AVP, and oxytocin. In particular, the female menstrual cycle as well as the use of oral contraceptives have been reported to affect secretion of free cortisol levels in response to stressors (Biondi & Picardi, 1999; Kirschbaum et al., 1999). In fact, it has been suggested that prolonged use of oral contraceptives may alter the reaction of the HPA axis to psychological stress (Biondi & Picardi, 1999). We suggest that the impact of such intra-sex variations on the development of PTSD should be more closely examined in future studies.

10. Conclusion

In conclusion, although numerous studies have been published on sex or gender differences in PTSD, most have focused on establishing and explaining sex differences in PTSD prevalence. There is general consensus that females are approximately twice as likely as males to be diagnosed with PTSD following a wide range of trauma types, although sex differences in the prevalence of PTSD following some trauma types (e.g. rape, CSA, combat) have not been fully established. Sex differences in the types of trauma that males and females are exposed to and in the risk factors associated with PTSD appear to account for at least some of the increased PTSD prevalence in females compared to males. However, more research is needed to establish the degree to which sex differences in PTSD prevalence and severity is mediated by trauma type and risk factors, which are more prevalent in females.

In this chapter we have gone beyond simply focusing on sex differences in the prevalence of PTSD and have examined how sex differences in the acute response to trauma may cause males and females to follow different pathways to PTSD. There is some evidence that whereas males tend to react to trauma with the well-known fight-or-flight response, females may be more prone to react with a tend-and-befriend response. These two distinct responses to stress are associated with marked physiological differences in SNS, PNS, and HPA activity. Dysregulation of these systems may lead to sensitisation of the fight-or-flight response in males and the tend-and-befriend response in females. This may result in males and females following separate pathways to PTSD. There is some support for the existence of such pathways, as preliminary findings suggest that sex may serve as a moderator on the relationship between certain risk factors and PTSD. In this chapter we have reviewed some support for the hypothesis that physiological arousal and possibly anxiety may be more closely associated with the development of PTSD in males compared to females. In contrast, some studies have found that social support and dissociation are more closely linked to the development of PTSD in females. Sex differences in the relationship between coping and PTSD may exist but are less well documented.

Despite sex differences in the initial response to trauma and risk factors associated with PTSD, there do not appear to be major differences in the core symptomatology of PTSD in treatment seeking males and females. However, there is some evidence that males may experience more physiological arousal and anger, whereas females report more dissociation and somatisation. Although the combined impact of multiple variables on the reactivity of the HPA axis makes it theoretically possible for males and females to primarily follow different pathways to PTSD, the end result appears to be similar, although sex differences in

posttraumatic symptomatology need to be studied further. Such research is particularly needed for symptomatology not covered by the PTSD diagnosis.
Finally, although different risk factors associated with PTSD as well as minor differences in symptoms such as anger and somatisation may call for different therapeutic approaches, sex differences in the efficacy of different PTSD treatments remain to be studied. However, the HPA axis appears to be influenced by multiple factors of both biological and social origin. This suggests that the core disturbances in PTSD may be treated through either biological, psychotherapeutic, or social interventions, regardless of the different pathways which may have caused it.
It is difficult to discern whether sex differences in PTSD are best accounted for by sex or by gender-based theories. We believe it to be most likely, that sex and gender differences work together to account for the increased prevalence and severity of PTSD in females compared to males, as well as the other sex differences in PTSD. It is therefore unlikely that scientists will ever be able to fully account for the unique influence of either. Many gender differences in society are likely to build on pre-existing sex differences related to differences in brain structure and functioning, physiological response to stress, and the influence of sex hormones on the different areas of human functioning. In contrast, the extent to which sex differences come to affect the actual behaviour of males and females may be affected by cultural factors, such as gender role expectancies, which can explain why the extent of sex differences in PTSD appear to vary between cultures (Norris et al., 2001).
As stated in a report by the Institute of Medicine on the biological contributions to human health: "sex matters and until the question of sex is routinely asked and the results - positive or negative - are routinely reported, many opportunities to obtain a better understanding of the pathogenesis of disease and to advance human health will surely be missed (Wizemann & Pardue, 2001).

11. References

Ahern, J., Galea, S., Fernandez, W. G., Koci, B., Waldman, R., & Vlahov, D. (2004). Gender, social support, and posttraumatic stress in postwar Kosovo. *Journal of Nervous and Mental Disease, 192*(11), 762-770.

American Psychiatric Association. (1980). *Diagnostic and statistical manual of mental disorders* (3rd ed.). Washington, DC: Author.

American Psychiatric Association. (2000). *Diagnostic and statistical manual of mental disorders* (4th ed., Text Revision). Washington, DC: Author.

Andrews, B., Brewin, C. R., & Rose, S. (2003). Gender, social support, and PTSD in victims of violent crime. *Journal of Traumatic Stress, 16*(4), 421-427.

Araya, M., Chotai, J., Komproe, I. H., & de Jong, J. T. V. M. (2007). Effect of trauma on quality of life as mediated by mental distress and moderated by coping and social support among postconflict displaced Ethiopians. *Quality of Life Research, 16,* 915-927.

Banyard, V. L. & Graham-Bermann, S. A. (1993). Can women cope? A gender analysis of theories of coping with stress. *Psychology of Women Quarterly, 17,* 303-318.

Biondi, A. & Picardi, A. (1999). Psychological stress and neuroendocrine function in humans: The last two decades of research. *Psychotherapy and Psychosomatics*, 68, 114–150.

Blain, L. M., Galovski, T. E., & Robinson, T. (2010). Gender differences in recovery from posttraumatic stress disorder: A critical review. *Aggression and Violent Behavior, 15,* 463-474.

Blanchard, E. B., Hickling, E. J., Galovski, T., & Veazey, C. (2002). Emergency room vital signs and PTSD in a treatment, seeking sample of motor vehicle accident survivors. *Journal of Traumatic Stress, 15(3)*, 199–204.

Bödvarsdóttir, I. & Elklit, A. (2004). Psychological reactions in Icelandic earthquake survivors. *Scandinavian Journal of Psychology, 45*(1), 3-13.

Breslau, N., Davis, G. C., Andreski, P., Peterson, E. L., & Schultz, L. R. (1997). Sex differences in posttraumatic stress disorder. *Archives of General Psychiatry, 54, 1044-1048.*

Breslau, N., Kessler, R. C., Chilcoat, H. D., Schultz, L. R., Davis, G. C., & Andreski, P. (1998). Trauma and posttraumatic stress disorder in the community: The Detroit area survey of trauma. *Archives of General Psychiatry, 55*, 626-632.

Bromet, E., Sonnega, A., & Kessler, R. C. (1998). Risk factors for DSM-III-R posttraumatic stress disorder: Findings from the National Comorbidity Survey. *American Journal of Epidemiology, 147*(4), 353-361.

Bryant, R. A. & Harvey, A. G. (2003). Gender differences in the relationship between acute stress disorder and posttraumatic stress disorder following motor vehicle accidents. *Australian and New Zealand Journal of Psychiatry, 37*(2), 226-229.

Bryant, R. A., Harvey, A. G., Guthrie, R. M., & Moulds, M. (2000). A prospective study of psychophysiological arousal, acute stress disorder, and posttraumatic stress disorder. *Journal of Abnormal Psychology, 109*(2), 341-344.

Cale, E. M. & Lilienfeld, S. O. (2002). Sex differences in psychopathy and antisocial personality disorder: A review and integration. *Clinical Psychology Review, 22*, 1179–1207

Cason, D., Grubaugh, A., & Resick, P. (2002). Gender and PTSD treatment: Efficacy and effectiveness, in *Gender and PTSD*, Kimerling, R., Ouimette, P., & Wolfe, J., pp. 305-334, The Guilford Press, ISBN: 1-57230-783-8, NY.

Christiansen, D. M. & Elklit, A. (2008). Risk factors predict post-traumatic stress disorder differently in men and women. *Annals of General Psychiatry, 7*, 24.

Chung, H. & Breslau, N. (2008). The latent structure of post-traumatic stress disorder: Test of invariance by gender and trauma type. *Psychological Medicine, 38*, 563-573.

Cloitre, M., Koenen, K. C., Gratz, K. L., & Jakupcak, M. (2002). Differential diagnosis of PTSD in women, in *Gender and PTSD*, Kimerling, R., Ouimette, P., & Wolfe, J., pp. 117-149, The Guilford Press, ISBN: 1-57230-783-8, NY.

Cottler, L. B., Nishith, P., & Compton, W. M., 3rd. (2001). Gender differences in risk factors for trauma exposure and post-traumatic stress disorder among inner-city drug abusers in and out of treatment. *Comprehensiv Psychiatry, 42*(2), 111-117.

Cromer, L. DM. & Smyth, J. M. (2010). Making meaning of trauma: Trauma exposure doesn't tell the whole story. *Journal of Contemporary Psychotherapy, 40*, 65–72.

David, D. H. & Lyons-Ruth, K. (2005). Differential attachment responses of male and female infants to frightening maternal behavior: Tend or befriend versus fight or flight? *Infant Mental Health Journal, 26*(1), 1-18.

Davis, M. & Emory, E. (1995). Sex differences in neonatal stress reactivity. *Child Development, 66*, 14-27.

De Bellis, M. D. & Keshavan, M. S. (2003). Sex differences in brain maturation in maltreatment-related pediatric posttraumatic stress disorder. *Neuroscience and Biobehavioral Reviews, 27*(1-2), 103-117.

Ehlers, A., Mayou, R. A., &, Bryant, B. (1998). Psychological predictors of chronic posttraumatic stress disorder after motor vehicle accidents. *Journal of Abnormal Psychology, 107*, 508-519.

Ehring, T., Ehlers, A., & Glucksman, E. (2006). Contribution of cognitive factors to the prediction of post-traumatic stress disorder, phobia and depression after motor vehicle accidents. *Behaviour Research and Therapy 44*, 1699–1716.

Elklit, A. (1996). Coping Style Questionnaire: A contribution to the validation of a scale for measuring coping strategies. *Personality and Individual Differences*, 21, 809-812.

Elklit, A. & Christiansen, D. M. (in press). Risk factors for PTSD in female help-seeking victims of sexual assault. *Violence and Victims.*

Elklit, A. & O'Connor, M. (2005), Post-traumatic stress disorder in a Danish population of elderly bereaved. *Scandinavian Journal of Psychology, 46*, 439–445.

Farhood, L., Zurayk, H., Chaya, M., Saadeh, F., Meshefedjian, G., & Sidani, T. (1993). The impact of war on the physical and mental health of the family: The Lebanese experience. *Social Science Medicine, 36*(12), 1555-1567.

Folkman, S. & Moskowitz, J. T. (2004). Coping: Pitfalls and promise. *Annual Review of Psychology, 55*, 745-774.

Fullerton, C. S., Ursano, R. J., Epstein, R. S., Crowley, B., Vance, K., Kao, T. C., Dougall, A., & Baum, A. (2001). Gender differences in posttraumatic stress disorder after motor vehicle accidents. *American Journal of Psychiatry, 158*(9), 1486-1491.

Galovski, T. E., Mott, J., Young-Xu, Y. N., & Resick, P. A. (2011). Gender differences in the clinical presentation of PTSD and its concomitants in survivors of interpersonal assault. *Journal of Interpersonal Violence, 26*(4), 789-806.

Gavranidou, M. & Rosner, R. (2003). The weaker sex? Gender and post-traumatic stress disorder. *Depression and Anxiety, 17*(3), 130-139.

Gavrilovic, J., Lecic-Tosevski, D., Dimic, S., Pejovic-Milovancevic, M., Knezevic, G., & Priebe, S. (2003). Coping strategies in civilians during air attacks. *Social Psychiatry and Psychiatric Epidemiology, 38*(3), 128-133.

Geary, D. C. & Flinn, M. V. (2002). Sex differences in behavioral and hormonal response to social threat: Commentary on Taylor et al. (2000). *Psychological Review, 109*(4), 745–750.

Gil, S. (2005). Coping style in predicting posttraumatic stress disorder among Israeli students. *Anxiety Stress and Coping, 18*(4), 351-359.

Goldberg, L. R. & Freyd, J. J. (2006). Self-reports of potentially traumatic experiences in an adult community sample: Gender differences and test-retest stabilities of the items in a brief betrayal-trauma survey. *Journal of Trauma and Dissociation, 7*(3), 39-63.

Griffin, M. G., Resick, P. A., & Mechanic, M. B. (1997). Objective assessment of peritraumatic dissociation: Psychophysiological indicators. *American Journal of Psychiatry, 154*(8), 1081-1088.

Grillon, C. (2008). Models and mechanisms of anxiety: Evidence from startle studies. *Psychopharmacology, 199*, 421–437.

Haines, V. A., Beggs, J. J., & Hurlbert, J. S. (2008). Contextualizing health outcomes: Do effects of network structure differ for women and men? *Sex Roles, 59*(3-4), 164-175.

Hapke, U., Schumann, A., Rumpf, H. J., John, U., & Meyer, C. (2006). Post-traumatic stress disorder: The role of trauma, pre-existing psychiatric disorders, and gender. *European Archives of Psychiatry and Clinical Neuroscience, 256*(5), 299-306.

Hawk, L. W., Dougall, A. L., Ursano, R. J., & Baum, A. (2000). Urinary catecholamines and cortisol in recent-onset posttraumatic stress disorder after motor vehicle accidents. *Psychosomatic Medicine, 62*(3), 423-434.

Hettema, J. M., Prescott, C. A., & Kendler, K. S. (2004). Genetic and environmental sources of covariation between generalized anxiety disorder and neuroticism. *American Journal of Psychiatry, 161*(9), 1581-1587.

Hovanitz, C. A. & Kozora, E. (1989). Life stress and clinically elevated MMPI-scales: Gender differences in the moderating influence of coping. *Journal of Clinical Psychology, 45*(5), 766-777.

Irish, L. A., Fischer, B., Fallon, W., Spoonster, E., Sledjeski, E. D., & Delahantya, D. L. (2011). Gender differences in PTSD symptoms: An exploration of peritraumatic mechanisms. *Journal of Anxiety Disorders 25*, 209–216.

Jakupcak, M., Osborne, T. L., Michael, S., Cook, J. W., & McFall, M. (2006). Implications of masculine gender role stress in male veterans with posttraumatic stress disorder. *Psychology of Men & Masculinity, 7*(4), 203-211.

Kaniasty, K. & Norris, F. H. (2008). Longitudinal linkages between perceived social support and posttraumatic stress symptoms: Sequential roles of social causation and social selection. *Journal of Traumatic Stress, 21(3)*, 274–281.

Kaplow, J. B., Dodge, K. A., Amaya-Jackson, L., & Saxe, G. N. (2005). Pathways to PTSD, part II: Sexually abused children. *American Journal of Psychiatry, 162*(7), 1305-1310.

Kendler, K. S., Kuhn, J., & Prescott, C. A. (2004). The interrelationship of neuroticism, sex, and stressful life events in the prediction of episodes of major depression. *American Journal of Psychiatry, 161*(4), 631-636.

Kessler, R. C., McGonagle, K. A., Zhao, S., Nelson, C. B., Hughes, M., Hughes, S., Wittchen, H., & Kendler, K. S. (1994). Lifetime and 12-Month Prevalence of DSM-III-R Psychiatric Disorders in the United States: Results from the National Comorbidity Survey. *Archives of General Psychiatry, 51*, 8-19.

Kessler, R. C., Sonnega, A., Bromet, E., Hughes, M., & Nelson, C. B. (1995). Posttraumatic stress disorder in the National Comorbidity Survey. *Archives of General Psychiatry, 52*(12), 1048-1060.

King, D. W., King, L. A., Foy, D. W., Keane, T. M., & Fairbank, J. A. (1999). Posttraumatic stress disorder in a national sample of female and male Vietnam veterans. Risk factors, war-zone stressors, and resilience-recovery variables. *Journal of Abnormal Psychology, 108*(1), 164-170.

Kirmayer, L. J., Robbins, J. M., & Paris, J. (1994). Somatoform disorders: Personality and the social matrix of somatic distress. *Journal of Abnormal Psychology, 103*(1), 125-136.

Kirschbaum, C., Kudielka, B. M., Gaab, J., Schommer, N. C., & Hellhammer, D. H. (1999). Impact of gender, menstrual cycle phase, and oral contraceptives on the activity of the hypothalamus-pituitary-adrenal axis. *Psychosomatic Medicine, 61*(2), 154-162.

Kivlighan, K. T., Granger, D. A., & Booth, A. (2005). Gender differences in testosterone and cortisol response to competition. *Psychoneuroendocrinology, 30*(1), 58-71.

Klein, L. C. & Corwin, E. J. (2002). Seeing the unexpected: How sex differences in stress responses may provide a new perspective on the manifestation of psychiatric disorders. *Current Psychiatry Reports, 4*, 441–448.

Koenen, K. C. & Widom, C. S. (2009). A prospective study of sex differences in the lifetime risk of posttraumatic stress disorder among abused and neglected children grown up. *Journal of Traumatic Stress, 22(6)*, 566–574.

Kudielka, B. M. & Kirschbaum, C. (2005). Sex differences in HPA axis responses to stress: A review. *Biological Psychology, 69*(1), 113-132.

Lilly, M. M., Pole, M., Best, S. R., Metzler, T., & Marmar, C. R. (2009). Gender and PTSD: What can we learn from female police officers? *Journal of Anxiety Disorders 23*, 767–774.

Lobbestael, J. & Arntz, A. (2010). Emotional, cognitive and physiological correlates of abuse-related stress in borderline and antisocial personality disorder. *Behaviour Research and Therapy, 48*, 116–124.

Lynn, R. & Martin, T. (1997). Gender differences in extraversion, neuroticism, and psychoticism in 37 nations. *Journal of Social Psychology, 137*(3), 369-373.

McNally, R. J., Bryant, R. A., & Ehlers, A. (2003). Does early psychological intervention promote recovery from posttraumatic stress? *Psychological Science*, 45-79.

Mollica, R. F., Caspi-Yavin, Y., Bollini, P., Troung, T., Tor, S., & Lavelle, J. (1992). The Harward Trauma Questionnaire: Validating a cross-cultural instrument for measuring torture, trauma, and posttraumatic stress disorder in Indochinese refugees. *The Journal of Nerveous ans Mental Disease, 180(2)*, 111-116.

Neumann, I. D. (2008). Brain oxytocin: A key regulator of emotional and social behaviours in both females and males. *Journal of Neuroendocrinology, 20*(6), 858-865.

Neumann, I. D., Wigger, A., Torner, L., Holsboer, F., & Landgraf, R. (2000). Brain oxytocin inhibits basal and stress-induced activity of the hypothalamo-pituitary-adrenal axis in male and female rats: Partial action within the paraventricular nucleus. *Journal of Neuroendocrinology, 12*(3), 235-243.

Norris, F. H., Foster, J. D., & Weisshaar, D. L. (2007). The epidemiology of sex differences in PTSD across developmental, societal, and research contexts, in *Gender and PTSD*, Kimerling, R., Ouimette, P., & Wolfe, J., pp. 3-42, The Guilford Press, ISBN: 1-57230-783-8, NY.

Norris, F. H., Friedman, M. J., & Watson, P. J. (2002). 60,000 disaster victims speak: Part II. Summary and implications of the disaster mental health research. *Psychiatry, 65*(3), 240-260.

Norris, F. H., Friedman, M. J., Watson, P. J., Byrne, C. M., Diaz, E., & Kaniasty, K. (2002). 60,000 disaster victims speak: Part I. An empirical review of the empirical literature, 1981-2001. *Psychiatry, 65*(3), 207-239.

Norris, F. H., Perilla, J. L., Ibanez, G. E., & Murphy, A. D. (2001). Sex differences in symptoms of posttraumatic stress: Does culture play a role? *Journal of Traumatic Stress, 14*(1), 7-28.

Nugent, N. R., Brown, N. C. C., Crow, J. P., Browne, L., Ostrowski, S., & Delahanty, D. L. (2010). The efficacy of early propranolol administration at reducing PTSD symptoms in pediatric injury patients: A pilot study. *Journal of Traumatic Stress, 23(2)*, 282–287.

O'Connor M. & Elklit, A. (2008). Attachment styles, traumatic events, and PTSD: A cross-sectional investigation of adult attachment and trauma. *Attachment & Human Development, 1* , 59–71.

Olff, M., Langeland, W., Draijer, N., & Gersons, B. P. R. (2007). Gender differences in posttraumatic stress disorder. *Psychological Bulletin, 133*(2), 183-204.

Ozer, E. J., Best, S. R., Lipsey, T. L., & Weiss, D. S. (2003). Predictors of posttraumatic stress disorder and symptoms in adults: A meta-analysis. *Psychological Bulletin, 129*(1), 52-73.

Pariante, C. M. & Lightman, S. L. (2008). The HPA axis in major depression: Classical theories and new developments. *Trends in Neurosciences, 31(9)*, 464-468.

Paris, J. J., Franco, C., Sodano, R., Freidenberg, B., Gordis, E., Anderson, D. A., Forsyth, J. P., Wulfert, E., & Frye, C. A. (2010). Sex differences in salivary cortisol in response to acute stressors among healthy participants, in recreational or pathological gamblers, and in those with posttraumatic stress disorder. *Hormones and Behavior 57*, 35–45.

Peirce, J. M., Newton, T.L., Buckley, T. C., & Keane, T.M. (2002). Gender and psychophysiology of PTSD, in *Gender and PTSD*, Kimerling, R., Ouimette, P., & Wolfe, J., pp. 177-204, The Guilford Press, ISBN: 1-57230-783-8, NY.

Perry, B. D., Pollard, R. A., Blakley, T. L., Baker, W. L., & Vigilante, D. (1995). Childhood trauma, the neurobiology of adaptation, and "use-dependent" development of the brain: How "states" become "traits". *Infant Mental Health Journal, 16(4)*, 271-291.

Pratchett, L. C., Pelcovitz, M. R., & Yehuda, R. (2010). Trauma and violence: Are women the weaker sex? *Psychiatric Clinics of North America, 33*(2), 465-474.

Ptacck, J. T., Smith, R. E., & Zanas, J. (1992). Gender, appraisal, and coping: A longitudinal analysis. *Journal of Personality, 60*(4), 747-770.

Punamäki, R-L., Komproe, I. H., Quota, S., Elmasri, M., & de Jong, J. T. V. M. (2005). The role of peritraumatic dissociation and gender in the association between trauma and mental health in a Palestinian community sample. *American Journal of Psychiatry, 162(3)*, 545-551.

Ray, S. L. (2008). Evolution of posttraumatic stress disorder and future directions. *Archives of Psychiatric Nursing, 22(4)*, 217–225.

Rosario, M., Shinn, M., Mørch, H., & Huckabee, C. B. (1988). Differences in coping and social supports: Testing socialization and role constraint theories. *Journal of Community Psychology, 16*, 55-69.

Saxe, G. N., Stoddard, F., Hall, E., Chawla, N., Lopez, C., Sheridan, R., King, D., King, L., & Yehuda, R. (2005). Pathways to PTSD, part I: Children with burns. *American Journal of Psychiatry, 162*(7), 1299-1304.

Schonfeld, W. H., Verboncoeura, C. J., Fifera, S. K., Lipschutz, R. C., Lubeck, D. P., & Buesching, D. P. (1997). The functioning and well-being of patients with unrecognized anxiety disorders and major depressive disorder. *Journal of Affective Disorders 43*, 105–119.

Shalev, A. Y., Freedman, S., Peri, T., Brandes, D., Sahar, T., Orr, S. P., & Pitman, R. K. (1998). Prospective study of posttraumatic stress disorder and depression following trauma. *American Journal of Psychiatry, 155*(5), 630–637.

Shumaker, S. A. & Hill, D. R. (1991). Gender differences in social support and physical health. *Health Psychology, 10*(2), 102-111.

Simmons, C. A. & Granvold, D. K. (2005). A cognitive model to explain gender differences in rate of PTSD diagnosis. *Brief Treatment and Crisis Intervention, 5*(3), 290-299.

Spindler, H., Elklit, A., & Christiansen, D. M. (2010). Risk factors for posttraumatic stress disorder following an industrial disaster in a residential area: A note on the origin of observed gender differences. *Gender Medicine 7*(2), 156-165.

Spitzer, C., Klauer, T., Grabe, H-J., Lucht, M., Stieglitz, R-D., Schneider, W., & Freyberger, H. J. (2003). Gender differences in dissociation: A dimensional approach. *Psychopathology, 36*, 65–70.

Stanton, A. L., Danoff-Burg, S., Cameron, C., & Ellis, A. P. (1994). Coping through emotional approach: Problems of conceptualization and confounding. *Journal of Personality and Social Psychology 66*(2), 350-362.

Stone, L. B., Gibb, B. E., & Coles, M. E. (2010). Does the hopelessness theory account for sex differences in depressive symptoms among young adults? *Cognitive Therapy Research, 34,*177–187.

Swaab, D. F., Fliers, E., Hoogendijk, W. J., Veltman, D. J., & Zhou, J. N. (2000). Interaction of prefrontal cortical and hypothalamic systems in the pathogenesis of depression. *Progress in Brain Research, 126,* 369-396.

Tamres, L. K., Janicki, D., & Helgeson, V. S. (2002). Sex differences in coping behavior: A meta-analytic review and an examination of relative coping. *Personality and Social Psychology Review, 6*(1), 2-30.

Taylor, S. E. (2006). Tend-and-befriend: Biobehavioral bases of affiliation under stress. *Current Directions in Psychological Science, 15*(6), 273-277.

Taylor, S. E., Klein, L. C., Lewis, B. P., Gruenewald, T. L., Gurung, R. A. R., & Updegraff, J. A. (2000). Biobehavioral responses to stress in females: Tend-and-befriend, not fight-or-flight. *Psychological Review, 107*(3), 411-429.

Taylor, S. E., Lewis, B. P., Gruenewald, T. L., Gurung, R. A. R., Updegraff, J. A., & Klein, L. C. (2002). Sex differences in biobehavioral responses to threat: Reply to Geary and Flinn (2002). *Psychological Review, 109*(4), 751-753.

Tolin D. F. & Foa, E. B. (2008). Sex differences in trauma and posttraumatic stress disorder: A quantitative review of 25 years of research, *Psychological Trauma: Theory, Research, Practice, and Policy, S(1),* 37–85.

Van der Kolk, B. A. (2007). The history of trauma in psychiatry. In Friedman, M. J., Keane, T. M., & Resick, P. A., pp. 19-36, Handbook of PTSD. Guilford: New York.

Veith, R. C., Lewis, N., Linares, O. A., Barnes, R. F., Raskind, M. A., Villacres, E. C., Murburg, M. M., Ashleigh, E. A., Castillo, S., Peskind, E. R., Pascualy, M., & Halter, J. B. (1994). Sympathetic nervous system activity in major depression: Basal and desipramine-induced alterations in plasma norepinephrine kinetics. *Archives of General Psychiatry, 51,* 411-422.

Wizemann, T. M. & Pardue, M.-L. (Eds.). (2001). Committee on Understanding the Biology of Sex and Gender Differences, Board on Health Sciences Policy. *Exploring the biological contributions to human health: Does sex matter?* The National Academies Press. Washington, DC.

Westenbroek, C., Ter Horst, G. J., Roos, M. H., Kuipers, S. D., Trentani, A., & den Boer, J. A. (2003). Gender-specific effects of social housing in rats after chronic mild stress exposure. *Progress in Neuro-psychopharmacology and Biological Psychiatry, 27*(1), 21-30.

Zeidner, M. (2006). Gender group differences in coping with chronic terror: The Israeli scene. *Sex Roles, 54(3/4),* 297-310.

Zetsche, U., Ehring, T., & Ehlers, A. (2009). The effects of rumination on mood and intrusive memories after exposure to traumatic material: An experimental study. *Journal of Behavior Therapy and Experimental Psychiatry, 40*(4), 499-514.

Zlotnick, C., Zimmerman, M., Wolfsdorf, B. A., & Mattia, J. I. (2001). Gender differences in patients with posttraumatic stress disorder in a general psychiatric practice. *American Journal of Psychiatry, 158*(11), 1923-1925.

7

Risk Factors and Hypothesis for Posttraumatic Stress Disorder (PTSD) in Post Disaster Survivors

Frank Huang-Chih Chou[1] and Chao-Yueh Su[2]
[1]*Department of Community Psychiatry, Kai-Suan Psychiatric Hospital, Kaohsiung,*
[2]*Department of Nursing, I-Shou University, Kaohsiung City,*
Taiwan

1. Introduction

Disasters, both natural and man-made, affect millions of people around the world every year. Natural disasters (e.g., earthquakes and hurricanes) and man-made disasters (e.g., traffic accidents, acts of terrorism and wars) can cause mental trauma with long-lasting consequences (Chou et al., 2005; Chou et al., 2007). The impact of a mass disaster or man-made trauma on the individual is a composite of two major elements: the catastrophic event itself and the vulnerability of those people affected by the event. To this end, post-disaster survivors need specific, systemic evaluation and management (Sapir, 1993).

2. The relationship between disasters and Posttraumatic Stress Disorder (PTSD)

Breslau et al. (1991) estimated that 6% to 7% of the US population is exposed to disaster or trauma every year, while Wang et al. (2000) showed that natural disasters affect an average of approximately 200 million people in China every year, several thousand of whom do not survive. In the aftermath of these catastrophic events, PTSD is one of the most common psychiatric diseases suffered by post-disaster survivors.

The prevalence of PTSD ranged from 3.0% to 34.3% in Taiwan after the 1999 earthquake (Chou et al., 2004a,b), it was approximately 25% in Turkey after the 1999 earthquake (Tural et al., 2004), and it was reported as 74% in Armenia after the 1988 earthquake (Armen, 1993). In a systemic review of the literature, Andrews, Brewin, Philpott, & Stewart (2007) found that delayed-onset PTSD in the absence of any prior symptoms was rare, whereas delayed onset that represented exacerbations or reactivations of prior symptoms accounted for, on average, 38.2% and 15.3% of military and civilian cases of PTSD, respectively. Generally, the lifetime and current prevalence rates for psychiatric disorders range anywhere from 1% to 74% (Breslau, Davis, Andreski, & Peterson, 1991; Carr et al., 1995; Chang et al., 2003; Chou et al., 2003; Tainaka et al., 1998), with women twice as likely as men to be affected. Furthermore, women report more symptoms of anxiety and depression than men (Chou et al., 2003; Chang et al., 2003).

3. The introduction of PTSD

Clinicians have recognized the juxtaposition of acute mental syndromes to traumatic events for more than 200 years. Observations of trauma-related syndromes were documented following the Civil War, and early psychoanalytic writers, including Freud, noted the relation between neurosis and trauma (Kaplan & Sadock, 1999).

The American Psychiatric Association (APA) (1952) published the "Diagnostic and statistical manual of mental disorders, first edition, DSM-I" and included in that edition gross stress reactions. However, the term PTSD was not included in the publications until the DSM-III in 1980 (Jones et al., 2003). It was then revised in the DSM-III-R (1987) and the DSM-IV (1994). According to the DSM-IV diagnostic criteria, PTSD has three core psychopathologies: (a) re-experience, (b) numbness and avoidance, and (c) hyper-arousal. The DSM-IV diagnostic criteria for PTSD allow clinicians to specify if the disorder is chronic, that is, the symptoms have lasted three months or more, or if the disorder exhibits delayed onset, that is, the onset of the symptoms was six months or more after the stressful event (Su, Tsai, Chou, et al., 2010). PTSD is an anxiety disorder that develops after a person has been exposed to a severe, life-threatening trauma. Its symptoms include a re-experiencing or reliving of the event, an avoidance or numbness toward the event, and/or hyper-arousal (American Psychiatric Association, 1994). Accordingly, PTSD is characterized by two special memory phenomena. The first is a facilitated memory of the traumatic event, including flashbacks and nightmares. The second is an inhibited memory involving the inability to voluntarily recall important aspects of the trauma (Hellawell & Brewin, 2002; Thomaes et al., 2009). These observations imply that emotional memory dysfunctions are key components in PTSD, and they include involuntary retrieval such as flashbacks and intrusions, exaggerated and context-independent fear, failure to integrate the trauma as a coherent episode into an autobiographical memory, and impaired fear memory extinction (Wolf, 2008).

4. PTSD with psychiatric co-morbidity

The majority of the research (Goenjian et al., 2000; Green, Lindy, Grace, & Leonard, 1992; Maj et al., 1989; McFarlane & Papay, 1992; Rubonis & Bickman, 1991) provides evidence of psychological sequelae that includes PTSD, major depressive episodes, sleep disorder, anxiety, and substance abuse after disasters. Furthermore, major depressive episodes and PTSD are the most common disaster-related psychiatric diagnoses and are strongly associated with one another (McFarlane & Papay, 1992; Goenjian et al., 2000; Green et al., 1992). Individuals confronted with disasters or major stressors exhibit greater psychological impairment and are more vulnerable to psychiatric diseases (Chou et al., 2005). The incidence of PTSD is higher than that of other major depressive episodes in the majority of the studies (Bromet & Dew, 1995; Chou et al., 2003; Chou et al., 2004a; Chou et al., 2004b; Chou et al., 2005; Davidson et al., 1991; Davidson 1995; Goenjian et al., 1994; Green et al., 1992; Sharan et al., 1996). In contrast to natural disasters, however, higher co-morbidity has been found with combat-related PTSD. Such co-morbidity includes drug and alcohol abuse, antisocial personality disorder, somatization disorder, and depression, and it is particularly prevalent when determined from an historical perspective (Green et al., 1992). PTSD can be triggered by a variety of traumatic events and is strongly associated with all other examined mental disorders (Brady, Killeen, Brewerton, & Lucerini, 2000; Goenjian et al., 2000;

Perkonigg, Kessler, Storz, Wittchen, 2000). For example, the combination of PTSD and panic and phobic disorders is an important predictor for PTSD chronicity (McFarlane & Papay, 1992; Ursano, Kao, & Fullerton, 1992). Furthermore, the rate of psychopathology is higher in post-disaster groups than in either the same groups prior to trauma or in control groups (Maj et al., 1989; Rubonis & Bickman, 1991).

5. Psychiatric studies of post-Chi-Chi earthquake survivors

Researchers focusing on survivors of the Chi-Chi earthquake in Taiwan (Su, Chou, Lin, Tsai, 2010) have found evidence of psychological sequelae that includes posttraumatic stress disorder (PTSD), major depressive disorder, sleep disorder, anxiety, and substance abuse (Chou et al., 2004a, 2004b, 2005, 2007; Chen et al., 2001; Chang et al., 2002; Lai et al., 2004; Hsu et al., 2002; Kuo et al., 2003; Liu et al., 2006; Tsai et al., 2007; Wu et al., 2006; Yang et al., 2003). The quality of life for survivors of traumatic events who develop psychiatric illnesses or impairments is worse than that for survivors without any psychiatric illness (Chou et al., 2004b; Tsai et al., 2007; Wu et al., 2006). In addition, rescue workers such as nurses, fire fighters, and soldiers may develop physical or mental impairments (Chang et al., 2008; Liao et al., 2002; Shih et al., 2002; Yeh et al., 2002). We used PubMed to identify Chi-Chi earthquake-related papers published through June of 2009. All of the Chi-Chi earthquake papers related to psychiatry are summarized in Table 1 (cited from Su, Chou, Lin, Tsai, 2011).

6. The risk factors of PTSD

Researchers who study risk factors for PTSD have identified aspects of demographic data, psychological factors, psychiatric symptoms, and post-trauma social resource factors as important factors that contribute to the development of the disease.

6.1 Demographic data

Some researchers who have examined gender differences suggest that females are more likely than males to develop PSTD (Chou et al., 2005; Helzer, Robins, & McEvoy, 1987; Johnson & Thompson, 2008; Lazaratou et al., 2008). A possible explanation for this is the specific reactions that result from feminine characteristics to a traumatic event (Chou, Tsai, Wu, Su, & Chou, 2006). Additionally, there are previous studies that have associated old age with an increased risk of developing PTSD (Goenjian et al., 1994; Lewin, Carr, Webster, 1998). However, a recent study has suggested contradictory results (Lazaratou et al., 2008).

6.2 Biological factors

Neuroendocrine data provide evidence of insufficient glucocorticoid signaling in stress-related neuropsychiatric disorders, while Nutt (2000) has suggested that individuals develop PTSD due to neuroendocrine dysregulation. Furthermore, impaired feedback regulation of relevant stress responses, especially immune activation/inflammation, may, in turn, contribute to stress-related pathology that includes alterations in behavior, insulin sensitivity, bone metabolism, and acquired immune responses (Raison & Miller, 2003). Because the hypothalamic-pituitary gland-adrenal axis (HPA) regulates hormone reactions during stress, PTSD severity seems to decrease when individuals exposed to traumatic

Authors	Year published	Study period after earthquake	Subjects	Purpose	Method	Conclusion
Chen et al.	2001	Within 1 month	525 residents	Screening for psychiatric morbidity and post-traumatic symptoms among early-stage survivors	Purposeful sampling	Approximately 11% of the subjects reported having thoughts of death or ideas of suicide; approximately 89.9% of respondents had psychological impairment
Chen et al.	2001	Within 1 year	210 residents	The Chinese version of the Davidson Trauma Scale, a practice test for validation	Translation, back-translation and concurrent validity	The sensitivity of the instrument was 0.9, specificity 0.81, positive likelihood ratio 4.74, and negative likelihood ratio 0.12
Chang et al.	2002	6 months later	171 pregnant residents	Psychiatric morbidity and pregnancy outcome in a disaster area	Purposeful sampling	The prevalence of minor psychiatric morbidity (MPM) was 29.2%.
Hsu et al.	2002	6 weeks later	323 student residents	Post-traumatic stress disorder among adolescent earthquake victims in Taiwan	Purposeful sampling	Of the 323 students, 21.7% had PTSD. Being physically injured and experiencing the death of a close family member with whom they had lived were identified as two major risk factors
Liao et al.	2002	2 months later	1,104 rescue workers serving in the area hit by the earthquake	Association of psychological distress with psychological factors in rescue workers.	Purposeful sampling	Prevalence of general psychological distress is high among rescue workers. Personality traits and pre-disaster life adjustment had a dominant predictive effect on psychological distress
Lin et al.	2002	1 year later	368 residents with 268 ≧65 y/ o	Geriatric survivors	Purposeful sampling	Lower quality of life in physical capacity, psychological well-being, and environment 12 months after the earthquake when compared to assessment prior to the earthquake were identified as risk factors

Authors	Year published	Study period after earthquake	Subjects	Purpose	Method	Conclusion
Shih et al.	2002	Within 1 year	46 nurses who worked in a hospital in the community	The impact of the 9-21 earthquake experience on Taiwanese nurses acting as rescuers	Purposeful sampling	Rescue experience strengthened most Taiwanese nurses' professional competency
Yeh et al.	2002	Within 16 days	187 young male military personnel who served as rescue workers	Characteristics of acute stress symptoms and nitric oxide concentration in young rescue workers in Taiwan.	Purposeful sampling	A significant inverse correlation observed between the severity of stress symptoms and plasma concentration of nitric oxide in rescue workers
Chang et al.	2003	5 months later	84 male fire fighters	Post-traumatic distress and coping strategies among rescue workers	Purposeful sampling	Study identified a 16.7% and 21.4% prevalence for general psychiatric morbidity and post-traumatic morbidity, respectively
Chou et al.	2003	21 months later	461 residents	Establishment of a disaster-related psychological screening test	Population survey	DRPST, administered in phase 1 of this two-phase study, may be used for effective and rapid screening for PTSD and MDE after an earthquake
Kuo et al.	2003	2 months later	120 bereaved survivors	To investigate the prevalence of psychiatric disorders and risk factors for PTSD and major depressive disorder among bereaved survivors	Purposeful sampling	The prevalence of PTSD was 37%,; that of major depressive disorder was 16%
Yang et al.	2003	3 months later	663 victims	To investigate the psychiatric morbidity and post-traumatic symptoms among earthquake victims in primary care clinics	Purposeful sampling	PTSD was 11.3%, partial PTSD was 32.0%

Authors	Year published	Study period after earthquake	Subjects	Purpose	Method	Conclusion
Chou et al.	2004a	21-24 months	461 residents	To investigate quality of life and related risk factors of Taiwanese earthquake survivors with different psychiatric disorders	Purposeful sampling	The prevalence of varied psychiatric disorders in earthquake survivors ranged from 3.3% to 18%
Chou et al.	2004b	4-6 months	4223 residents	To investigate the relationship between quality of life and psychiatric impairment	Purposeful sampling	PTSD: 7.6%, suspected PTSD: 26.7% and poor quality of life with psychiatric impairments in these respondents
Guo et al.	2004	1 month	252 rescue workers	To investigate the prevalence of post-traumatic stress disorder (PTSD) among professional and non-professional rescue workers involved in the 1999 Chi-Chi Earthquake in Taiwan	Purposeful sampling	Professional and non-professional rescue workers showed prevalence of 19.8% and 31.8%, respectively. Disaster rescue work is associated with a high level of stress, even for highly trained professionals. This work may, therefore, lead to mental health problems.
Lai et al.	2004	10 months	252 residents	Full and partial PTSD among earthquake survivors in rural Taiwan	Randomly selected from two rural communities	Prevalence rates for PTSD (n=26) and PTSS (n=48) were 10.3% and 19.0%, respectively
Chou et al.	2005	4-6 months	442 residents	To assess the development of psychiatric disorders among residents after earthquake	Population survey	Females had s ignificantly higher rates for most psychiatric disorders compared to males
Yang et al.	2005	During a 7-year period	---	To examine the time trends of increased suicide rates	Time-series analysis	Mean monthly suicide rate for earthquake victims was higher; indicates the need for providing strengthened psychiatric services during first year following major disasters

Authors	Year published	Study period after earthquake	Subjects	Purpose	Method	Conclusion
Sepalki et al.	2006	Before and after the earthquake	1160 older individuals	To investigate variability in resilience to depressive symptoms in aftermath of 1999 Taiwan earthquake	Longitudinal survey with interviews	Persons of low socioeconomic status (SES), socially isolated individuals, and women reported higher levels of depressive symptoms than respective counterparts, as did persons who experienced damage to their homes; Psychological effects of damage were strongest among those aged 54-70
Wu et al.	2006	33-36 months	405 residents	To investigate quality of life and related risk factors in earthquake survivors diagnosed with different psychiatric disorders	Population survey	Prevalence range for psychiatric disorders in earthquake survivors was 0.2% to 7.2%; persistence of long-term economic problems was one of many important factors affecting quality of life
Chen et al.	2007	2 years later	6412 earthquake survivors whose houses were destroyed	To examine prevalence and risk factors for post-traumatic stress symptoms and psychiatric morbidity	Purposeful sampling	Estimated rates of PTSD and psychiatric morbidity were 20.9% and 39.8%, respectively; severe earthquakes can cause long-term psychological impact in survivors
Chou et al.	2007	0.5, 2, 3 years later	442, 461, 405 residents	To survey dynamic population for risk factors for PTSD and major depression and assess prevalence of different psychiatric disorders 6 months, 2 years, and 3 years after the earthquake.	Population survey	PTSD prevalence significantly decreased 3 years later; suicidal tendency and drug abuse/ dependence significantly increased

Authors	Year published	Study period after earthquake	Subjects	Purpose	Method	Conclusion
Kuo et al.	2007	1 years later	272 victims from temporary housing units	To investigate the incidence of PTSD and psychological health status among earthquake victims one year after quake	Purposeful sampling	Posttraumatic stress symptoms and psychological problems more prevalent among women (22.2% and 64%, respectively) than men (9.2% and 47.9%, respectively)
Tsai et al.	2007	3 years later	1756 respondents	To evaluate, prospectively, the relationship between clinical course of PTSS and quality of life (QOL)	Fixed cohort follow-up	Three years after earthquake, estimated rate of posttraumatic stress symptoms declined; survivor quality of life varied according to survivor's progression of PTSS
Chang et al.	2008	---	193 fire fighters	To investigate modification effects of coping strategies on relationships between rescue effort and psychiatric morbidity in earthquake rescue workers	Purposeful sampling	Older age and longer job experience (>3 years) were associated with both general psychiatric and post-traumatic morbidities

Cited from Su et al. (2011) under permission

Table 1. Summary of Chi-Chi earthquake papers related to psychiatry (PubMed search through June 2009) (cited from Su et al., with permission)

events experience decreased stress levels. Thabet & Vostanis, (2000) and Gurvits et al. (1997) found more positive soft neurological signs in PTSD participants than in participants who experienced similar trauma but did not develop PTSD. Many trauma victims complain of memory impairment, such as difficulty remembering daily activities, frequent compulsive recall of the traumatic event in detail, memory gaps, island-like memory, difficulties with declarative memory, and intrusive memories. Anderson et al. (2004) used functional magnetic resonance imaging (MRI) to identify the neural systems involved in keeping unwanted memories out of one's awareness. Controlling unwanted memories is associated with increased dorsolateral prefrontal activation (DLPF), reduced hippocampal activation, and impaired retention of those memories. Both prefrontal cortical and right hippocampal activations predicted the magnitude of forgetting. These results confirm the existence of an active forgetting process and establish a neurobiological model for guiding inquiry into motivated forgetting.

There are still gaps in our understanding of the genetic underpinnings of PTSD. For example, while Stein et al. (2002) have found moderate hereditary factors in individuals with PTSD symptoms, no single gene that causes PTSD has been identified.

6.3 Psychological factors and psychiatric symptoms

Meyer et al. (1999) indicated that some psychiatric symptoms and disorders are risk factors for PTSD (Meyer, Taiminen, Vuori, Aijälä, Helenius, 1999). For example, certain personality traits, such as neuroticism and introversion, are associated with an increased risk of PTSD (Lewin, Carr, & Webster, 1998; McFarlane, 1988) while some studies indicate that certain psychiatric disorders may be predictive of chronic PTSD (Engdahl, Dikel, Eberly, & Blank, 1998; McFarlane & Papay, 1992). Then again, other studies have examined the long-term course of PTSD. A longitudinal analysis of the mental health of school children after the great Hanshin Awaji earthquakes indicated that some survivors' psychological reactions emerged early and disappeared early (i.e., within two years after the disaster); however, this is contrary to findings from other studies (Shioyama et al., 2000). Lazaratou et al. (2008) have found that greater numbers of PTSD symptoms emerged during the first 6 months after the earthquake and were associated with a greater impact on the victims' lives 50 years after the event. Uemoto et al. (2000) posited that the best predictor of recovery from chronic PTSD was the initial level of post-traumatic reaction immediately after the accident. However, few data are available on the long-term effects caused by a disaster (Chou et al., 2007).

6.4 Post-trauma social resource factors

Inadequate social support after the trauma adds to the risk of developing PTSD (Chou et al., 2004a; Wang et al., 2000). Not surprisingly, higher levels of post-disaster life events are also related to the risk of developing PTSD (Chang, Connor, Lai, Lee, & Davidson, 2005). Similarly, social stressors such as economic or marital issues or a disruption of one's daily life, including relocation, the death of an intimate partner, or other significant loss problems are associated with a greater risk for developing PTSD.

7. Hypothesis for PTSD

Hobfoll's conservation of resources (COR) model has been well supported by previous studies on natural disasters (Sumer, Karanci, Berument, & Gunes, 2005). According to

Hobfoll's conservation of resources stress theory (Hobfoll, 1989; Chou et al., 2007), resource loss is an important determinant of individual stress and physical and mental health, including PTSD. Brewin et al. (2000) also found that the effect sizes of all risk factors were modest. Factors operating during or after the trauma (e.g., trauma severity, lack of social support, and additional life stress), however, had somewhat stronger effects than did pre-trauma factors.

Consequently, multiple risk factors constitute a network that results in psychiatric illness. According to Hobfoll's conservation of resources theory, resource loss is an important determinant of individual stress and physical and mental health, including PTSD. Our hypothesis states that an individual reaches a sub-threshold of psychiatric illness and then develops the illness due to a decreasing availability of resources, an accumulation of risk factors, and/or a major stressful event. Furthermore, unresolved, sub-clinical psychiatric symptoms caused by a disaster or major life event may increase a survivor's sensitivity to future stresses. When faced with stress, frustration (e.g., life events), or traumatic events (e.g., brain damage or deprivation of internal or external resources) individuals, either suddenly or gradually, become more vulnerable to psychiatric impairment and diseases such as PTSD. An individual might reach a sub-threshold of PTSD and then develop the illness due to a decreased availability of resources, an accumulation of risk factors (personality traits, poor social interactions, etc.) or a major stressful event. Furthermore, unresolved, subclinical psychiatric symptoms caused by a disaster may increase a survivor's sensitivity to future stresses. Other factors that tend to increase an individual's vulnerability to psychiatric problems include brain damage, heredity, personality traits, life events, and social interactions.

8. The treatment and rehabilitation of PTSD

Treatment or rehabilitation efforts should concentrate not only on severe psychiatric symptoms, emotional disturbances and personality traits or disorders, but also on interpersonal and social-environmental interactions. To treat PTSD, clinicians only use drugs and do not provide psychosocial treatment; thus, they cannot meet the true needs of the survivor. Based on the bio-psychosocial causation model of psychiatric disease as it applies to public health, we propose a model of the causation of PTSD. Issues related to PTSD that are most in need of further study include biological causation, psychosocial recovery, and long-term evaluation of psychological rehabilitation.

9. Conclusion

Although changes in emotional, cognitive, behavioral, and biologic states are transitory for most individuals after a catastrophe or major trauma, psychological trauma may persist much longer in some victims. While the psychological profiles of these victims are often altered, given their vivid and repetitive recollection of the traumatic events (Chou et al., 2004b; Chou et al., 2005; Lin et al., 2002), Wang et al. (2000) determined that prompt and effective post-disaster intervention might mitigate the impact of initial exposure and reduce the probability of PTSD occurrence. Issues related to PTSD most in need of further study include biological causation, psychosocial recovery, and long-term evaluation of psychological rehabilitation.

10. References

American Psychiatric Association (1994). *Diagnostic and Statistical Manual of Mental Disorders-Fourth Edition.* Washington, DC: American Psychiatric Association.

Anderson MC, Ochsner KN, Kuhl B, Cooper J, Robertson E, Gabrieli SW, et al. (2004). Neural systems underlying the suppression of unwanted memories. Science. 303: 232-235.

Andrews B, Brewin CR, Philpott R, Stewart L (2007). Delayed-onset posttraumatic stress disorder: a systematic review of the evidence. Am J Psychiatry. 164:1319-26.

Armen G. (1993). A mental health relief programme in Armenia after the 1988 earthquake: implementation and clinical observations. Br J Psychiatry. 163: 230-239.

Brady KT, Killeen TK, Brewerton T, Lucerini S. (2000). Comorbidity of psychiatric disorders and posttraumatic stress disorder. J Clin Psychiatry, 61(Suppl)7: 22-32.

Breslau N, Davis GC, Andreski P, Peterson E. (1991). Traumatic events and post-traumatic stress disorder in an urban population of young adults. Arch Gen Psychiatry. 48: 216-222.

Brewin CR, Andrews B, Valentine JD. (2000). Meta-analysis of risk factors for posttraumatic stress disorder in trauma-exposed adults. J Consult Clin Psychol. 68: 748-766.

Carr VJ, Lewin TJ, Webster RA, Hazell PL, Kenardy JA, Carter GL. (1995). Psychosocial sequlae of the 1989 Newcastle earthquake: I. Community disaster experiences and psychological morbidity 6 months post-disaster. Psychol Med. 25: 539-555.

Chang CM, Connor KM, Lai TJ, Lee LC, Davidson JR. (2005). Predictors of posttraumatic outcomes following the 1999 Taiwan earthquake. J Nerv Ment Dis. 193: 40-46.

Chang CM, Lee LC, Connor KM, Davidson JR, Jeffries K, Lai TJ. (2003). Posttraumatic distress and coping strategies among rescue workers after an earthquake. J Nerv Ment Dis, 191:391-398.

Chang CM, Lee LC, Connor KM, Davidson JR, Lai TJ. (2008). Modification effects of coping on post-traumatic morbidity among earthquake rescuers. Psychiatry Res. 158(2): 164-171.

Chang HL, Chang TC, Lin TY, Kuo SS. (2002). Psychiatric morbidity and pregnancy outcome in a disaster area of Taiwan 921 earthquake. Psychiatry Clin Neurosci. 56: 139-144.

Chen CC, Yeh TL, Yang YK, Chen SJ, Lee IH, Fu LS, et al. (2001). Psychiatric morbidity and post-traumatic symptoms among survivors in the early stage following the 1999 earthquake in Taiwan. Psychiatry Res. 105: 13-22.

Chou FH, Chou P, Lin C, Su TT, Ou-Yang WC, Chien IC, et al. (2004b). The relationship between quality of life and psychiatric impairment for a Taiwanese community post earthquake. Qual Life Res. 13: 1089-1097.

Chou FH, Chou P, Su TT, Ou-Yang WC, Chien IC, Lu MK, et al. (2004a). Survey of quality of life and related risk factors for a Taiwanese village population 21 months after an earthquake. Aust N Z J Psychiatry. 38: 358-364.

Chou FH, Su TT, Chou P, Ou-Yang WC, Lu MK, Chien IC. (2005). Survey of Psychiatric Disorders in a Taiwan Village Population Six Months after a Major Earthquake. J Formosan Med Association. 104: 308-317.

Chou FH, Su TT, Ou-Yang WC, Chien IC, Lu MK, Chou P. (2003). Establishment of a disaster-related psychological screening test. Aust NZ J Psychiatry. 37(1): 97-103.

Chou FH, Tsai KY, Wu HC, Su TT, Chou P. (2006). Disaster and Posttraumatic Stress Disorder. Taiwanese J Psychiatry. 20: 85-103.

Chou FH, Wu HC, Chou P, Su CY, Tsai KY, Chao SS, et al. (2007). Epidemiologic Psychiatric Studies on Postdisaster Impact among Chi-Chi Earthquake Survivors in Yu-Chi, Taiwan. Psychiatry Clin Neurosci. 61: 370-378.

Davidson JRT, Hughes DL, Blazer DG, George LK (1991). Posttraumatic stress disorder in the community: An epidemiological study. Psychol Med. 21:713.

Davidson JRT. (1995). Posttraumatic stress disorder and acute stress disorder, in Comprehensive Textbook of Psychiatry, 6th ed. Edited by Kaplan HI, Sadock BJ. Baltimore, Williams & Wilkins, 1227-1236.

Engdahl B, Dikel TN, Eberly R, Blank A Jr. (1998). Comorbidity and course of psychiatric disorders in a community sample of former prisoners of war. Am J Psychiatry. 155: 1740-1745.

Goenjian AK, Najarian LM, Pynoos RS, Steinberg AM, Manoukian G, Tavosian A, et al. (1994). Posttraumatic stress disorder in elderly and younger adults after the 1988 earthquake in Armenia. Am J Psychiatry. 151: 895-901.

Goenjian AK, Steinberg AM, Najarian LM, Fairbanks LA, Tashjian M, Pynoos RS. (2000). Prospective study of posttraumatic stress, anxiety, and depressive reactions after earthquake and political violence. Am J Psychiatry. 157: 911-916.

Green BL, Lindy JD, Grace MC, Leonard AC. (1992). Chronic posttraumatic stress disorder and diagnostic comorbidity in a disaster sample. J Nerv Ment Dis. 180: 760-766.

Gurvits TV, Gilbertson MW, Lasko NB, Orr SP, Pitman RK. (1997). Neurological status of combat veterans and adult survivors of sexual abuse PTSD. Ann N Y Acad Sci. 821: 468-471.

Hellawell SJ, Brewin CR (2002). A comparison of flashbacks and ordinary autobiographical memories of trauma: cognitive resources and behavioural observations. Behaviour Research & Therapy, *40*(10): 1143-1156.

Helzer JE, Robins LN, McEvoy L. (1987). Posttraumatic stress disorder in the general population : findings of the Epidemiologic Catchment Area Survey. N Engl J Med; 317:1630-4.

Hobfoll SE(1989): Conservation of Resources: a new attempt at conceptualizing stress. The American Psychologist 44: 513-524.

Hsu CC, Chong MY, Yang P, Yen CF. (2002). Posttraumatic stress disorder among adolescent earthquake victims in Taiwan. J Am Acad Child Adolesc Psychiatry. 41: 875-881.

Johnson H, Thompson A. (2008). The development and maintenance of post-traumatic stress disorder (PTSD) in civilian adult survivors of war trauma and torture: a review. Clin Psychol Rev. 28: 36-47.

Jones E, Vermaas RH, Mccartney H, et al. (2003). Flashbacks and post-traumatic stress disorder: the genesis of a 20th-century diagnosis. Br J Psychiatry. 158-63.

Kaplan, Sadock's (1999). Comprehensive Textbook of Psychiatry, 7th ed.

Kuo CJ, Tang HS, Tsay CJ, Lin SK, Hu WH, Chen CC. (2003). Prevalence of psychiatric disorders among bereaved survivors of a disastrous earthquake in taiwan. Psychiatr Serv. 54: 249-251.

Lai TJ, Chang CM, Connor KM, Lee LC, Davidson JR. (2004). Full and partial PTSD among earthquake survivors in rural Taiwan. J Psychiatr Res. 38: 313-322.

Lazaratou H, Paparrigopoulos T, Galanos G, Psarros C, Dikeos D, Soldatos C. (2008). The psychological impact of a catastrophic earthquake: a retrospective study 50 years after the event. J Nerv Ment Dis. 196: 340-344.

Lewin TJ, Carr VJ, Webster RA. (1998). Recovery from post-earthquake psychological morbidity: who suffers and who recovers? Aust N Z J Psychiatry. 32: 15-20.

Liao SC, Lee MB, Lee YJ, Weng T, Shih FY, Ma MH. (2002). Association of psychological distress with psychological factors in rescue workers within two months after a major earthquake. J Formosan Med Association. 101(3): 169-176.

Lin MR, Huang W, Huang C, Hwang HF, Tsai LW, Chiu YN. (2002). The impact of the Chi-Chi earthquake on quality of life among elderly survivors inTaiwan:Abefore and after study. Qual Life Res. 11: 379-388.

Liu A, Tan H, Zhou J, Li S, Yang T, Wang J, et al. (2006). An epidemiologic study of posttraumatic stress disorder in flood victims in Hunan China. Can J Psychiatry. 51(61): 350-354.

Maj M, Starace F, Crepet P, Lobrace S, Veltro F, De Marco F, et al. (1989). Prevalence of psychiatric disorders among subjects exposed to a natural disaster. Acta Psychiatr Scand. 79: 544-549.

McFarlane AC, Papay P. (1992). Multiple diagnoses in posttraumatic stress disorder in the victims of a natural disaster. J Nerv Ment Dis. 180: 498-504.

McFarlane AC. (1988). The longitudinal course of posttraumatic morbidity. The range of outcomes and their predictors. J Nerv Ment Dis. 176(1): 30-39.

Meyer H, Taiminen T, Vuori T, Aijälä A, Helenius H. (1999). Posttraumatic stress disorder symptoms related to psychosis and acute involuntary hospitalization in schizophrenic and delusional patients. J Nerv Ment Dis. 187: 343-352.

Nutt DJ. (2000). The psychobiology of posttraumatic stress disorder. J Clin Psychiatry. 61(Suppl): 24-9; discussion 30-2.

Perkonigg A, Kessler RC, Storz S, Wittchen HU. (2000). Traumatic events and post-traumatic stress disorder in the community: prevalence, risk factors and comorbidity. Acta Psychiatr Scand. 101: 46-59.

Raison CL, Miller AH. (2003). When not enough is too much: the role of insufficient glucocorticoid signaling in the pathophysiology of stress-related disorders. Am J Psychiatry. 160: 1554-1565.

Rubonis AV, Bickman L. (1991). Psychological impairment in the wake of disaster: the disaster-psychopathology relationship. Psychol Bull. 109: 384-399.

Sapir DG. (1993). Natural and man-made disasters: the vulnerability of women-headed households and children without families. World Health Status Quartly. 46: 227-233.

Sharan P, Chaudhary G, Kavathekar SA, Saxena S. (1996). Preliminary report of psychiatric disorders in survivors of a severe earthquake. Am J Psychiatry; 153:556-558.

Shih FJ, Liao YC, Chan SM, Duh BR, Gau ML. (2002). The impact of the 9-21 earthquake experiences of Taiwanese nurses as rescuers. Social Science of Medicine. 55: 659-672.

Shioyama A, Uemoto M, Shinfuku N, Ide H, Seki W, Mori S, et al. (2000). The mental health of school children after the Great Hanshin-Awaji Earthquake: II. Longitudinal analysis. Seishin Shinkeigaku Zasshi. 102: 481-497.

Stein MB, Jang KL, Taylor S, Vernon PA, Livesley WJ. (2002). Genetic and environmental influences on trauma exposure and posttraumatic stress disorder symptoms: a twin study. Am J Psychiatry. 159: 1675-1681.

Su CY, Chou FH, Lin WK, Tsai KY. (2011). The Establishment of a Standard Operation Procedure (SOP) for Psychiatric Service after an Earthquake. Disasters. 2011;35 (3):587-605.

Su CY, Tsai KY, Chou FH, Liu RY, Lin WK. (2010) A Three-year, Follow-up Study of the 8 Psychosocial Predictors of Delayed and Unresolved PTSD in Taiwan Chi-Chi, Earthquake Survivors. Psychiatry Clin Neurosci 2010; 64(3):239-48.

Sumer N, Karanci AN, Berument SK, Gunes H. (2005). Personal resources, coping self-efficacy, and quake exposure as predictors of psychological distress following the 1999 earthquake in Turkey. J Trauma Stress. 18: 331-342.

Tainaka H, Oda H, Nakamura S, Tabuchi T, Noda T, Mito H. (1998). Workers' stress after Hanshin-Awaji earthquake in 1995—symptoms related to stress after 18 months [in Japanese; English abstract]. Sangyo Eiseigaku Zasshi. 40: 241-249.

Thabet AA, Vostanis P. (2000). Post traumatic stress disorder reactions in children of war: a longitudinal study. Child Abuse Negl. 24: 291-298.

Thomaes K, Dorrepaal E, Draijer NP, de Ruiter MB, Elzinga, BM, van, BAJ, et al. (2009). Increased activation of the left hippocampus region in Complex PTSD during encoding and recognition of emotional words: a pilot study. *Psychiatry Res, 171*(1): 44-53.

Tsai KY, Chou P, Chou FH, Su TT, Lin SC, Lu MK, et al. (2007). Three-year follow-up study on the relationship between posttraumatic stress symptoms and quality of life among earthquake survivors in Yu-Chi, Taiwan. J Psychiatr Res. 41(1-2): 90-96.

Tural U, Coskun B, Onder E, Corapçioğlu A, Yildiz M, Kesepara C, et al. (2004). Psychological consequences of the 1999 earthquake in Turkey. J Trauma Stress. 17(6): 451-459.

Uemoto M, Shioyama A, Koide K, Honda M, Takamiya S, Shirakawa K, et al. (2000). The mental health of school children after the Great Hanshin-Awaji Earthquake: I. Epidemiological study and risk factors for mental distress. Seishin Shinkeigaku Zasshi. 102: 459-480.

Ursano RJ, Kao T, Fullerton CS. (1992). PTSD and meaning: Structuring human chaos. J Nerv Mental Dis. 180: 756-759.

Wang X, Gao L, Shinfuku N, Zhang H, Zhao C, Shen Y. (2000). Longitudinal study of earthquake-related PTSD in a randomly selected community sample in north China. Am J Psychiatry. 57: 1260-1266.

Wolf OT (2008). The influence of stress hormones on emotional memory: relevance for psychopathology. *Acta Psychol,* 127(3):513-531.

Wu HC, Chou P, Chou FH, Su CY, Tsai KY, Ou-Yang WC, et al. (2006). Survey of quality of life and related risk factors for a Taiwanese village population 3 years post-earthquake. Aust N Z J Psychiatry. 40: 355-361.

Yang YK, Yeh TL, Chen CC, Lee CK, Lee IH, Lee LC, et al. (2003). Psychiatric morbidity and posttraumatic symptoms among earthquake victims in primary care clinics. Gen Hosp Psychiatry. 25: 253-261.

Yeh CB, Leckman JF, Wan FJ, Shiah IS, Lu RB. (2002). Characteristics of acute stress symptoms and nitric oxide concentration in young rescue workers in Taiwan. Psychiatry Res. 112: 59-68.

Part 3

African Perspective

8

War in Côte d'Ivoire and Management of Child's Post Traumatic Stress Disorders

A. C. Bissouma[1], M. Anoumatacky A.P.N[2] and M. D. Te Bonle[3]
[1]*Child Guidance Center-National Institute of Public Health and National Mental Health Programme Côte d'Ivoire,*
[2]*Psychiatric Hospital of Bingerville and National Mental Health Programme,*
[3]*Child Guidance Center-National Institute of Public Health, Côte d'Ivoire*

1. Introduction

For two decades, Cote d'Ivoire has gone through a number of crises that undermined the Ivoirian national cohesion. In September 2002, almost two years after the 1999 putsch and the stabilization of the social, military and political situation, with a president and a Government recognized by all and sundry, as everybody looked forward to leading a normal life, the suddenness of the war, which broke out in the night of 18th -19th, shook the foundation of the nation. Côte d'Ivoire was under attack, towns were besieged, populations were running up and down for dear lives…The death toll, the mass displacement of people both internally and externally, the medico-psychological trauma brought about by this situation provoked a real trauma among the populations with its consequences in terms of social disorganization.
The country was divided into three zones: the governmental zone, the trusted held by the French forces "*Licorne*" and the U.N forces "*ONUCI*", and another zone called CNW (Central, North and West) held by the rebels.
In December 2002, the conflict initially intensified in the western region, in the *Moyen Cavally* region, especially on the *Guiglo-Toulepleu* road where people witnessed the birth of the phenomenon of child soldiers. The conflict was deadly there and the fight lasted 3 years in that region. From 2006 to 2008, these child soldiers were attended to.
Psychiatrists were put in place. Unfortunately, the socio-psycho-medical interventions, which started in 2002, have not been supported by a formal organization to ensure sustainability.
Our objective is twofold: to describe on the one hand a unique experience in Côte d'Ivoire, that of a field care of child soldiers by an Ivoirian psychiatrist and on the other hand, analyze the intervention in order to better prepare for future interventions.

2. Background

What is the context of psycho trauma in Côte d'Ivoire? It is rather difficult to trace back the first actions on psycho trauma.
Although the country had previously passed through a number of situations and catastrophes with a number of traumatic experiences, no significant interventions to improve people's mental health in situations of mass trauma had been instituted.

2.1 Côte d'Ivoire geographic and sociodemographic context

Côte d'Ivoire is situated in West Africa in the sub-Saharan area. It covers an area of 322,462 square kilometers. It is bordered in the North by Burkina Faso and Mali, in the West by Liberia and Guinea, in the East by Ghana and in the South by the Gulf of Guinea.

The political capital of the country is Yamoussoukro, located in the heart of the country, some 248 km from Abidjan (in the South), and the economic capital. The official language is French. It is a country of immigrants, on account of being a crossroad of economic and cultural exchange. It has witnessed an urban growth since independence. The country probably has the best urban centers in Africa south of the Sahara.

On the sociopolitical level, Côte d'Ivoire is a democratic republic led by an executive President.

The population of Côte d'Ivoire was estimated in 2008 at 20,179,602 inhabitants. Forty three percent of the population is less than 15 year-old, and 49% are female among whom 51% are within the active reproductive age.

The Ivoirian population is characterized by its ethnic diversity. There are more than 60 ethnic groups divided into 4 main groups: the *Malinkés* in the northwest, the Voltas in the northeast, the *Krous* in the southwest, and the *Akan* sin the southeast.

Ivoirians are essentially religious-minded people, and the freedom of worship is guaranteed by the Constitution. The main religions are Christian faith, Islam and Animism.

As a rule, the Ivoirian population is diversified, young, barely literate and highly fertile; which constitutes a strong pressure on health agents who are over worked most of the time, especially in the situations of crises.

2.2 The different wars

Since the death of the Founding Father, His Excellency Felix Houphouet-Boigny in 1993, the country has always been prey to many uprisings. The climax was reached on the eve of Christmas in December 1999. The country knows its first putsch and a transition military take-over that lasted around 11months. At the end of the military confrontations linked to the putsch, people were traumatized and a few actions were taken against this traumatic experience. In 2000, a controversial election, urban confrontations and a military and political crisis, brought President Laurent Gbagbo to power. A number of initiatives, such as Reconciliation Days were organized in order to reunite the nation, as well as a few attempts of psychosocial actions. Despite this, on 19 September 2002, an armed rebellion cropped up that attempted to topple the Government. The failure of this attempt saw the partition of the country. The northern part fell in the hands of the rebels, while the Southern part remained under the control of government forces. A third zone, the trusted zone in the hands of the international forces (*Licorne* and *ONUCI*) representing the intervention forces separated the two warring forces.

In November 2004, the French army based in Abidjan, the economic capital in the South, attacked the Ivoirian army. People took to the streets and many casualties were recorded.

In August 2006, people were, once again, shaken by the problem of toxic waste damped into a number of sites in Abidjan. People concluded that, after the failure of the military coup, it was the time of chemical and bacteriological war.

On the political level, union governments came into existence, but their operations were once again hampered by internecine, partisan and political war. The country remained divided into two, even though on 31 July 2007, the reunification was announced. Despite of all these difficulties, the country lived on.

2.3 The 2002 war

In the night of 18 to 19 September 2002, a number of towns were attacked simultaneously: Korhogo in the North, Man in the West, Bouake in the central region, and Abidjan in the South.

The military and political crisis, facing Côte d'Ivoire at that moment would give birth to a humanitarian catastrophe, without precedent in the history of the country, and that would result in loss of human life among the civilian population affecting manily women and children. Around 1,500,000 persons were forced to leave the theater of war, (OCHA, 2004) either to seek refuge in areas under government control, or to seek shelter in neighboring countries (400,000 Ivoirian refugees).

More than 2,600 teachers and 704,800 students were displaced, including about 59,000 who were able to resume classes in the institutions labeled as relay schools in the free zone. But, only few of those displaced children did enjoy psychological health.

This mass movement of traumatized people and the disorganization of the social structure yielded dramatic health, social and psychic consequences yet to be investigated and addressed.

All these displaced persons had many difficulties to readapt because few of them received psychological support as part of handling war trauma. The absence of medico-psychological and social intervention due to lack of qualified health personnel was visible.

The phenomenon of child soldiers actually appeared in December 2002, on the occasion of the outbreak of a new tension source in the western region. The interethnic conflicts took around three years, in an area where many factions, including those from Liberia, a neighboring country fought the battles. This tension source developed in a region that has been receiving traditionally a number of Ivoirians and foreigners for years, and also, since the Liberian war, some refugee camps.

2.4 Actions on behalf of traumatized people

Few documents have reported the intervention undertaken since 1999, and few research works have explored this issue. No one can deny that some activity reports did exist, but they are yet to be known by the public. It is only in the course of the 2002 war that we discovered traces of the humanitarian interventions undertaken. Theses and dissertations carried out at National Institute of Public Health (INSP). Unit Taking over an Integrated of Abidjan (UPECI), at the psychiatric hospital of Bingerville (in the district of Abidjan) as well as at the *Centre Mie N'Gou of Yamoussoukro* (Bissouma& al, 2005; Kouadio, 2004; Kouakou, 2003), documented the psychopathological facts and disorders and confirmed the data of the international literature on the social, economic, psychological and medical consequences of the Ivorian war. These consequences may be categorized mainly into an increase in unemployment rate of 87.73% in Yamoussoukro after the war versus 21.82% before the war; an increase in psychological disturbances in the form mainly of anxiety, depressive and psychotic disorders; an increase insomnia, aggressive behavior and psychosomatic disorders, such as high blood pressure and diabetes mellitus; and the non-adaptation of the civilian populations to their host environments manifested by loss of interest in productive activities.

The consequences of this war have been dominated by post-traumatic stress and co-morbid signs with 93% of the victims experiencing sleeplessness and loss of appetite, depression, and an increase in consumption of toxic products(alcohol, tobacco, drug), and a certain degree of loss of social values for most of the population.

3. Situation analysis

In 2004, we took interest in the issues of psycho-trauma. At that time, despite the importance of the problem, there was insufficient interest in the issue and inadequate research had been carried out on it Cote d' Ivoire. The few studies that had come out were done in preparation for a doctorate of medicine dissertation and these studies were conducted among adults. Researches were about psychiatric disorders (depression, anxiety, and psychosis), but the issue of psycho trauma was not tackled. The situation of children was ignored.

A number of factors account for the relative lack of research on psycho-trauma among children and adolescents.

As a rule, in African societies, the child and adolescent mental disorders are completely neglected. Instead, people are more concerned with the mental well being of parents and guardians. Some reform to improve the mental health of adults has occurred though more needs to be done. In African societies, to resort to psychiatric care means madness. Psychiatry is still stigmatized and a number of people refuse to accept the value of psychiatric care. The Brazzaville WHO conference held in 2005 came up with solutions for a better follow-up of mental health issues among children and adolescents.

When, finally, the issue of child soldiers emerges, it was more or less unwelcome, more or less unrecognized and not readily accepted by the Government. People working within a national non-governmental organization (NGO) to rehabilitate former child soldiers expressed their desire to be backed up in rehabilitation efforts by qualified mental health specialists. Convinced that these children were at risk of the traumatic effects of war and were at risk of developing post-traumatic stress disorders, it became urgent, for this NGO, to evaluate the mental health of the former child soldiers in preparation to help the children.

In June 2006, an assessment mission went to the western region of the country to assess the real needs in the field of psychosocial reform program for former child soldiers. The terms of reference of this consultancy were to assess the mental health situation of 400 child soldiers and their families living in five (5) villages on the *Guiglo-Toulepleu* road (*Kaade, Behoue, Ke-Bouebo, Pehe and Pantrokin*) and to assist them psychologically.

This activity ran from June 2006 to August 2008. In that period, a number of working sessions in *Guiglo*, a western region city, were conducted in order to work with the NGO staff in assisting the former child soldiers, but secondarily with Liberian young refugees (we will not discuss here). Our additional task was to support psychologically the personnel and to help them in their psychosocial activities.

In Côte d'Ivoire, that experience was unheard of because children's psycho traumatic experiences were not well appreciated and were ranked as second-class activity. Ivoirian psychiatrists did not invest in this area, leaving the field to humanitarian agencies. Similarly, Jézéquel (2006) underlines that, in the case of conflicts in Africa, the issue of child soldiers has been initially the prerogative of aid agencies. For him, child soldiers have become the symbols of an African continent on the decline, a "heart of darkness" alien to European culture. It becomes the object of a new "aid agency crusade", a western neo-interventionism paralleling the civilizing missions of the past.

At the beginning, this mission proved difficult on account of the nonexistent consensus national structure to provide various forms of assistance for victims of traumatic stress in Côte d'Ivoire, including the availability of competent human resource and material means, difficulties to access the area to carry out research (distances, situation and insecurity on the road), the impossibility to undertake a regular follow-up, and the persistent instability and insecurity in the zone

My initial survey carried out during the draft of my specialization in psychiatry dissertation entitled: "*The war and the medico-psychology situations of children received at CGI and colliged case in the community*" formed the background literature for our intervention in the region.

This map shows the division of Cote d'Ivoire after the outbreak of the War of 2002. The country was divided into three zones (government, CNW zone under the control of the rebels and trusted zone). Cote d'Ivoire has been reunited July 30, 2007. The red line materializes the axis Guiglo-Touleupleu.

Fig. 1. Map of Côte d'Ivoire

As a young psychiatrist leading the team, we were doomed to venture on the slippery field of psycho-trauma without reference or theory.

We had to cross the country from South to West (around 600km) under difficult conditions. We used to leave Abidjan at sunrise to reach *Guiglo* at sunset, most of the time after an endless journey on a car or on a "*gbaka*" (a dilapidated 18-20 seat mini-car). We had to go there at our own expenses. On site, the NGO personnel organized both activities and sojourn.

At that time, there were instances of insecurity in the area and armed bands were still operating in spite of the program of disarmament set up by the Government.
At the same time, in these villages, there was no health center, and the people's somatic problems were difficult to solve (people had to go either to *Guiglo* or to *Toulepleu*); finally the NGO had to hire a male nurse. Schools were closed down and teachers had not come back. Some villages had no electricity and no telephone.
The traces of the war were visible everywhere: houses destroyed, walls riddled with bullet impacts, faces mirroring unspeakable suffering; misery and poverty seemed to be the daily companion of the population.
In such a situation, the implementation of this far-reaching project that consisted in rebuilding human lives, in giving back a meaning to life and to raising children psychically by healing their invisible wounds named traumas, proved to be an arduous but inspiring task. We needed to face a huge undertaking, that of children requesting care, not always psychic, but often somatic, that of parents for whom we were all doctors and who were begging for assistance, that of participants who were most of the time overworked, psychically suffering sometimes from the burden of the task, under the tough conditions of the mission. The question was to bring answers, a little satisfying to everybody and to each one.

4. Methodology

The implementation of such a project required a methodology with a clear-cut description, feasible and doable but at the same time flexible enough to adapt to unforeseen field events in order to let us plan our actions on a daily basis.

4.1 Population

The population in this project comprised of two groups: child soldiers or those associated with the battle and the untrained field assistants to tackle the issues of psychological consequences of traumatic stress (10 NGO "Social workers" and 10 organizers from the village community).
The structure, which organized the field activities, had listed 500 children, but, finally 400 were recruited into the project. The sample retained for the study of psychopathological disorders was made up of 345 children broken down as follows: 93 children from *Kaadé,* 80 from *Béoué*, 118 from *Ké-Bouébo*, 67 from *Péhé* and 67 from *Pantrokin.*
Local NGO workers and workers from the village community, many of whom had no experience, assisted children with psychic suffering. Most local agents were either veterans or inhabitants of the village who had lived the events themselves and who had not been assisted psychologically. As for the local agents, they were condemned to live with permanent anguish, in the midst of the villagers under precarious sanitary and security conditions.

4.2 The project itself

The project, which took two years to complete, progressed through many phases:
an initial assessment phase before intervention, with an initial assessment of the psychopathological situation of child soldiers and a definition of the intervention to be taken for social reintegration of those children;

an action phase with psycho medical consultations, educative assistance by the organizers of the structure (group therapy, literacy, re socialization), term assessment seminars.
Children's psychiatric consultation took place every three months. Thus, for a week, the organizers made an assessment of the situation of children and the stock of global assistance.
A final assessment phase of the children's psychopathological situation was conducted after intervention in the year 2008.

4.3 Means

We set up a data bank in order to study the psychopathological characteristics of the former child soldiers.
The collected data concerned the socio demographic (sex, age) and psychological characteristics (sexual activity, symptoms). This data gave us the opportunity to gather indicators to better plan our actions.
We asked for educational and playing equipment: balls, toys, pencils and felt-tip pens, building in games, paint and drugs: haloperidol, chlorpromazin, levomepromazin, and trihexyphenidyl (an antiparkinsonian) to palliate the side effects of neuroleptics.

4.4 Project implementation

The NGO project started at the end of the year 2005. Our fieldwork kicked off in June 2006 with a series of consultations with 400 former child combatants. The objectives of the initial consultations were to assess the mental health needs of the children, to train the community mobilizers to be attentive to the needs of the children, to initiate and supervise the administration of drugs as indicated, and finally to provide psychotherapeutic services. We received children either individually or in groups depending on their needs.
We planned and conducted field activities in series, and each activity lasted seven days; field activities began with an assessment of the psychological needs of the children followed by the training of the community mobilizers.
We conducted community awareness campaigns to convince parents on the harmful effects of war on the psychological health of individuals. The campaigns were conducted at market places, and an average of 100 persons per village (300 persons were present in *Kaadé)* attended the sessions.
At the end of the awareness campaigns, 177 persons asked for medical assistance though only 30 people eventually came for consultation.
Some major signs of psychic disorder were identified among parents including depression, psychosis and fear with a feeling of suspicion, demotivation with low performance at work and an accentuation of poverty situation.
Following our assessment of children's mental health needs, we engaged them in a variety of therapeutic activities such as drawing (draw your house and your family, before, during and after the war), and we engaged the older children in income generating activities to determine how the children would adapt to work situations. In the course of implementing our fieldwork we wanted to observe how the children behaved, how the village and family environment might be a limiting factor, how the weather changes and seasons influenced the engagement of children and how the children's social and cultural factors might influence the future of the children. Our fieldwork ended in August 2008, four months after the NGO project ended in April 2008. We now present the results of our medical assistance to the former child soldiers.

5. Results

5.1 The socio demographic characteristics

More than half of the children were male (60.3%) against 39.7% female.
Just below 1% (0.9%) of the children born during the war were aged less than 5 years. Figure 2 below shows the distribution of the children by age category.

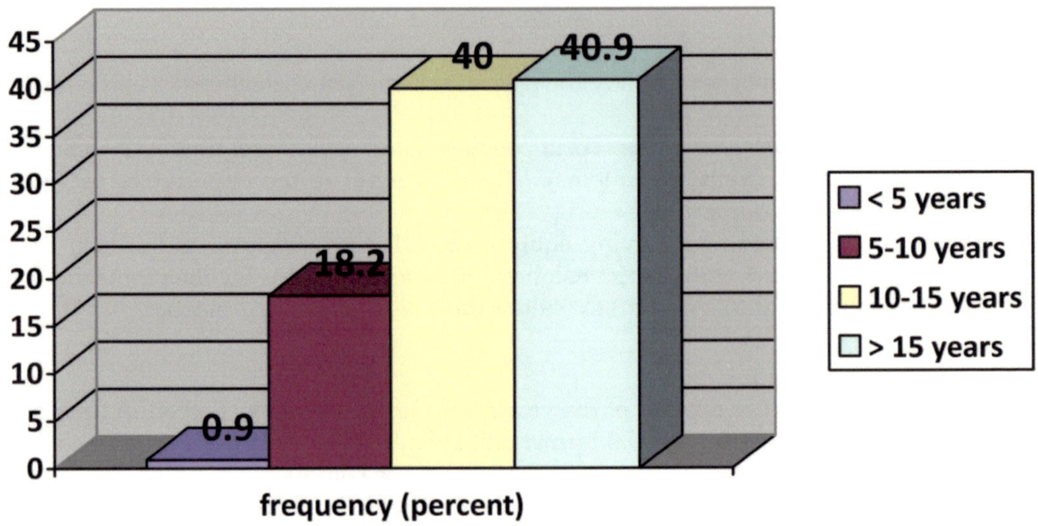

Fig. 2. Distribution of age

Ninety-two (92.2%) of the children did not go to school: Fifty-eight percent (57.7%) of them dropped out of school due to the war (0.3% at Kindergarten, 52.8% at primary school and 4.6% at secondary school) , 34.5% have never gone to school and only 7.8% attended school (4.6% at primary school and 3.2% at secondary school). Among these children, 86.8% had no activity while 13.2% were farmers, fishers or breeders.

5.2 The experience of the war

The consultations carried out and the reports made by the community mobilizers gave us the opportunity to trace the war experiences of those children. Eighteen percent (17.68%) of the children actually fought in the war and 61.74% worked either as cooks and cleaners in houses, or carried goods or worked as security guards. Information on the exact nature of involvement was available for 20.58% of the children.
A number of dramatic stories were told; like the one of this 17-years-old teenager from *Kébouébo*. He attended the 8th Grade in another town where he was sequestered for two weeks at the time the town was attacked by the rebels. In the course of his captivity, he was sodomized on a regular basis by a group of six armed men. Upon escaping, he returned to his village where he met a group of armed men from Liberia. He told that *" when they asked me to come with them, it was to avenge my parents…but I thought of what these people did to me, I saw red and wanted to kill…they did not force me to take up arms."*
We met in the same village a pregnant female teenager of 14. She had been captured in the bush. Her family had been massacred under her eyes, she was forced to cook the dead body

of her mother as food for her tormentors, and one of them desired to take her as his wife. At the time we met her, she had being living for over one year in another village without any link with her family and was pregnant with her second child.

Many children told their war experiences, how they stood up to defend their villages and their region; many of them witnessed atrocities and some of them carried the physical marks of their contribution to the war.

According to these children, their involvement in the war was motivated by revenge (13.91%); defense of the country/village (4.35%); liberation of their farms (3.19%); solidarity (2.32%); imitation of others (0.8%); and no reason (75.36%).

Ten percent of the children (10.43%) joined the conflict in 2002; 29.27% in 2003; 4.06% in 2004; and 1.74% in 2005. Over fifty percent of the children (54.5%) of the children did not specify the period of their involvement in the war; 4.35% were engaged in the war below the age of 10 and 49.86% between 12 and 15.

Over fifty percent (56.81%) saw a man being killed but 20.87% did not witness such a scene. Eight percent (8.41%) of the children said they did not have any reaction to a person being killed, 25.51% were afraid, 9.27% were upset, 4.93% were delighted, 2.32% felt pity, 1.45% revolted, and 3.78% said they felt traumatized. Just under ten percent (9.86%) joined deliberately the armed group to make war and 36.52% were inspired by someone they knew and 26.38% followed a parent. A third (30.43%) had learned how to manipulate weapons and 27.64% were trained in a camp. Among those who had been trained in a camp, one could list 46.8% of children recruited by an armed Liberian group called *LIMA*, 33.0% were recruited by *ZAKPRO* (an Ivoirian militia), 12.8% were recruited by *FLGO* (an Ivoirian militia for the liberation of the western region), 3.2% were recruited by *APWE* (the Alliance of Wê Patriots), and 2.1% were recruited by an unidentified special force. Forty percent (41.16%) said there were children in camps during the war. Their number varied from less than five to more than thirty. All of them talked about the presence of girls among the child soldiers whose size reached sometimes 20, according to the groups. They served as cooks, (20%), fighters (5.22%) cleaners and maids (3.19%), security guards (2.32%), and as porters (2.03%). The number of girls used for sexual purposes was not specified. Forty percent (38.26%) of the children reported to have an affective and physical proximity with someone among the rebels.

5.3 Evidence of psychopathology

Over two percent (2.61%) of the participants had already had problems with the law (arrests by the police for offence). None had a previous record of psychiatric illness; 52.17% were sexually active with, at times, several partners (63.33% had 1, and 36.67% had as many as 6 sexual partners). Fewer than ten percent of the children (9.57%) had been victims of sexual abuse and violence (70% for girls against 30% for boys); many of the girls had served, as sexual slaves for the rebels. Twenty seven percent of girls involved in the research were teenage mothers. The teenage mothers justified these early motherhoods on the basis that the traditional *Guéré* cosmogony (the ethnic group of the region) required girls to give birth to prove their capacity to give life, which similarly gives them the status of being woman. Just over fourteen percent(14.49%) of the children reported the use of cannabis, gunpowder; 37.39% the use of various brand of locally brewed alcohol (*distilled cane sugar* or *Koutoukou (distilled palm wine*, or some adulterated alcohol); 16.52% smoked tobacco; and 0.3% inhaled solvents (glue). All the children showed evidence of mental health problems among which insomnia ranked first. There were a group of traumatized children presented with clear memories of war events: the war was still present in their mind and the eventuality of its resumption was not warded off; the children frequently formulated this eventuality. The children's personalities were characterized by narcissistic fragility and failures. They openly expressed their anger, their aggressiveness and their tendency to revolt. One could note, at times, a great pessimism associated with a feeling of a future blocked and a social disinvestment. The children moved about in the village in groups. For example, in August 2006, in the course of the parade of the children of *Ké-bouébo*, they arrived in battle order chanting war songs wielding tanks, rock launchers, Kalashnikovs, and pistols sculpted in bamboo fiber. An important fact is to be noted: in *Pantrokin*, the former child soldiers surveyed showed evidence of psychiatric disorders linked to the war. When we asked them about their mental condition, they reported that they had received some form of traditional treatment (plant-made medicine). However the nature of this treatment has not been revealed.

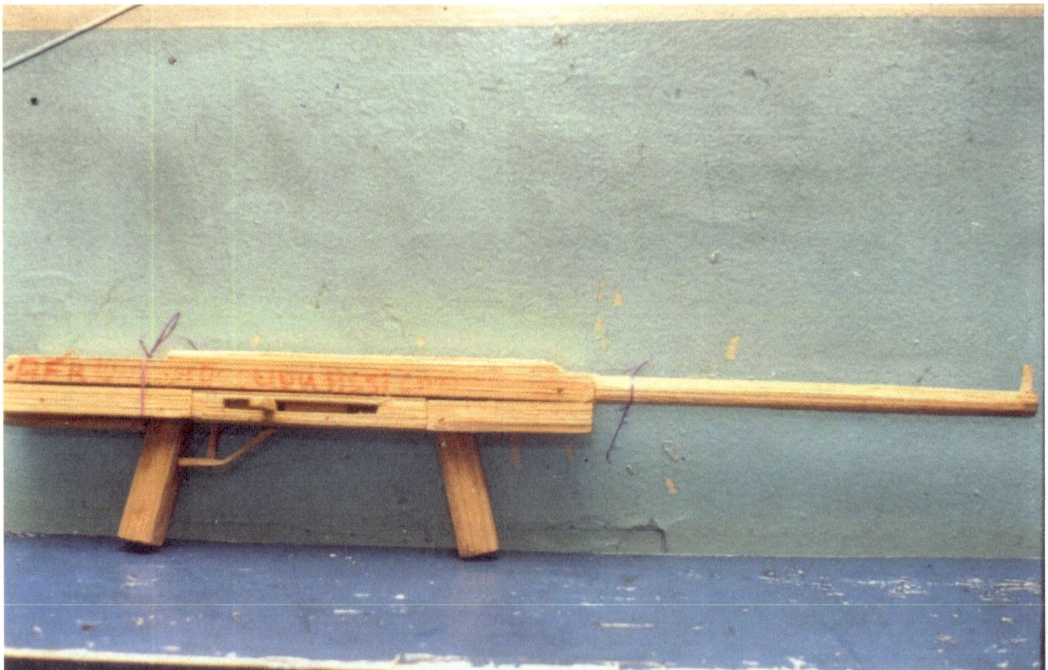

Fig. 3. Bamboo rifle made by a child-Ke-bouébo

The children were showing different psychic disorders *(one child could show one or more symptoms*).

Mental Health Problem (Symptom)	Frequency	Percent (%)
Behavior disorder (theft)	281	81.45
Insomnia	265	76.87
Anxiety	261	75.8
Disorder of character	249	72.24
Easily moved to tears	143	41.6
Sadness	114	33.0
Social isolation, withdrawn	98	28.4
Delirium or hallucinations	54	15.6
Memory disturbance	47	13.8
Suicidal gestures or behavior	40	11.6
Logorrhea	39	11.3

Table 1. Symptom patterns of mental health problems among former child soldiers

Psychiatric diagnosis was possible among 81.45% of the children with the following (CIM 10) diagnoses present in table 2 below.

probable psychiatric disorders among former child soldiers	Frequency	Percent (%)
Post-traumatic stress disorders (F43.1)	150	53.38%
Depressions (F32.11, F32.8	57	20.29%
Acute psychosis (F22.0, F23.31)	36	12.81%
Anxiety (F41.1, F41.2)	34	12.10%
Schizophrenia (F20.0, F20.1)	4	01.42%

Table 2. Frequencies of probable psychiatric disorders among former child soldiers

5.4 The somatic problems

A number of the children had various forms of health problems (hypo gastric pains among children who had been raped, dermatological diseases, sequelae of head injury, deafness, ear infections, and lumbar pains).

The somatic problems were dominated by74.75% headaches, 67.97%palpitations, 29.10% feeling of suffocation, 22.30% loss of appetite and 06.10% enuresis

5.5 The socio-cultural aspects

It is important to underline here some socio cultural characteristics observed in that region, as a minimum knowledge of the cultural environment in which one wants to act significantly influences one's ways of life.

A latent conflict existed between children and adults. Precarious economic situation and poverty were important in the region. The involvement of the children in the project, financed by national and international structures was backed by a food donation (rice, oil). Parents had been complaining sometimes because they were willing to receive themselves the provisions or they claimed the provisions of their children, when the latter had given up their activities with the project. The children were used as foils by parents, but there existed between them a conflict of authority reflecting the reversal of social order from the war.

Income generative activities have been impaired by the way children perceived them. The activity which did not kick off well enough was truck farming. Boys were saying that " *cultivating okra, egg-plant was sauce*" meaning that they are ingredients to cook sauces, and that it is women who cook, therefore this activity could not be exercised by men. In a number of regions in Côte d'Ivoire, truck gardening, especially those destined to daily consumption are the prerogative of women.

Poultry also did not fare well. Chick distribution created jealousy in some villages, but children themselves had problems in adapting. Without wasting time, either they ate up the chickens or some children embezzled the products of the sale.

The activities, which were the most successful, were sewing and carpentry because the children were put in apprenticeship. The boss himself monitored their vocational project.

An important rate of early pregnancies was observed. Girls explained that to be considered as a woman, one has to prove one's fertility. To be a mother is to achieve the status of a woman.

Excision of the young women was still carried out in this region; girls and boys put it that it is important to do it, if a woman was not excised this would bring bad luck to her husband.

6. Strategy of intervention

In order to develop a meaningful relief program we first took stock of existing interventions in the region. Next we trained the NGO staffs in the management of traumatic stress symptoms. In the course of fieldwork we constantly modified what information was useful to adopt in the management of psycho trauma symptoms.

Activity follow up was made through telephone calls with the field teams who used to call as soon as a need arose or whenever problems cropped up.

6.1 Activities already implemented

A number of humanitarian agencies had implemented a number of activities including songs, funny stories, code of living, and promotion of children's rights, literacy campaigns and sports. Those activities, although useful, did not always meet the real needs of the children or led to the relief of required mental health problems of affected children. But those children were preys to anguish, psychic disorders and aggressiveness. Adults were not equipped to receive and accept such brutal and violent emotions.

6.2 Medical prescriptions

Psychotropic drugs prescription was dominated by antidepressants (amitryptillin). Generally speaking, these drugs were prescribed in small dose. Drugs prescribed appear in table 3 below.

Number of Children	Medication	Daily Dosage
34	Amitriptilin	35-75 mg
18	Haloperidol	5-7.5 mg
14	Chlorpromazin	150-200 mg
7	Bromazepam	3-4.5 mg
30	Trihexyphenidyl	5 mg
5	Carbamazepin	400-1,200 mg

Table 3. Psychotropic medications prescribed in the management of mental health problems

The range of drugs used was voluntarily restricted in order to promote their rational use by the field agents who have no medical background. Dosage has been at their minimum to facilitate handling and limit the risk of side effects e.g. chlorpromazin, which is usually prescribed in dosages of 200 to 300 or 400 mg per day, was given in with maximum dosages ranging from 150 -200 mg in our sample . This permitted, in most cases, drug taking without difficulty.

The molecules recommended in international literature in the treatment of post-traumatic stress disorder and co-morbid states, such as paroxetin and hydroxyzin were not used, especially because of their high price and the difficulties involved in their availability in rural areas.

Once the first positive effects of the treatment were reported, the organizers discontinued further use of the drugs. Reports indicated that some parents even took the drugs with the view to enjoying their sedative effects. Thus, with one chlorpromazine tablet in the evening, they slept well and they could the next day return to their farms without effort.

We asked the organizers to take back the drugs to avoid drug misuse, and to administer the drugs to the children.

The NGO had a package of drugs available to field workers based on drug prescriptions. Living in the community, they distributed a sufficient quantity every week and ensured good treatment compliance.

6.3 Additional examinations

In order to provide comprehensive health care to the children, we referred some of them to other specialists as follows: the children who needed surgical consultation (2), gynecological consultation (1), general medicine consultation (2), ophthalmological consultation (1), ENT consultation (2), and urological consultation (2). It proved important to carry out electroencephalogram (EEG) for 3 children and an x-ray of the lumbar vertebrae for one child. Consultations required that children be sent down to *Guiglo*. As for the EEGs, (which were disrupted later), the children had to come down to Abidjan.

6.4 Psychotherapeutic action

Beside the activities implemented by the volunteers, such as literacy campaigns, animal breeding, agriculture, training of volunteers and the populations about post-traumatic disorder, a number of therapeutic activities were initiated: family drawing, game, and therapeutic workshops.

Concerning drawing, the instructions were "*draw your house and your family before, during and after the war*". All the children took up this activity, even those who had never gone to school. Their drawings were full of memories of their trauma. The lines were strong, violent testifying to an internal aggressiveness and violence. The dominant colors were black, red and orange. Few children were imagining a return to normal life after the war. They had drawn their house destroyed by the war and which remained the same after the war; not rebuilt but over grown with weeds.

It seemed important to initiate some psychotherapeutic activities. The children had come off, in this context, with objects from their environment, most of the time it was bamboo or raffia. Children with the most important psychic disorders had built objects recalling the war, while those who overcame their problem had drawn houses, churches and cars.

Working tools and activity comment cards were difficult to use because there were no specialized education officers among the field teams. Even if the officers showed interest, their expected involvement in psychotherapeutic work represented an extra work, as they had other activities to attend to (identification of children in need of birth certificates, HIV/AIDS awareness activities). So that, psychotherapy was necessary for a number of children, an activity based on talking with the children, proved impossible for field workers.

7. Intervention results

We made two assessments of our intervention. In September 2007, 15 months after the beginning of our intervention the number of children showing disorders had decreased considerably. 74.20% of the children showed a positive mental health. By linking this rate to the 281 children suffering at the beginning, one noted an improvement rate of 89.32%. The children had a better health condition; they were dynamic and jovial. They carried with ease their activities despite would-be internal conflicts. Concerning this, they had implemented a system of justice relying on the oldest among them. Those who went back to school had good academic results.

The breakdown of the children in line with persistent symptoms after our intervention appears in table 4 below.

Behavioral disorder (theft, running away, aggressiveness, instability)	26	07.47%
Insomnia	13	03.91%
Anguish/ Fear	10	03.04%
Headaches	12	03.56%
Emotional disturbance	10	03.04%
Amnesiac disturbance	4	01.20%
Enuresis	5	01.52%
Sadness	1	00.30%
Isolation/withdrawn attitude	3	00.90%
Delirium/ Hallucinations	2	00.60%

Table 4. The proportion of children showing symptoms of mental health problems 15 months after the initiation of our intervention

Thus, residual symptoms were dominated by behavioral disorder, insomnia and headaches. None of the children showed signs of palpitations, fits of crying, suffocation, loss of appetite, logorrhea, or suicide.

After the first phase of intervention, the diagnosis assessment showed:

- PTSD ranging from 53.38% to 2.90% (8 children)
- A rate of depressed children ranging from 20.29% to 01.45% (5 children)
- Anxiety disturbance ranging from 12.10% to 00.58% (2 children)
- Acute psychotic disturbance ranging from 12.81% to 04.35% (12 children)
- Schizophrenic disturbance remained stable (01.42%, 4 children)

During this 2007 September mission, we noted an improvement in social behaviors. An inter-village sport and cultural event were organized. The children presented sketches, dances and they played football. User-friendliness and brotherhood were the orders of the day.

The field agents organized also awareness campaigns on the dangers of circumcision. The topic was rather delicate because, at the same time, five girls involved in the project had retired in the bush to be excised. A meeting was organized and the matron responsible for that activity empowered us to talk to the girls. When we asked them about their motivation for genital mutilation and pointed that the Government had prohibited such practices, and that some were already mothers (like two of them), one of them argued that excision favors marriage because a non excised woman is a source of evil to her husband. Even if they lose some sensibility during coitus and their libido will be negatively affected, excision is worth being carried out. Our questions seemed to disturb some of the girls, and a dispute even cropped up.

Before such a practice, the field agents were powerless and villagers barely listened to them. One of them explained laughing that all the talks made by the officer was meaningless because if the woman is not excised she could bring evil to her husband (confirming thus the opinion expressed by the young girl). What could be added if ancestral beliefs are so strong?

At the final assessment of our intervention in August 2008, only 58 children were assessed. Many were the children who had left the area before the project came to an end without an assessment of the impact of the actions undertaken and without a real reintegration. It was hoped that any improvement in the mental health of the children would permit the children to view their future with hope and resume normal life activities for survival after the war.

The children who better rebuilt their life were those who returned to school or those who learned sewing: the young girls of *Pantrokin* who enlisted in the sewing project built their own sewing shop; they bought new machines and were receiving customers. One of the boys, Joel, who demonstrated leadership capacities, was handling alone truck farming and breeding. As for the other boys, they had, either left the village or abandoned the project.

In *Ké-bouébo* and in *Béoué*, breeding and farming were abandoned. Only the vestiges of a promising project remained (abandoned hen houses, fallow ground).

The overall appreciation of that mission is the following:

In general, the children were better off on the psychopathological point of view though of them showed a reactivation of psychiatric symptoms (psychotic disturbances, depressions).

The community volunteers seemed to have given up their commitment toward these children possibly for a variety of understandable reasons including lack of funds (therefore no salary),lack of food donation and perhaps because the NGO local agents were no longer there to provide the services they did before.

8. Reflexions from this experience

8.1 Who are the Ivoirian children soldiers?

8.1.1 They are boys and girls also

Most of them were boys (60.3%). Girls made up 39.7%. The children revealed the presence of girls among the child soldiers. The girls were used as cooks (20%), fighters (5.22%), dish and clothes washers (3.19%), security guards (2.32%) and porters (2.03%).

The number of girls used for sexual purposes was not specified. It is difficult to establish a ratio between the numbers of girls out of a total size of children present with the armed groups. In some countries, girls represent up to 40% of the child soldiers, like, for example, within the *Tigres de Liberation de l'Eeclam in Shri Lanka.* Those children are also brought to carry a number of functions as we have seen (Huyghebaert, 2009; Ayissi& Maia, 2004). Girls play different roles on the same day; they are fighters, cooks, messengers, spies, nurses, sexual slaves, even "captive wives", as it was the case of a young girl we met at *Ké-bouébo.*

In some countries, according to Ayissi & Maia (2004), women fighters are also used as suicide bombers and in delicate tasks like the security guards of warlords or in spy missions and infiltrations of enemy troops because of their efficiency and fidelity to "their" men. The same authors argue that, even if boys are not saved from these troubles, it is girls and teenagers who pay a heavy cost in rape and sexual abuse. Those abuses are followed by serious physical injuries, sometimes painful and disabling, as a result of unplanned pregnancies followed by high-risk abortions. This corroborates the data on the sexual violence that we were talking about.

Be they victims of the barbarism of fate, those young girls and all the children in general, remain profoundly traumatized both physically and psychologically through the hardships that they endured in their early age.

Most children that we saw were aged between 5 and 15 years (58.2%). In countries like Sudan, the Democratic Republic of Congo, Sierra Leone or Liberia, child soldiers were enrolled between 7 to 18 years of age?(Baingana&Bannon, 2004). Though Huyghebaert(2009), citing an ILO publication (2006) concerning the child soldiers' enrollment age, places the age of teenagers at15 or more at the time of their enrolment, the young boy enrolment at 7 to 8 years tends to become more and more frequent.

The children that were the object of our study found themselves in a situation of war at an age when education was an important element of life, at a period of great psychological vulnerability when a human being is growing, where the child is socializing and where he develops psychologically. In that part of the country, sending children to school is difficult on the one hand because of lack of classrooms, and on the other hand because of the general poverty levels of the populations. The conflict has played a role in this context, leading thus to mass removals from schools (60.50%). One of the NGO addressed this problem by introducing literacy classes and a strategy of school resumption by providing school equipment and by helping families in sending back children to school, either in the village or in the nearby town. Children who returned to school had good results. That situation in *Guiglo* is different from what is seen in general as underlined by Tomkiewiez (1997) who cites extreme difficulties in having the Ugandan civil war children back to school.

8.1.2 Did they take or were they given arms?

Some of those children from *Ké-bouébo* were recruited, or enrolled forcibly. Only 17.68% handled or manipulated weapons. The use of non-combatant children mirrors our case like in many other countries as evidenced by Mouzayan (2003), for whom those children were enrolled for dangerous and alienating activities (fights, chores, spying, messengers, and sexual slaves). Whatever children are involved directly or indirectly in they are in danger (Anwo, 2009). Many reasons motivated those children of *Moyen-Cavally*. They put it bluntly that revenge was the N° 1 motive (13.91%) though as Honwana (2006), argues, in some conflicts, a variety of reasons, including coercion, poverty, or sheer violence turns young men into assassins before they are able to understand the complexities of morality. As seen on the field, all the children were not enrolled by force, or by constraint. In fact, some became "willing" members of a gang or an armed force to protect their family or themselves, changing thus their status: from children who must be protected by parents, they become those on who lies the survival of the family. For others whose family members became victims of the conflicts a desire for revenge (Huyghebaert, 2009) motivated them to take up arms as we found it. For Ayissi& Maia (2004), children are above all vulnerable physically, mentally and emotionally, and therefore more docile and more malleable than the adults, while if their engagement is "voluntary", it comes from a desperate strategy for survival. To boot, when the war gets stuck and decimates whole families, a number of children become orphans, with no perspective of subsistence than joining the armed groups where to bear a weapon will give them the feeling of existing and to be protected. If it is true that those children are the real victims, Jézéquel (2006), cites researchers like Paul Richards, who, while denouncing the violence on children during the war, shows that children are the real actors capable of displaying their own tactics in a field of constraints imposed by war dynamics.

8.2 Ravaged lives in a ravaged region

8.2.1 Pain or psychic suffering?

The war attacks children and destroys them as Bertrand (1997) puts it; those who survive will carry almost irreversible marks. Even if this argument carries a rather violent character, it is however the sad reality.

The war generates psychic wounds with which children are doomed to live with because they are part of their history. Those "invisible wounds" are visible, identifiable, through their own expressions in children's behavior and through their relation to others and to the outside world.

The encounter with those child soldiers was followed by a direct contact with an important psychic pain. This psychic pain is different from psychic suffering. Citing Ferenczi, Bertrand (1997)recalls that this pain may have extremely serious effects; destroyers of the individual. The risk may be the sudden decomposition under the form of a delirium or traumatic cleavage, but also as depression, somatizations, and disabling chronic pains with no detectable lesions. The idea of the existence of this pain in the children's psyche could explain the various clinical charts that we mentioned. Anguish and pain are limits to experiences, at the border of our being; limit of both the possibility of existence and of the possibility of subjectivities.

With those children, this brutal and raw pain, assimilated with difficulty because the trauma could not possibly be comprehended by their young minds, and therefore still carried (even after a period of four years) of its emotional impact, was palpable at the beginning of our fieldwork. Is it not an important obstacle to the possibility of being fully human for those children? For Tomkiewicz (1997), psychic disturbance could be the must link between traumatic stress and its psychological consequences.

Around 56% of the children had seen a man killed under their own eyes. Dapic & Coll (2002) have studied this parameter with primary school children (Grade 5) victims of the Bosnia-Herzegovina war, in a Sarajevo district. 90.7% of them had seen war wounded and 74.3% had seen a dying man.

Among our sample, before the "show" of execution of a person, some were said to be abreactive (8.41%), scared (25.51%), and shattered (9.27%); other emotions had been described as joy (4.93%), pity (2.32%), revolt
(1.45%). 3.78% said to have been traumatized.

The issue of social support in the course of events was also brought forward. Most of the time, the children had accompanied someone they knew into the war (36.52%) and 26.38% said that they had followed a parent. Others have certainly experienced solitude, but few talked about it. On this subject, Tomkiewicz (1997) explains that solitude may be considered as an aggression, because it engenders or increases suffering. It pervades children when they have lost their families, their bearings, their friends and their dearest around them;

when they find nobody with whom to share their fear, their anguish, and their hope. The absence of schools and of any educational and socializing institution participates to this solitude. The question in this context is how the children who lost everything including social networks (27.6% who joined armed groups out of solitude) could ever be helped to find solace in their lives.

8.2.2 Psychiatric and social consequences

Our data revealed that 9.6% of the children were victims of sexual violence and they are forever marked by this aggression; 52.17% were sexually active and among them 36.67% had many partners, perhaps as a method of coping with their experiences and aggression from the war as, a child told us: "when I think of all that happened and of what I saw during the war, I don't sleep and consequently, I could go with five or six girls a day."

In the course of the conflicts, cases of child abuses are probably countless, sometimes increasing the risk of exploitation and sexual abuse. Cases of abuse apparently continue into the "post-conflict" period: chores becoming servitude, recrudescence of child trade, and sexual violence and exploitation in refugee camps (Huyghebaert,2009).

However, the issue of violence is pushed into the background, as no direct allusion to sexual abuse transpired in the course of our discussions with the former child soldiers. This observation may be explained by the customs of the people of the western region of Côte d'Ivoire where teenagers have been raised with the idea that having a child is a sign conferring the status of adult and challenges mental health professionals' concerns over child sexual abuse as an explanation for psychological problems in times of conflict and war. Since early active sexual involvement is socially accepted in Cote d'Ivoire, 27% of the girls were mothers at the 1st assessment and at the 2nd assessment, and this rate rose to 34%. Moreover, despite awareness campaigns, one had more than 30% of the children had evidence of STI recurrence.

Though the early involvement of the children in Cote d'Ivoire exposes them to the risk of health problems and crime (early and unwilling pregnancies, STI/HIV/AIDS, infanticide, deserting children) the link with mental health problems is not explainable as early sexual activity is socially and culturally sanctioned. Some studies indicate HIV sero-positivity in post-conflict period of 60% nationwide (Leblanc, 2004). Concerns exist concerning the welfare of children born during the period of violence to mothers who were themselves victims of violence. However Tomkiewicz (1997) explains that most of the time, the children who survived conflicts reach a better social adaptation than that could have been predicted. The same author claims that more or less rapidly, those children succeed in integrating into an "after war" society, and very few become really marginalized. As Tomkiewicz (1997), our results allow us to reach such a conclusion, as from one village to another, "survivors" were more or less integrated

As a matter of fact, most children surveyed had been integrated in the villages because they used to live there and were able to resume, for most of them, an activity: returning to school, farming or breeding. That's how most of the children interviewed had been integrated into the villages because they lived there and were able to resume, most of them, an activity: back to school, farming or breeding ... This is especially important that the integration (re-integration) refers to a social return be inserted individually in the community clean.Our data agrees with observations made in Mozambique where families of former child soldiers rejected the children who faced social integration problems as a result (Green&Honwana,

2001). Explanations for the rejection of former child soldiers in Mozambique resides in the enrolment motivation in Mozambique where children were fighting outside their communities, where as those of *Moyen Cavally* in Cote d'Ivoire took up arms to defend their communities. In return, the community fully supported them. The Ivoirian children of *Moyen Cavally* were motivated in great extent by the desire of revenge or the liberation of their village, which has a great community connotation. As Gannagé puts it, parents must act as protecting filter and as pare-incitement to the child. The capacity of the children to dominate and to memorize trauma depends on their parents' capacities of elaboration and implementation of trauma, of figuration and representation, to think and to communicate to the child about the event.

The best social reintegration (in the sense of being able to contribute anew to the development of one's community) is perhaps that of the *Péhé* and *Pantrokin* children. In saying it, we have in mind the idea that some factors linked to the recuperation environment have positively influenced this normalization of social life. In fact, *Péhé* is a sub prefecture endowed with a number of commodities like *Pantrokin* the nearby village. This standard of living as well as the advantages, which go with it in terms of employment, contributes to reintegration. Yet, even here, as in *Kébouébo*, a village with no drinking water and electricity as well as in *Béoué* and *Kaadé,* located on a tarred road, provided with water and electricity, a good number of children had hard time to readapt and to invest in the activities proposed to them. Even if all children of the project initially had the same motivation to participate to the war, the socio-cultural post-conflict reintegration to promote better for most of them.

As for the conditions of girls, Huyghebaert (2009) argues that it is difficult for girls who have been kidnapped and who, during their captivity, gave birth to babies to return to their families and communities. Reintegration therefore proves to be a complex process of re-adaptation and, at times, of community expiation, as well as negotiation with the families to convince them to accept to take back their children and, the children of their children. Concerning the girls that we met and their progenitor, the cultural context encouraging maternity has probably contributed to facilitate their reintegration and the acceptance of their children's babies born to the enemy or the invaders. .

Thirty-seven percent of the children in our sample used alcohol, and fourteen percent used drugs such as gunpowder and cannabis (14.49%). The use of psychoactive substances is a misuse because there are no data to discuss here an insulated handles abuse. But the children have told us to use these substances to overcome the atrocities they saw, lived and committed. They sought out the effects of these mind-altering substances.

The misuse of alcohol and other psychoactive substances enhanced the children's ability to act and endure the hardships, psychic pain and anguish, insomnia, and physical pain of war (Douville, 2007). Alcohols *(cane juice, koutoukou*) that children took came primarily from local manufacturers and were sold at low prices and were easily accessible to the poor.

In our sample the war was still present in the children's mind and the possibility of the resumption of war was not brushed aside. Personalities were marked with fragility and by narcissistic flaws. The children expressed their anger, their aggressiveness and their revolt. One could notice, at times, a great pessimism associated with a feeling of a blank future and a social disinvestment. The children were moving about in the village in groups. The need to reform them was evident and urgent: it seemed imperative to set up a reform program in order to neutralize "the many bombs of aggressiveness" in the children.

Many of the children (81.45%) showed evidence of mental disturbance before intervention. Our estimates are higher than those of Cordahi et al (2002) who found 62.5% of Lebanese children and teenagers who went through the 1996 "*raisins of wrath.*" were psychologically affected one year after the events of war and loss of a father/mother. We might explain the high rates of psychiatric disturbance in our sample of children and teenagers by the duration and number of traumatic events, the scarcity of communal resources, social disorganization and the threat of Cote d' Ivoire been split into two countries at the time of our intervention in the border region with Liberia, a country that had itself been in conflict for years.

As observed in other war areas, the commonest diagnoses in our study were post-traumatic stress disturbance (53.38%) and depression (20.29%) four years after war. However the rate of post-traumatic stress disorder among children one year after war in a study conducted by Schwarzwald et al in 1994 and cited by Jolly (2000), was 12%with those whose house had been bombed showing a more significantly prevalence: 23.8 *versus* 9.1% (Green &Honwana, 2001). In 1994, in the course of the conflicts between Muslims, Croats and Serbs, a study of displaced Bosnian children aged 6 to 12 years old, showed a PTSD rate of 93.8% and the rates of associated disorders were observed: sadness (90.6%), anxiety (95.5%), feeling of guilt (66.6%) and anorexia (59.7%) (Green &Honwana, 2001).These and our study show that a traumatic event provokes fear, horror, a feeling of disarray and desperation.

The experience or commission of violence creates a staggering level of depression and a melancholic behavior that could lead to suicide (Douville, 2007).

As in other studies, the physical pain that participants in our study experienced was palpable, the violent emotions were, so intense that words could not tell, the narration being interrupted the liberating power of words sometimes became incommunicable because, in the first place, the traumatized people were unable to talk about them for a number of reasons including the low literacy of the children, the lack of opportunity for the children to talk about their experiences, and the relative lack of skills among our organizers to enhance liberating cathartic communication from the children about their experiences.

Tomkiewicz (1997) asserts that, if one wants to save a child victim of war, it is not enough to heal his body, his wounds, his lesions; it is not enough to give him something to eat, to vaccinate him; we need to caress him, to smile to him, to talk to him. To be able to talk to him and to listen to him, it is necessary to associate him to a local team, whose members could infuse confidence in him. We could not wholly act according to this principle for a number of reasons including the fact that our own humanness was too upset by this human catastrophe, by this dehumanizing catastrophe at the beginning; the organizers' emotional equilibrium, and ours was seriously too inadequate to restore a minimum of decent psychic and social life among our participants.

The local organizers were themselves, either veterans, or war victims and were ill equipped to help our participants to talk about their emotional and psychic pain. The relations of the organizers with those children were most of the time conflicting.

8.3 The status of children in African societies

African societies have developed and they retain at times their own perceptions of childhood (Jézéquel J.H, 2006; Ferme, 2001). It transpires from the works of anthropologists such as Ferme (2001) that childhood is, in sub-Saharan Africa, sometimes assimilated to a time of ambiguity, an unstable and hybrid situation. In Ivory Coast a number of children were involved in the conflict, most of the time by taking up arms, to defend their village, in the place and along side adults. They fought and they gained (at an early age) a status of adults, of defenders, of liberators; this status conferred upon them a special position in the village. Thus the children are not denied any capacity as adults in society (Howana, 2000). But, in reality, it is a precarious position and psychologically unbearable situation to be no longer a child and not to be a real adult. The young freedom fighters hold interstitial social spaces, between the adult and juvenile worlds. In Côte d'Ivoire it appears that the place of the child is ill defined, ill conceptualized, moving from a traditional conception of the child (submissive, usable and stooge for the parents) to a more modern conception that parallels what is going on in the West. Even if the relation between children and parents is changing, the perception of children as potential labor force remains still very strong (Berry, 1985). The issue of child soldiers appears to parallel a long history of child labor force in colonial and postcolonial African economies.

8.4 The shake-up of social order

Jézéquel (2006) explains that war situations are marked by inversion phenomenon through which elders lose their authority over the youngest, whole towns are conquered by bands of teenagers not always controlled by their leaders. This is due to the mass deterioration, loss of links to reference and to the ancestry caused by the war. Conflicts lead children and teenagers to roam about feeling that they have been destroyed in their actual humanness. There is a fragmentation of the society. The possible role of family running over social running is destroyed, sometimes inexorably so that the common space limit is often narrowed to extremely precarious clans (Douville, 2007). Douville goes on to say that there is a destruction of the image of the other and that when is war over, children and teenagers have much difficulty in entering an ordinary social link. In addition to generation gap, war aggravates the erosion of parental and adult authority and family links are disrupted and destroyed (Bertrand, 1997). As war dismantles and destroys the civil organization of a country, schools, justice system, police and post office services, children run the risk of losing all their marks of common life and fail to conform to what is permitted or what is good or forbidden and bad (Tomkiewicz, 1997).As Tomkiewicz puts it, the apparent ignorance of the law and the loss of bearing constitutes the main expression, or at least the most visible psychological consequences of the war on short term.

8.5 When culture invites itself, we should not avoid it

Most of the time, post conflict recovery programs aim to demobilize and reintegrate children associated with force or armed groups. For children affected by the conflict, there are usually no programs of social reintegration, of education (private lessons, for example), of vocational training, of cultural activities, of sport and/or income generating activities (for example, setting up small businesses). Destined to all the children affected by the conflict and not exclusively targeting child " soldiers" or child victims of sexual violence, programs, while privileging a communal reintegration that should be as inclusive as possible, constitute sound programs of preventing (re) enrolment. To this extent, children and community's involvement in the choice of programs of reintegration finds its place in order to optimize the latter (Huyghebaert, 2009).

To have children and communities get involved at a certain level of the project implementation relies on taking into account the socio-cultural considerations in order to foster the success of social re-integration programs. It is important, not only to take care of investigating the psychological and societal aspects peculiar to the host milieu, but also to insure the approval of the parents and community members. The passion of some adults, in the villages where we worked, to thwart, even to destroy the projects and the difficulties met by our agents in overcoming this difficulty leads us to argue that it is imperative to prepare field through a good alliance with the villagers and the parents of the beneficiaries of post-conflict reintegration programs. The question was to offer the possibility of the villagers to accept the activity through an understanding of the project and a better knowledge of the short and long term benefit for the children, but also for them and for the whole community.

8.6 Which efficiency, during and after those actions?

In September 2007, we noted an improvement rate in the health condition of 89.32% of the participating children and teenagers. Residual symptoms were dominated by behavior disorders, insomnia and headaches. Most studies have reported an improvement in symptoms or a decline in the level of psychopathology in the course of time (from some months to a number of years) with the majority of the children. The duration and the degree of remission depends, on the degree of exposure to traumatic stress, of the manner in which families react and of the social support for families (Cordahi & al, 2002).

To back our psychotherapeutic action, we implemented medical treatment, especially some amitryptillin (35-75 mg/day).

The premature giving up of the drugs parallels a general observation marked by the difficulty for Ivoirians to take drugs on a long term basis, which could explain the short period of medication and the lack of compliance. Or the medical treatment, especially anti-depressive may permit to attenuate the symptoms and favor thus the verbalization of the psychic pain and allow a psychotherapeutic access (Deniau& Cohen, 2011).

The continuous psychosocial help permits the improvement of emotional disorders in children and teenagers. Supported by a psychosocial environment, the children had seen their situations improve. This is particularly the case when antidepressant medication is used to support psychotherapeutic and psychosocial intervention. However the reluctance of Ivorians to take medication on a long-term basis often undermines the potential clinical outcomes of post-conflict interventions for psychological consequences of traumatic stress. In line with African tradition, which argues that a child belongs to the community before belonging to his parents, child orphans always found someone to accommodate them in our study settings. The fact that the children returned to their village after the war, that they found a home, a house, and a place in the community was certainly comforting for children who returned from war. In our view reintegration of former child soldiers into accepting homes gave the children new meaning to the children against the background of chaos brought about by the war and its associated trauma. Finding a home gives the child a new sense of being human, hope as a member of the human community where war dehumanizes, to find parental comfort and love where solitude and acts of violence enslaves human beings.

9. Conclusion

This project, which lasted 4 years after the conflicts in the western region of Côte d'Ivoire during the 2002 war aimed at assessing the magnitude of post-traumatic stress disorder.

Tomkiewicz (1997) argues that, if we want to save child soldiers, it is not enough to heal their bodies, their wounds, their lesions; it does not suffice to give them food, to vaccinate them; we must caress them, smile to them; we need to talk to them. In order to be able to

talk to them and to listen to them, it is absolutely necessary to associate them to a local team whose members could bring confidence in them. Even if we find this idea judicious, this seemed difficult to implement in our context for several reasons.
On the one hand our own humanness was too upset by this human catastrophe, and by this dehumanizing catastrophe. At the beginning, the organizers emotional equilibrium, and ours was seriously required by anything to be done in order to restore a minimum of decent psychic and social life to permit us help the traumatized former child soldiers.
On the other hand, the local organizers were themselves, either veterans, or war victims and their relations with the children were, most of the time conflicting.
Our results indicate that 89.32% of the children achieved some degree of improvement in their mental health situation.

10. References

Anwo J. Conscription and use of child soldiers in armed conflicts. Journal of Psychology in Africa, Vol 19(1), 2009. Special issue: Violence against children in Africa. pp. 75-82.

Ayissi A.e, Droits et misères de l'enfant en Afrique. Enquête au cœur d'une « invisible » tragédie. *Etudes* 10/2002 (Tome 397), p. 297-309. Ayissi Anatole et Maia Catherine « Les filles-soldats », *Etudes* 7/2004 (Tome 401), p. 19-29

Baingana F., Bannon I. Intégrer les interventions psychosociales et la santé mentale dans les opérations de prêt de la Banque mondiale pour les populations touchées par des conflits : Un ensemble d'outils. Document Banque Mondiale, Septembre 2004

Bertrand M. Du trauma au récitin *les enfants dans la guerre et les violences civiles, approches cliniques et théoriques. Sous la direction de BERTRAND M. eds L'Harmattan, coll. Espaces Théoriques, 1997, pp 117-132*

Berry S., *Fathers Work for their Sons. Accumulation, Mobility, and Class Formation in an Extended Yoruba Community*, Berkeley, University of California Press, 1985.

Classification Internationale des Maladies 10ème édition (CIM-10) ; chap. V : Troubles Mentaux et Troubles du Comportement.

CordahiC, KaramE G, NehmeG, FayyadJ, MelhemN, RashidiN. Les orphelins de la guerre-expérience libanaise et méthodologie d'un suivi prospectif. Revue francophone du stress et du trama, 2002, 2 (4), 227-235

Dapic, R., Sultanovic M., JahicH.s., Cerimagic D, Bajramovic I. et Lomigora A. Polytraumatismes de guerre chez les enfants de Dobrinja. L'Esprit du Temps *Champ Psychosomatique*. 2002/4 - n° 28, pages 23 à 36

Deniau E, Cohen D. prescription d'antidépresseurs chez les enfants et adolescents. EMC (Elsevier Masson SAS, Paris), Psychiatrie/Pédopsychiatrie, 37-209-A-30, 2011

Douville O. « Enfances en guerre », *La lettre de l'enfance et de l'adolescence* 4/2007 (n° 70), p. 93-98.

Gannage M. Après la guerre, quel type de prise en charge pour l'enfant et l'adolescent ? Revue francophone du stress et du trama, 2002, 2 (2), 91-95

Green E.C., Honwana A. Des thérapeutiques autochtones pour soigner les enfants traumatisés par les guerres en Afrique-2001/03/30. http://www.irenees.net

Honwana A.« Innocents et coupables : les enfants soldats comme acteurs tactiques », *Politique africaine*, 80, 2000, p. 58-78]

Honwana A., Child soldiers in Africa. Baltimore, MD, US: University of Pennsylvania Press, 2006. 202 pp.

Huyghebaert P. « Les enfants dans les conflits armés : une analyse à l'aune des notions de vulnérabilité, de pauvreté et de "capabilités" », *Mondes en développement* 2/2009 (n° 146), p. 59-72. Jézéquel J-H.« Les enfants soldats d'Afrique, un phénomène singulier ? », *Vingtième Siècle. Revue d'histoire* 1/2006 (nº 89), p. 99-108. Jolly A. Evénements traumatiques et Etat de stress post-traumatique, une revue de la littérature épidémiologique .Annales Médico-psychologiques ; 2000 ; 158 ; 5.

Jolly A. Epidémiologie des PTSD (Post-traumatic Stress Disorder). Journal International De Victimologie (JIDV) Année 2, Octobre 2003, 1.

Leblanc H., administrateur - UNICEF France. La situation mondiale des enfants soldats en 2004. *www.unicef.fr/mediastore/7/2075-4.pdf.*

Mariane C. - The Underneath of Things. Violence, History and the Everyday in Sierra Leone. Berkeley-Los Angeles-London, The University Press of California, 2001, 287 p.,

Mouzayan Osseiran-Houbballah. L'Enfant-soldat, 2003, 240p, édition Odile Jacob

Perlemeuter L., Perlemeuter G., Guide de thérapeutique, Nouvelle édition actualisée, Masson, Paris 2001, 2002, 2003, 2006: 4e édition; p 1608-1618. Rapport UNICEF-statistique Côte d'Ivoire. http://www.unicef.org/french/infobycountry/cotedivoire_statistics.html

RousseauC. Développement moral et santé mentale en situation. *Neuropsychiatrie de l'enfance et de l'adolescence 56 (2008) 199–205*

Tomkiewicz S., l'enfant et la guerre, in *les enfants dans la guerre et les violences civiles, approches cliniques et théoriques. Sous la direction de BERTRAND M. eds L'Harmattan, coll. Espaces Théoriques, 1997, pp 11-43*

9

Post Traumatic Stress Disorder - A Northern Uganda Clinical Perspective

Emilio Ovuga[1] and Carol Larroque[2]
[1]*Faculty of Medicine, Gulu University,*
[2]*University of New Mexico,*
[1]*Uganda*
[2]*USA*

1. Introduction

Post-Traumatic Stress Disorder (PTSD) is a psychiatric condition, which develops after a person experiences, witnesses, is confronted with or hears about emotionally stressful and painful experiences beyond what a human being can bear. The traumatic event may be life threatening; threatens body integrity and causes considerable fear, horror and a sense of helplessness in the affected individual (APA, 1992). Traumatic events are psychologically wounding to the individual and leave deep scars (Anonymous, 2009; and Tonks, 2007) on trauma victims; they are dehumanising, demoralising and humiliating, and may put an abrupt end to the hopes and plans of an otherwise enterprising individual, as the individual loses the sense of the future (Bardin, 2005) as one of the clinical features of post-traumatic stress disorder.

The experience of traumatic stress in the history of human kind is perhaps not new and was probably limited to the processes of survival in pre-historic times. However with civilization and modernization the nature scope and experience of traumatic stress has become more complex and sophisticated in terms of clinical significance, individual perception and interpretation of the traumatic experience, and public health importance. In Uganda the nature of traumatic stress ranges from natural events including road traffic accidents, industrial accidents, domestic accidents, floods, landslides and occasional earth tremors to manmade traumatic events such as orchestrated domestic violence, child abuse and neglect, and organized violence and war, the most recent of which took place from 1986 to 2009. In recent times there has been an upsurge of child abductions for human sacrifice in Uganda, and this has been extremely traumatizing to affected families and relatives.

War is of particular significance as it is manmade, is associated with significant mental health problems (Murthy & Lakshminarayana, 2006), causes more suffering, deaths and disability within the same time unit than an epidemic, and imposes considerable economic and social burden on communities (Murray et al, 2002). The psychological and psychiatric consequences of traumatic experience have been the subject of initial disagreement and debate in the international literature. However the personal accounts of victims of violence and war as detailed by Judith L. Herman (1997) in her book on trauma; and client accounts in clinical settings and to journalists (Anonymous, 2007) leaves no doubt as to the clinical and public health significance of traumatic stress experience.

Traumatic events are often sudden and overwhelming irrespective of their origin or nature though certain traumatic experiences last for a short time while others take a protracted or repeated course, particularly if they are politically motivated or occur in the hands of hostage takers or domestic abusers. With almost no exception, traumatic experience seems so unreal, horrible and unimaginable to most victims that its experience leaves victims helpless with a serious challenge to the human sense of omnipotence over the environment. Manmade traumatic events cause intense fear, a systematic weakening of the struggle for freedom, the break-up of victim's self-control fabric, and a total dependence on the perpetrator of the traumatic experience for survival. In most cases trauma victims may hold society as accomplices in their experience with the development of a sense of abandonment and loss of basic trust in the social order. Further more trauma victims develop self-blame, guilt feelings, loss of self-confidence and self-esteem. Emotional numbness that accompanies the traumatic experience causes severe loss of control over personal routines and dignity with a pervasive loss of sense of the future with the victim living by the day (Herman, 1997).

The bulk of published research data on post-traumatic stress disorder concerns adults compared to children. However isolated published research data highlights the magnitude and psychological effects of traumatic stress, physical effects and long-term social consequences of conflict and war among child populations in conflict affected areas of Africa (Anonymous, 2007; Bardin, 2005; Betancourt et al, 2010; Betancourt et al, 2008a, b, c; Mock et al, 2004; Ovuga et al, 2008; and Pham et al, 2009). In an exception to current emphasis on providing care to adult clients, Onyut et al (2005) describe the potential value of narrative exposure therapy for war-affected children in two camps in Uganda.

Most available published data from war zones of Africa pay little attention to the clinical features of post-traumatic stress disorder, and most cases of probable post-traumatic stress disorder presenting at primary care units are misdiagnosed and mismanaged. In this chapter we describe the complex settings and clinical presentation of post-traumatic stress disorder in Northern Uganda. We supplement the chapter with material from our own assessment of mental health needs among children and adolescents in one district of northern Uganda that was the epicentre of Uganda's most protracted and brutal armed conflicts since the country attained independence from Britain in 1962.

2. Historical background of Post-Traumatic Stress Disorder

Historically, the awareness of PTSD as a clinical syndrome followed wars such as the American Civil War, world wars I and II, and the Vietnam and Gulf Wars. Modern wars have become sophisticated and assume the form of guerrilla wars that take place in cities and directly affect civilians exposing adults and children alike to the senselessness of humans killing humans with little regard to the sacred value of human life. In Africa, active wars and organized violence have recently affected millions of civilians in Ivory Coast, Tunisia, Egypt and Somalia. Recent acts of ethnic violence, and organized violence following elections and religious fanaticism have affected hundreds of civilians in previously stable and peaceful communities in Rwanda, Democratic Republic of Congo, Kenya, Tanzania, and Uganda's capital city, Kampala on July 11, 2010.

2.1 PTSD among refugees and in Internally Displaced Persons (IDP) camps in Northern Uganda

Uganda is a country in sub-Saharan Africa that suffered from the ravages of war, poverty and the consequences of the HIV/AIDS epidemic. Gulu in northern Uganda was hit especially hard by recent political upheaval and guerrilla warfare. While safety is no longer an issue in this region, grave social concerns remain arising from the deep wounds the war inflicted on the children in northern Uganda (Tonks, 2007). During the war thousands of children had been abducted: boys to serve as child soldiers; girls to be both killers and sometimes as officers' "wives". Many of the young girls became pregnant and bore children of their own. While the region may now be peaceful many of the young people are not at peace with themselves or their community. For those youngsters who were not abducted they grew up in a shadow of war and lived in extreme fear. Many lost their parents, siblings and friends to the war. Their lives and education were disrupted. In addition, in this developing country many children experience the stress of extreme poverty on a daily basis. They are often hungry and have experienced the loss of loved ones because of medical conditions such as malaria, tuberculosis and HIV/AIDS. In early 2000 Gulu district was hit with the deadly hemorrhagic fever, Ebola that killed many including 21 health care providers in St Mary's Hospital Lacor, the only mission hospital serving many parts of the country. In the year 2008 at least three districts in Northern Uganda were hit by hepatitis B epidemic that killed many pregnant women, adding to the troubles of the region.

More than an estimated three hundred thousand refugees fleeing the civil war in Southern Sudan lived in various districts of the West Nile Region of Uganda. Kanarukana et al (2004) and Neuner et al (2004) reported high levels of mental health problems among the refugees and nationals including post-traumatic stress disorder, alcohol abuse and suicide. Following the fall of dictator Idi Amin of Uganda in 1979 wanton acts of violence against civilians in the West Nile region exposed nearly every family to horrible events of traumatic stress. The Northern Uganda war between government forces and the Lord's Resistance Army of Joseph Kony displaced more than two million civilians from their homes to internally displaced persons' camps in the entire Acholi, Lango and Teso sub-regions of Northern Uganda. Recent surveys have demonstrated significantly high rates of PTSD in the camps (Roberts et al 2008). Published data among various population groups from Northern Uganda suggest high levels of mental health problems including depression, alcohol abuse, anxiety and suicide (Ovuga, 2005; Ovuga et al, 2005a; Ovuga et al, 2005b; Roberts et al, 2008; and Ovuga et al, 2008; Roberts et al, 2009).

While poverty, personal loss and war trauma can produce devastating effects on children, not all children in a community will be impacted to the same degree or in the same manner. In fact, some very resilient children flourish in spite of severe adverse experiences (Betancourt and Khan, 2008). While some studies have been done on the emotional well being of specific groups in war affected areas, information about children is scarce, especially information about children less than 12 years of age. Most of the work that has been published was carried out at a time when there remained significant insecurity in the region of Gulu and many individuals feared for their wellbeing. Studies to date that have examined the emotional well being of individuals in northern Uganda have focused on two primary groups: 1) Internally displaced adults living in camps because of the war and 2) former child abductees of the Lord's Resistance Army (LRA). Research in these populations revealed a very high prevalence of PTSD and depression.

Roberts et al. (2008) and Vinck et al. (2007) separately conducted studies on adult Ugandans living in camps for internally displaced persons (IDPs). Roberts and associates used the Harvard Trauma Questionnaire in 2006 to study traumatic exposures and PTSD symptoms in 1,210 participants while Vinck and his colleagues used the PTSD Checklist-Civilian Version in 2,585 adults. Both sets of investigators used the Hopkins Symptom Checklist-25 to assess for levels of depression in the study participants. The data of Roberts and Vinck each showed very high rates of PTSD (54% and 74.3% respectively) as well as high rates of depression (67% and 44.5% respectively). The prevalence of PTSD was high even compared with other groups with post conflict PTSD (de Jong 2001). The high rates of PTSD in the Ugandan IDPs may be explained by the long duration of exposure (almost 2 decades) of highly traumatizing events including mutilation, abduction, abductees forced to commit violent crimes and displacement of approximately 2 million people. Furthermore, the conditions in the camps contributed to ongoing trauma and deprivation.

Performing a cross-sectional study of 2,875 individuals, selected through a multi-stage stratified cluster design in 8 districts of northern Uganda, Pham and associates (2009) reported that one-third of subjects experienced abduction and more than half of the respondents and greater than two-thirds of abductees met criteria for PTSD. Factors that increased the risk for former abductees experiencing PTSD were: female gender, being a member of the Acholi ethnic group (not surprising as the war began in the Acholi sub-region), witnessing or participating in a number of traumatic events and experiencing difficulty upon re-entry into their communities. Increased risk for depression was associated with an older age of males at time of abduction, lower score on a social relationship scale, high incidence of exposure to traumatic events, high incidence of forced acts of violence and difficulty with re-entry into their communities.

Ilse Derluyn and colleagues (2004) confined their research to 301 former child soldiers in Gulu and Lira towns. The researchers used a semi-structured interview format to learn of past experiences. Additionally, 71 of the children were randomly selected to complete the Impact of Events Scale Revised (IESR). The age span of participants was from 12 to 28 years. Close to one third of the children were orphans. On the average, each child experienced six traumatic events during abduction. The rate of PTSD in the group was extremely high at 97%. The age of the child, the length of abduction, and period of time between escape and research did not affect the rate of PTSD.

Interested in the psychological and social rehabilitation of former child soldiers Bayer et al (2008) performed a cross-sectional field study of 169 former child soldiers in rehabilitation centres in Uganda and the Democratic Republic of the Congo. At the time of this 2005 study the former soldiers ranged in age from 11 to 18 years (mean age 15.3 years). The purpose of this study was to investigate the association between PTSD symptoms and feelings of openness to reconciliation as well as revenge in the study subjects. The investigators used a sample specific events scale and the Child Posttraumatic Stress Disorder Reaction Index. To study openness to reconciliation and feelings of revenge structured questionnaires were utilized. Data indicated that the child soldiers were exposed to high levels of trauma. Over 90% witnessed a shooting and more than half reported having killed someone. Close to 35% of the youngsters scored significantly for PTSD symptoms. Those with more PTSD symptoms were significantly less open to reconciliation and had more feelings of revenge.

The work of Ovuga et al (2008) most clearly demonstrates the need to screen all former child soldiers for PTSD and depression and to provide psychological interventions as a component of rehabilitation and reintegration of these children into their homes and community. Their research was prompted by the emergency psychiatric admission of 12 former child soldiers of the LRA in 2006 because of mass psychotic symptoms. Ovuga and colleagues studied a total of 102 children aged 6 to 18 years. This included the 12 children who were hospitalized and 90 schoolmates of those children. The 58 girls and 44 boys in the study were attending a rehabilitative boarding school for former abductees in northern Uganda. Data on posttraumatic stress disorder, depression, physical disabilities, social-demographic variables and children's war experiences were collected by using the Harvard Trauma Questionnaire, a modified Hopkins Symptom Checklist and a 15- item War Experience Checklist. Results indicated a very high percentage of children had serious emotional symptoms. This group of youngsters had been severely traumatized with 87.3% reporting that they experienced 10 or more war related events. The data indicated that 55.9% of the children reported symptoms of posttraumatic stress disorder; 88.2% depressed mood and 21.6% had various forms of physical disability. A high percentage of the children (42.2%) reported a family history of severe mental illness. It was the clinical opinion of the authors of this study that the school environment may have contributed to the exacerbation of emotional symptoms. The "ultra-modern" school had limited resources and teachers; yet, the children's learning curriculum was significantly accelerated to enable them to "catch up" for educational time lost while in the bush placing significant pressure on these youngsters. Those older than 16 were limited to vocational training which in reality promised a future of hard work and poverty and left the children with little hope. In addition, the children were in a confined, structured environment away from home and family which may have recreated for them their days in rebel captivity. Finally, the school viewed some of the youngsters as being possessed by demons (Ovuga et al 2008).

2.2 Complications of traumatic experiences in Northern Uganda

Victims of PTSD may suffer a variety of complications that may take the form of physical injuries or the psychological impact of the traumatic event per se. In addition war and violence have destabilizing effects on the social and individual lives of members of affected communities. In northern Uganda former child soldiers are called stigmatizing and criminalizing names that makes it difficult for the affected individuals to be reintegrated into their communities. The children of former female child soldiers who returned from the bush war are often not accepted by the communities of the child mothers, thus essentially uprooting the former child mothers from their communities and social roots. Additionally communities in northern Uganda still face the prospects of ethnic conflicts with potential for further trauma, as war has major impacts on children's development (Bardin, 2005).

Physical injuries include fractures, brain damage, seizures, sexually transmitted diseases and unplanned pregnancies. Psychological impact of traumatic events as seen in Northern Uganda include post-traumatic stress disorder, panic disorder, alcohol abuse and psychosis. At social and community levels, the Northern Uganda war has contributed significantly to lack of education, poverty, early marriages especially among girl children, family breakup among older persons, fear among male youths about getting married and the disempowerment of men as heads of households. Trans-generational effects and conflicts between neighbouring communities (Volkan, 2004) remain serious threats to social security and stability as renewed cycles of violence, war, prejudice, revenge motifs and lack of social development remain significant issues in the region.

3. Clinical features

Four main features characterize post-traumatic stress disorder; namely: intrusive memories of the traumatic event, flashbacks, re-experience of the traumatic event and avoidant behaviour. Intrusive memories may take the form of bad dreams while re-experience of traumatic events and avoidant behavior may take the form of loss of interest in social contact; fear of visiting places where individuals experienced initial traumatic events such as farm fields; depersonalization or aggressive outbursts triggered by conversations; and or bad dreams. The Fourth Edition of the Diagnostic and Statistical Manual of Mental Disorders of the American Psychological Association (APA, 1992) gives the following clinical diagnostic criteria.

- History of having experienced, witnessed or been confronted with events or an event that involved actual bodily harm or serious bodily injury or threatened death or threat of integrity to the person's body.
- During the event the person responded with fear, sense of helplessness or horror in adults, or agitated disorganized behaviour in children.
- The affected person experiences recurrent painful recollections of the traumatic event including images, thoughts or perceptions among adults, or frightening dreams among children.
- Acting or feeling as though the traumatic event were reoccurring manifested by re-living the event, illusions, hallucinations and flash-backs among adults, or plays involving actual traumatic events among children.
- Marked experience of emotional distress whenever the individual is exposed to situations that resemble an aspect of the traumatic event.
- Experience of physical symptoms whenever the person is exposed to situations that resemble the traumatic event.
- Active efforts to avoid thoughts, feelings or conversations about the traumatic event; the client might actively attempt to avoid visiting places that might evoke memories of his/her traumatic experience.
- Active efforts aimed to avoid activities, places or people that arouse recollections of the trauma.
- Inability to remember important aspects of the traumatic event despite the individual's preoccupation with his/her traumatic experience.
- Significant loss of interest or participation in pleasurable activities and daily chores.
- A feeling of emotional detachment or estrangement from friends and other social groups.
- Inability to experience loving warmth for others.
- Loss of hope in the future, e.g. feeling that one has no chance in marriage, raising children or that one might die before long.

In addition the individual might report the following symptoms that might have been absent before the experience of the traumatic event.

- Difficulty falling asleep or staying asleep
- Being easily upset or getting angry
- Difficulty concentrating at tasks involving mental effort
- Being attentive and alert to spot signs of danger
- Exaggerated response to stimuli in the surrounding such as sudden loud noise.

The recognition of post-traumatic disorder in northern Uganda is, however, not simple due to widespread beliefs in witchcraft and supernatural powers in rural areas, and many individuals with the disorder do not receive the intervention they need for a number of considerations. Children in rural areas and in schools receive severe corporal punishment almost routinely as a strategy by adults, teachers, and older children to instil discipline in them. Sometimes children are denied access to basic necessities of life including food in retribution for wrongs they might commit. Thus the nature and scope of traumatic stress in rural communities in northern Uganda is diverse and may pass as normal in the eyes of the ordinary individual. Rural communities in the region are more likely to somatise their ailments and to explain psychological distress in terms of witchcraft; spirit possession and or the non-performance of rituals to appease displeased ancestral spirits. As large communities were exposed to the traumatic events in the region, most people are inclined to underrate the psychological impact of their experiences in their lives, and to consider their psychological experiences as universally normal responses to their traumatic experiences. Informal social support exists at community level, which offers some degree of protection against psychological distress at least at superficial level (Betancourt and Khan, 2008), and most child soldiers (and adults) appear to adjust remarkably well to their traumatic experiences (Betancourt et al, 2010). As a result most investigators who are not accustomed to the social and cultural life of the communities mistakenly believe that post-traumatic stress disorder is rare in northern Uganda and that the communities in the region do not require any form of psychological intervention.

Indeed some individuals may not in fact recover fully from their traumatic experiences due to the delayed onset of post-traumatic stress symptoms in some individuals (Jones, 1987), and the long-term effects of traumatic experiences such as rape (Shanks and Schull, 2000) despite appearing to function well in daily activities. Unpublished work from northern Uganda also indicates that poor parental mental health evidenced by previous history of traumatic stress, depression, suicidal behavior and alcohol abuse may predispose children to poor mental health either independently or arising secondary to children's own traumatic experience. Thus at least in the context of northern Uganda, despite the availability of ubiquitous social support networks children and adults alike may or may not be resilient to the effects of war experience in the region. This thus highlights the importance of routine screening for depression, suicidal behavior, anxiety disorders and post-traumatic stress disorder symptoms among patients attending primary care.

3.1 Common symptom patterns of PTSD in northern Uganda

Post-traumatic stress disorder usually presents with vegetative symptoms of depressive and anxiety disorders or alcohol use disorder symptoms. Patients may complain about poor sleep due to dreams involving the dead beckoning them unto death. Direct inquiry about probable history of exposure to a traumatic event is required as dreams about the dead may be a significant sign of depressive disorder, anxiety disorder or PTSD representing intrusive thoughts. Sometimes patients may complain about having many thoughts or thinking too much. Too many thoughts may mean being worried, and signify depressive disorder or an anxiety disorder, particularly in association with frightening dreams in which the dreamer is visited by dead relatives, is chased by enemies/armed men, or is involved in battle. However too much thoughts may be an idiom for intrusive thoughts seen in post-traumatic stress disorder. Individuals may be described as preferring to be alone, and this description is the equivalent of loss of interest in social contact and pleasurable activities as in depressive disorder or post-traumatic stress disorder; it is not uncommon for post-traumatic stress disorder and depression to co-exist in the same patient. Such individuals are usually intolerant to conversations that might remind them of their traumatic experiences, and may exhibit considerable levels of irritability and may therefore not wish to participate in conversations with family and friends. Individuals who prefer to be alone following exposure to traumatic events also exhibit episodes of depersonalization with aggressive outbursts. The triad of social isolation, depersonalization and aggressive outbursts is so characteristic of former rebel soldiers in northern Uganda that some communities readily recognize the psychological instability in affected individuals and often arrange a quiet room for the victims to rest before they can rejoin their peers in social activities.

Vignette 1

The following vignette about a 19-year old former child soldier in northern Uganda perhaps illustrates the complex manner in which post-traumatic stress disorder presents sometimes. The patient was referred to the psychiatric unit in a general hospital in northern Uganda by a humanitarian agency. The patient presented with severe cognitive impairment suggesting severe brain damage; he was disoriented to time, place and person and he had poor attention span and poor concentration with poor short-term and recent memories. The young man had no idea as to where his home was claiming that he came from a location in Okokoro county in northern Uganda; however his name suggested that he came from the West Nile region of Uganda. He claimed he had graduated from a university in central Uganda (non-existent) and that he came from the Congo-Somali-Ethiopia border. Clinical examination, and laboratory and radiological investigations revealed no physical abnormality. This case illustrates the psychological consequences of brainwashing and indoctrination that the rebel captors used to keep control of the young children they abducted and trained as members of their forces. However the cognitive impairment in this case may also be explained on the basis of memory impairment that accompanies the clinical features of post-traumatic stress disorder.

Vignette 2

A 40-years-old married man and father of four children was admitted with severe psychotic symptoms and features of alcohol dependence to the psychiatric unit of a general hospital in northern Uganda. The man had been violent toward his wife for her failure to respond to his sexual advances, as he seemed to her not stable psychologically. Additionally the man's admission was prompted by his unusual behavior of watching a line of ants as they moved into and out of an anthill. The man interpreted the line of ants as government soldiers tracking rebel forces in northern Uganda. In a systematic order he crushed and killed some of the ants that he believed were government soldiers while sparing the ones that he thought were rebel soldiers. In therapy the man lamented the extent to which people in northern Uganda had suffered from the effects of the northern Uganda war and he wished that he were able to prevent a return of war in the region. The man reported repeated dreams of him hiding up in very tall trees to avoid being spotted by helicopter gunships, or looking down on government soldiers who would stare up helplessly at him from down below after escaping from them. Despite the location of his residence in northern Uganda, and the symptoms of post-traumatic stress disorder, the man denied any history of traumatic experience or links with either rebel forces or government troops.

Vignette 3

A 72-years-old man who had been involved with religious work in northern Uganda was admitted to the psychiatric unit of a general hospital in northern Uganda with what seemed to be unclear to the junior mental health professionals in the hospital. The man had features of early dementia as well as major depressive disorder, and he complained that his memory was poor and that his mind kept going blank as he held conversations; indeed as he narrated his traumatic experiences, the man seemed hesitant, paused frequently to recollect his thoughts and repeatedly asked for questions or sentences to be repeated to him. The man had been exposed to repeated traumatic experiences both in Southern Sudan and northern Uganda for over four decades and he reported several episodes of witnessing torture, killings and human suffering; he himself reported at least three occasions of near escape from death, making him wonder as to why it was always him who had to go through the sort of traumatic experiences that he repeatedly encountered. While in hospital the man always ran out from his hospital room to spend the night outside the room in the open, claiming that attackers had come for him. Out of his several dreams related to traumatic situations, the man reported one example of him leading a group of five men who had attacked a refugee camp and killed many of the refugees before the authorities sent in reinforcement to rescue the refugees. When the reinforcement arrived they informed him and his men that since they were the ones that started the fighting, they would all be killed. Though he ordered his men to retreat, his men were all killed and he woke up from sleep just in time before he was himself killed.

Vignette 4

A four-year-old male child was admitted to a large mental hospital in a former homeland in South Africa with scanty history of his psychiatric problem. By the time one of the authors (EO) saw him, the child's parents had returned to their remote rural village. The available information indicated that the child had been attended to at the rural district hospital and at a traditional healer's shrine without benefit. Each of the child's hands was firmly and securely crepe-bandaged into a fist. He had a combination of recent and old healing wounds and scratches in both temporo-frontal areas of the head. The child was otherwise of good nutritional status and there was no immediate evidence of child neglect. The reason for bandaging the child's hands became immediately obvious when the child suddenly began to hit himself with both hands in the injured areas of the head. The blows were so strong and fierce that an onlooker would feel sympathy and pity for the child. Efforts to prevent the child from his self-injurious behavior only led to resistance and even stronger blows to the child's head. Whenever the child got tired from hitting himself he would sometimes hold the hand of the nearest adult and beckon the adult to hit him. Routine laboratory test results were normal. Report of an electroencephalography indicated non-specific occasional spikes and waves in the temporal lobes, particularly in the right temporal lobe. The child was treated with an initial low dose of haloperidol, followed by ethosuximide but with no clinical benefit.

Nursing report indicated, however, that the child was attracted to two female nurses both of who responded appropriately as surrogate mothers. The reports further indicated that each time the primary surrogate mother lifted the child into her arms he would reach out for and pull out her breasts though he did not attempt to breast-feed. While in the company of the surrogate mothers, the frequency and severity of the child's blows to his head decreased. This observation led us to believe that there was a problem with loss of attachment and to develop and break down a set of the child's self-injurious behavior pattern from the most complex to the smallest units for purposes of drawing up a behavior modification strategy based on appropriate rewards if the child refrained from self-injurious behavior and the withholding of attention or reward if the child engaged in any form of behavior considered by the surrogate mothers or other nurses as unacceptable. Each time the child's behavior was considered positive he was praised and occasionally presented with a personal toy, but each time his behavior was unacceptable this was indicated to him in a clear simple language promptly. The rules of the therapy were typed out and pinned on the notice boards on the children's ward for all nurses to follow in support of the two surrogate mothers. Though the author (EO) was from another culture, the behavior modification strategy was planned carefully with the nurses, written out in simple language and explained before its implementation. As part of the therapy, the bandages were removed from the child's hands as the initial reward for non-injurious behavior. All medications were also withdrawn and the child was left free to do whatever he wanted within the provisions of the behavior modification strategy. Using this strategy, the child's self-injurious behavior progressively and eventually resolved completely within two weeks. The child's clinging behavior on either of the surrogate mothers stopped; he became social and interacted freely with all nursing staffs on the ward and began to play with other children.

When the parents eventually came to take him home after six months, the mother narrated the history of the child's mental health problem as follows. The child was the first-born in the family and received the full attention of his mother. When he was two-and-half years old, a sibling came in between him and his mother. The child reacted with intense rivalry with his infant sibling who the child attempted repeatedly to push off from their mother's lap. When he failed in his efforts to push the infant from the mother's lap, he became more and more vicious in his attacks on the infant sibling. In a final effort to stop the child's hatred toward the infant sibling his mother confessed hitting the boy so hard that he stopped pushing the infant from her lap. In reaction the child turned his hatred toward himself and started to slap and scratch himself. As observed in the hospital whenever he got tired he would come to the mother and beckon her to slap him in the face as she had done. A full explanation was made to the mother as to the probable origin of his self-injurious behavior, which the parents accepted, and the mother believed the explanation would help in her future relations with the little boy, who we shall call Sipho in this chapter.

It is possible that this child suffered from two episodes of traumatic stress; first his loss of his first love object, the mother, and secondly the physical attack on his physical integrity by the mother. Though young, the child apparently drew the correct relationship between his hateful feelings toward his sibling and the punishment that he received from the mother. In order to protect himself and his infant sibling, the child took a middle option; self-punishment that in adult term would have led him to suicidal behavior, which is a common occurrence in post-traumatic stress disorder. One might interpret his never-ending urge for punishment as an obsession, and the self-injurious behavior as a compulsive disorder. It is therefore not surprising that a program of response prevention that aimed to modify his behavior into a healthy lifestyle in the face of unavoidable challenge in life worked for him. The child's mental health problem that we might refer to basically as an obsessive-compulsive disorder probably qualifies to be intrusions and attempts to re-experience his traumatic experience in the hands of his mother. Further his behavior interfered significantly with his social functioning to the extent that it interrupted his normal relations with his parents and sibling resulting in hospitalization. Given the history this was a case of post-traumatic stress disorder co-morbid with obsessive-compulsive disorder.

4. Mental health problems of children in Northern Uganda

Methods of data collection

In this section we summarize the findings of our research on the patterns of mental health problems of children in northern Uganda. The findings highlight the diverse nature of traumatic experiences and their associated psychological distress symptoms the children aged 4-17 in the region experience. We conducted a cross-sectional survey of children in Gulu district using both qualitative and quantitative research methods. We used stratified cluster sampling strategy to select two urban and two rural villages in Gulu District. We randomly selected the participating villages from 2 sub-counties (one rural and one urban) in Gulu District. We estimated that 100-150 children would participate in the study. The parent or caretaker of each child or adolescent selected was also requested to participate in the study.

Participant selection involved community leaders in each village who helped the research team to discuss the research in general terms with the identified children and their caregivers and gave them the opportunity to ask questions and to think about possible participation. We explained the research project; and gave the participating children and their caregivers the opportunity to ask questions. A simple consent/assent form was explained to each potential participating child and caregiver. If they still wished to participate we asked that they sign the consent/assent form or place a thumbprint in the case of those who could not write. (In some of the studies referred to above only verbal consent was obtained). Throughout the interviews participants were asked if they were okay in participating and given the opportunity to stop if they chose to. At the end of each interview participants were asked how they felt about having participated and if they had any questions about the project at the end. Each caregiver was given a phone number to call or a person they could contact (they may not have access to a phone) who could contact one of our team members (EO or CL) if any concerns or questions should arise in the future about the research interview.

Using a semi-structured interview we collected demographic information, descriptions given by the children of their personal experiences and their reactions to events in their lives. We covered areas of strengths as well as difficulties. A principal investigator (CL) or a trained assistant conducted the interviews. When indicated, an interpreter asked the questions in Luo (the primary language spoken in Gulu district) and translated the answers for the primary interviewer who spoke English. The primary interviewer clarified answers with the research subjects through the assistance of the interpreter. The primary interviewer wrote down answers to the questions on each questionnaire. We also used a semi-structured interview with parents/caregivers of the children to determine how well the children functioned emotionally and behaviorally. The investigators of this research project constructed the questions for the interview with the assistance of community members in order to be sure that the concept of how an individual functioned in daily life was consistent with cultural expectations. We examined the child's ability to function in 3 domains: 1) the home 2) in peer relationships and 3) at school, job (such as farming), or age appropriate activity.

In order to gather the information required we developed three questionnaires to obtain information about the emotional well being of children in northern Uganda. In order to be culturally and linguistically accurate each questionnaire was developed with input from professionals and community members in the region. The questionnaires were first written in American English and converted into Ugandan English to assure accuracy in communication. The questionnaires were then translated from Ugandan English to Luo. In order to be sure that the original meaning of each questionnaire was not lost in translation the questionnaires were translated from Luo back to American English to check for accuracy.

1. *Questions for Caregivers*

Parents, guardians or other caregivers of children participating in the study were asked questions about the children in the study. One interview took place using a semi-structured format to determine how the children functioned at home, school or work, and with their peers. The interview took approximately 30 minutes to complete.

2. *Interview of Children and Adolescents (Ages 9 years to 17 years)*

This semi-structured questionnaire was administered to children and adolescents aged 9 years to 17 years of age. It was administered in 2 parts. Part one was administered during a first meeting in order to establish rapport. Part two was administered during a separate meeting during which time questions related to feelings; reactions and functioning were more personal. Each interview took approximately one hour.

3. *Interview with Young Children (Ages 8 Years and Younger)*

To date studies related to the mental health of children in Uganda have focused on older children, primarily adolescents. There is little information about children 8 years of age and younger. This semi-structured questionnaire was designed to engage younger children by using puppets and giving stories about the puppets. After hearing about the puppets the children were asked questions about themselves in a qualitative approach using a semi-structured interview. The questionnaire was administered in two parts on two separate occasions. Part one was administered during a first meeting in order to establish rapport. Part two was administered during a separate meeting during which time questions related to feelings, reactions, and functioning were more personal. Each interview took approximately 30 minutes.

To participate in the study we included a) children or adolescents who participated in the study were aged between 4 and 17 years and were willing and able to answer our questions and b) those that agreed to be in the study and an adult responsible for the child (parent/caregiver/guardian) also willingly consented to the child's, and their own participation in the study. We excluded from the study a) children who could not speak English and there was no appropriate interpreter to interpret for the subject related to the study and b) children that were unable to communicate due to a medical or severe psychological problem such as mutism, catatonia, and severe mental retardation. Children under 18 years of age, who were able to answer our questions, as well as their guardians / caregivers, were interviewed. We took special care to be sure the children and their caregivers knew that participation was voluntary; that there would be no negative consequences for not participating, and that any benefits they might receive from the community, university or hospital would not change if they decided not to participate. We received informed consent from the caregivers and assent from the children in the study. Because our study did not offer specific interventions and because mental health resources are limited in rural areas of the region we interviewed the children in a non-direct manner asking about their life – what they enjoyed, what annoyed them, what they routinely ate and or how they slept, what they would like to see be different, etc. Such an approach allowed the children to disclose information while not putting them in a position of forcing them to talk about things that are emotionally very upsetting for them. If we were to notice some children (or caregivers) who were of serious concern to us, e.g. severely depressed, suicidal etc. we attempted to help them utilizing any resources that might be available such as Hospital. In addition, CL who is child psychiatrist provided supervision for mental health workers referring them to the mental health unit at Gulu Regional Referral Hospital.

Pre-coded numerical data from the semi-structured questionnaires were entered and analyzed with SPSS version 10.0. Chi-squared test was used to determine statistical significance levels between groups. One-way ANOVA multivariate and logistic regression analyses were used to determine factors associated with emotional disorders and impaired social functioning among the participants. Significance levels were set at 0.05 and 95% Confidence Intervals. Prose accounts from the questionnaires were analyzed manually according to emerging themes emotional disorders and psychosocial functioning of participants. In this chapter we provide only qualitative material to present the nature of post-traumatic stress disorder among children aged 4-17 years in Gulu district, northern Uganda. We conducted our interviews in a private room or outdoor space chosen by the caregivers of every subject in the comfort of their own homes. In general, the adolescent participants were interviewed alone without an adult caregiver present; children, especially those younger than 9 years, were interviewed in the presence of their caregiver. However, children and adolescents were given the opportunity to determine whether they wished to have an adult caregiver present or not during the interview. We received ethical clearance for the study from the Institutional Review Committee of Gulu University and the Uganda National Council for Science and Technology.

We analyzed the interviews to determine themes and patterns that were expressed by each child. We then worked with selected community members and faculty members of the school of medicine to determine how certain themes and patterns such as somatic complaints, visitation by spirits, feelings of abandonment, etc. might compare with western constructs of such disorders as PTSD, depression and anxiety. We hoped that the analyzed information would give us a percentage of the children in each village who were experiencing significant emotional difficulty and those who were not functioning adequately. We also hoped that we would have qualitative and descriptive data, which would give us information that would take into consideration the culture and context of the participating children and their caregivers.

5. Results

Ninety-eight families from four separate randomly selected villages in Gulu district in northern Uganda participated in a Fulbright-supported qualitative study to determine the mental health needs of one child per family in January to March in 2010. The study related to the mental health of the children and the children's functioning, their general concerns, attitudes and coping strategies of each child who was aged between four and seventeen years. This review highlights the complex situation of children in northern Uganda where they not only cope with the day-to-day problems of poverty, the aftermath of war and conflict but also troubled relationships within their own families. We summarize our findings under seven themes; namely: stress related to difficulties paying school fees, aggression/violence, fear, sleep disturbance, emotional problems, spirit possession, and coping strategies. Coping strategies are particularly significant as they relate to the resilience described by Betancourt and Khan (2008), Betancourt et al (2010), and Akello et al 2010, 2011). For purposes of clarity we group the children in this study as younger children aged between 4 and 8 years, and older children aged from 9-17 years.

5.1 Significant stress related to the families' inability to pay school fees with resulting lack of education or inconsistent education for the children

Fifty-eight caregivers (59%) expressed concerns that lack of school fees was a source of stress for them or their children. Many of the children expressed feeling worried about being able to pay fees as well as sadness when "chased from school" because of an inability to pay. Though education is free in government-aided primary and secondary schools in Uganda, many parents in rural areas are sometimes unable to meet small amounts of fees levied by schools to meet the welfare needs of their children; the experience of a child being chased away from school is therefore traumatic for affected children. For some children the stress was very severe with one youngster reporting that when his family failed to pay his school fees he felt that he would never be happy again; another teen reported that when school fees are not paid she thinks of committing suicide.

5.2 Aggression/violence

Children and caregivers reported that much aggression occurred in the children's lives. The aggression was experienced directly by many children and many also witnessed it. The aggression occurred in the home, in the community, at school and with peers. At times children were victims and at times they participated in the aggression.

Of the 98 youngsters in the study 66 (65%) experienced physical aggression that was directed towards them. Almost 31% of the children experienced physical aggression toward them at school by teachers and or other students; almost 29% of the children experienced physical aggression at home by caregivers and or siblings and almost 28% of the children experienced physical aggression in the community.

The percentage of children who reported that they experienced physical aggression in the home was much higher in the rural communities as compared to the urban communities. In the two rural communities: 11 out of 24 youngsters from one community and 11 out 25 youngsters from the second community, for a total of 22 out of 49 children (almost 45% from both rural communities) experienced aggression at home. In the urban communities 3 out 25 youngsters from the first community and 2 out 24 from the second community, or a total of 6 out of 49 (12%) from both urban communities experienced aggression at home. Five of the 98 children reported experiencing physical aggression in all three places; i.e.: home, school and community. Three of the five were from the same rural village, and there was one child from each of the two urban villages.

Fifteen of the 98 children were reported to be aggressive. Out of the 15, eight children came from the rural village in which 3 children experienced physical aggression in all three sites examined. The children in the study were more vocal about the aggression they experienced than any other topic. They reported fear, sadness and anger related to the behaviour directed toward them.

5.3 Fear

Many of the younger children expressed that they were fearful. While it is not uncommon for young children to express fear the content and extent of the fear was troubling as it suggested significant stress and at times trauma. 36 out of 47 children from the younger age group (76%) reported that they had experienced fears from at least one f the three categories; 1) some one or something 2) illness and 3) going to sleep. Eight out of 47 children (17%) feared something from all 3 categories; 15 children (almost 32%) feared something from 2 categories and 13 (almost 27%) from one category. Twenty-five children (53%) feared people or things. Some of the people or things the children feared were "strangers who chop off children's heads for human sacrifice" (witchcraft), thieves, snakes, cats, mad people, darkness, elephants, dogs, rats, kidnappers, and other children who beat them. Twenty-eight of 47 children (59%) reported that they feared getting ill. Many feared an illness that they had in the past; some children feared they would die or never get well. Fifteen of 47 children (almost 32%) feared going to sleep. The reasons for fearing going to sleep were fear of darkness; fear that someone such as a thief, someone to cut off their heads and other scary people would come to them. Some feared animals such as snakes, hyenas or rats would come and harm them while asleep. Two children reported being bitten by a rat in the past; one child reported that a snake bit his younger brother.

5.4 Sleep disturbance

Sleep disturbance was common with nightmares being the most prevalent problem. Many nightmares were triggered by actual experiences. Sixty-seven out of 98 youngsters (68%) reported a sleep disturbance. For the younger children the sleep disturbances were: difficulty falling to sleep; fear of falling asleep, and bad dreams. For the older children sleep disturbances were: worrying at night; trouble falling asleep; waking up at night; and nightmares. Fifty-two out of 98 children (54%) reported having bad dreams. The content of the dreams varied some between the younger and older groups as well as between villages. However, the most common themes of the nightmares of the younger group of children were; death, or being harmed or chased by animals or bad people. The most common themes in the nightmares of the older children were: death, fighting, being attacked or abducted. In the first rural community, the predominant themes of the nightmares of younger children were death and being chased by ghosts. The content of the bad dreams of the older group of children was primarily of fighting, or being chased or attacked. In the second rural community, the themes of nightmares of younger children were: abandonment, being in a life-threatening situation, and being harmed and or killed by bad people. In regard to the older children almost 86% reported nightmares. For the older children, in general the nightmares were traumatic, with the predominant themes of death or life threatening experiences. In 66% of the older children who reported nightmares they described someone dying in their dream. Some nightmares were recurrent and some were reported as occurring every night. In the third (urban) community, the themes of the bad dreams reported by younger children were that of bad animals coming to harm them or the death of someone close. Almost 79% of the older children in this community reported nightmares. The content of the nightmares were significant. The themes were of death and dying, killing and fighting and abduction by the LRA. In the fourth (urban) community, the prominent sleep disturbance of the younger children was bad dreams. The themes of the nightmares were; animals or a kidnapper coming to harm them, a collision of motor vehicles, and falling in a lake. Of the older children only 2 reported nightmares and the theme was of death - one of them being killed by a witch; the other was of the deceased father coming to him (child) to take him away to die.

5.5 Emotional problems: sadness/ isolation/ anger/ worries

Boxes 1 and 2 below illustrate various reasons that study participants gave for children feeling sad, angry or worried. Twenty-two of the younger children reported that there were times when they were sad; 39 of the older children reported episodes of sadness/ isolation and/or anger. A total of 61 youngsters out of 98 (62%) reported feelings in the spectrum of

Children's reasons for feeling sad, angry or isolative	Caregivers' reasons for their children feeling sad, angry or isolative
• 9 year old female feels sad when she arrives at school; she fears being caned by the teacher • 10 year old male feels very sad when he is sick; he feels he will not get better; he feels very angry when friends beat and hurt him one time per week • 13 year old female reports that she is lonely, sad and likes isolating because friends gossip; she is angry with herself after quarrelling at home • 15 year old male is sad when he quarrels with friends • 13 year old female fees sad when she is lacking food and when she asks for something and is denied; she is angry when school asks for fees and there is no money; she thinks about and plans suicide when relatives compare her to her late father • 14 year old female is sad often when there is no money for school fees and when there is no food in the house about twice a week; she often thinks of committing suicide by collecting pills and overdosing; she feels suicidal because of failure to pay school fees • 17 year old female reports that she is sad when she lacks school fees or when she is sick; she is sad when she thinks about her deceased father and he is not here; she is angry when relatives disturb her because her father is dead; she is angry when she is all alone • 14 year old female says she is angry when she is delayed by being asked to do something at school and because of the delay she is denied food at home and gets no lunch • 11 year old female is lonely because she likes to stay away from boy neighbors; she feels bad when friends say she is stupid • 16 year old female is sad after she refuses to do something for her parents; she isolates when someone gossips about her • 16 year old male feels that he will never be happy again because of a lack of school fees • 16 year old female whose both parents are deceased is lonely and sad when friends are not around	• Lonely; wants to see her dad/ wants to go to school • Fears other children will fight him • Lonely; friends like beating him • Thinking about mother; parents are dead • Stomach pain; "many thoughts"; does not like father because he beats the mother • Sad one time per week; caregiver does not know why • Sees other children going to school and cannot go because of fees • "Many thoughts"; not going to school because of fees; isolates when being refused something asked for • He has "many thoughts " 2 or 3 times per month and is sad because of the death of his father • He is sad and lonely for reasons unknown to caregiver; "many thoughts" about how he will make a living • She is sad and lonely when she thinks about her deceased parents; sad and has "many thoughts" when she is chased from school because she lacks fees • He is sad but caregiver does not know the reason • 9 year old male is sad; caregiver does not know why

Box 1. Reasons why children felt sad, angry or isolative

sadness and anger. Sixty-two caregivers (63%) reported that their child had periods of being sad, lonely, isolative and/or angry. Reasons youngsters gave for being sad or angry were: physical abuse by caregivers, teachers or peers; death or illness of parent or parents with child being neglected or being deprived of food or school because of no money; being overworked; illness; relationship issues with peers and missing someone. Nineteen of the younger children and 26 of the older children, a total of 45 out of 98 (45%) reported that they worried. Fifty-five caregivers (56%) reported that their youngsters worried. Children reported that they worried about the following things: a sick parent or sibling who they fear will die or who would transmit disease to them; physical abuse by teachers, caregiver or peers; well being of family members when they travelled or went to school; being turned away from school because of no fees or not passing grades; fear of nightmares; fear of abandonment; lack of food; poor living conditions; becoming burned when starting a fire; and alcohol problems of caregiver.

Children's reasons for being worried	Caregivers' reasons for their children being worried
• 9 year old male worries and will not eat at school because he fears the teacher will beat him • 17 year old male worries about the living conditions of his mother • 16 year female worries when her brother is delayed that he might be in an accident; worries when boys slap her and say bad things to her • 15 year old female worries when her brother lacks school fees; sometimes she goes out digging to earn school fees for the brother • 16 year old female worries about getting bad dreams about a man slashing her throat for human sacrifice • 16 year old male worries that his aunt (caregiver) has alcohol problems • 16 year old female whose parents are deceased worries about her brother when he has no money	• No food; friends fight him • When friends or children at school try to take her belongings • Lack of school fees • Thinking about father's death last year • When hungry; when friends steal her books or pens; when friends fight over toys • Lack of school fees • Lack of fees; being beaten by teachers

Box 2. Reasons children and caregivers gave for children being worried

5.6 Spirit possession

No caregiver from urban villages reported their child as having experienced spirit possession currently or any time in the past. However three caregivers from rural villages felt that their youngster might have spirit possession. In one case the caregiver thought this because the child would awaken from sleep 2 to 3 times per month shouting, "something is coming to hurt me". Another caregiver reported that a teenager experienced spirit possession at the beginning of each month. Rebels abducted the child when he was very young. He has had episodes of spirit possession since returning from the bush. A third caregiver felt that her child experienced possession during the previous year. The nine-year-old girl reported that she is sad most of time and has repetitive dreams of her deceased grandmother coming to take her away. She cries because her father was killed by the rebels and there is no one to help her mother care for her. The previous year when her mother was hospitalized, the mother's sister-in-law cared for the child. The child did poorly in school due to the hardships she went through during that period. One day during the same year the child ran away to the bush for just 1 day. Her mother saw her rolling on the ground and her mother heard spirits making sounds over the girl's head and the mother thought the girl was possessed or cursed. Two additional caregivers initially reported that their children did not experience spirit possession. Each caregiver then added that their child does get nightmares and, therefore, they wondered if the child had spirit possession.

5.7 Coping strategies

The children in the study discussed various strategies that they used to help themselves when they were sick, sad or angry.

5.7.1 Religion

Almost universally (94 out of 98 youngsters) of both the older and younger age groups reported both believing in a religion and also found religious beliefs and praying helpful when coping with illness and emotional difficulties. The children came from Catholic, Protestant and Muslim backgrounds.

5.7.2 The use of alcohol or drugs (Asked only of the older children ages 9 through 17)

No youngsters reported using street drugs. Only 2 children reported the use of alcohol and neither described it as a method of coping. One 10-year-old female from a rural village reported drinking 250 mgs of local brew with sugar "for enjoyment". One 10-year-old male from an urban village reported drinking alcohol on one occasion only – Christmas. He drank alcohol because he saw adults drinking it.

5.7.3 The use of traditional medicine/ceremonies (Asked only of older children 9-17)

Seven youngsters from the 2 rural villages reported that they used traditional medicine or had used it in the past. Traditional medicines were described as tree bark, herbs or roots that were used to treat such ailments as cough, rash, vomiting and convulsions. Two of the youngsters who had used traditional medicine were not sure if it had been useful. From the 2 urban villages none of the youngsters reported using traditional medicines or ceremonies. However, one teenager felt that traditional beliefs and ceremonies were sometimes useful. Another youngster reported that some traditions are helpful but that some are not correct.

5.7.4 Coping when Ill (Asked only of the older children ages 9 through 17)

Almost universally, 50 out of 51 older children used medication that they received from a hospital or elsewhere when they were ill and found the medication useful. The medications that were specifically mentioned were: Panadol (Paracetamol), quinine and painkillers. Bathing in cold water to relieve fever or putting cold water on the head to relieve headache was also mentioned. Foods and beverages were also taken by some children when ill. All the remedies were thought to be helpful.

5.7.5 Use of food to feel better when ill (Asked only of the older children ages 9-17)

Out of 51 older youngsters, 30 (59%) reported that they drank specific beverages or ate specific food to help them feel better and for most this was helpful. More of the young people who lived in the urban villages ate or drank items for their wellbeing than did those in the rural villages. Some of the beverages the youngsters drank were juice, which gave energy and strength and improved appetite; tea, especially to reduce shivering: water which "added blood", and soda to relieve headaches. Some of the food items eaten were: peas to relieve headaches, beans, vegetables, rice, *posho* (bread made from maize or corn flour), eggs, porridge and fish.

5.7.6 Coping when sad (Asked only of the older children ages 9 through 17)

Of the 51 older children 35 (69%) reported using specific strategies when sad; 13 specifically reported that they did not do anything when feeling sad; 1 child reported that he did not get sad and for 2 coping strategies are unknown. Of the children who used strategies there were several approaches to coping.

a. Going to another person such as a friend or relative (mother, brother or sister) to get advice, to play, or to listen to a song or stories. Twenty-two youngsters chose this strategy making it the most popular. Friends were chosen more often than family members
b. To participate in an enjoyable activity. Six youngsters chose this approach and went to play; played football; sang or enjoyed music or a movie.
c. To engage in a quiet or contemplative activity. One child bathed or relaxed; 1 prayed and 1 just sat.
d. To isolate. Two children reported that they isolated and 1 "moved away".
e. To correct the reason for sadness. One young lady would go to those who offended her and ask for their apology.

5.7.7 Coping when angry (Asked only of the older children ages 9 through 17)

Thirty-four of 51 youngsters (66%) reported using specific strategies when angry and for most the strategies worked. However, 1 teen coped very poorly. He would cry and then he thought of running in front of a car. One child reported that he did not get angry. Fifteen reported doing nothing when angry. For 1 child the coping approaches were unknown. For the young people who used strategies there were several approaches to coping:

a. Going to another person such as a friend or relative for advice, to talk, to play, to tell stories and laugh or get away and sit with people. Thirteen young people chose this strategy. Twelve went to friends or other people; 1 went to an older sister.
b. To participate in a useful or enjoyable activity. Three children chose this strategy and cleaned utensils; ate; and played by concentrating on a game.
c. To engage in a quiet or contemplative activity. Ten youngsters chose this strategy and would go and lay down; sat and did nothing; slept; kept quiet; and prayed.
d. To move away from the situation or to isolate. Five young people chose this approach and would move away from home or the situation; stay or sit alone; and isolate to think of new things.
e. Aggression. Two youngsters reported that they would fight.
f. To wait for the situation causing anger to be corrected - to seek fairness. One child would ask her parents to give her the same things given to her siblings and wait for them to do so.
g. Self-destruction. One child would cry and think of killing himself by running in front of a vehicle.
h. Some of the youngsters chose more than one strategy to help them overcome feelings of anger.

6. Comment

The children of Uganda are at high risk for emotional problems yet there is limited information about children's mental health in the country. This is especially true of very young children (8 years of age and younger) for which information is scarce. Our study was an exploratory mental health needs assessment that could facilitate the development of mental health programs in northern Uganda. The study focused on the children in the community at large and not one targeted group of children. It included a representative sample of all children between 4 and 17 years of age who were willing and able to participate in a semi-structured interview. The study took place at a time when the region of Gulu was secure, and without war related violence or ongoing abductions. In a secure environment selected by children's parents, guardians or caretakers in the comfort of their homes, we obtained a different assessment of mental health problems and needs of children than that obtained from research conducted at a time of political upheaval, warfare and unrest despite the documented long-term effects of war (Jones, 1987); in this way much of our results do not include cases of acute stress disorder among children related to war and organized community violence though some of our study participants appeared to live under the threat of aggression in their homes, school or community almost on daily basis. Our study gathered information that might be useful in guiding mental health interventions for children and adolescents. It is necessary to understand the types of emotional difficulty children experience, as well as the degree to which they are affected, in order to provide appropriate treatment interventions and resources that might prevent further psychological damage.

By sampling a broad spectrum of children from several communities, by not limiting our research to one specialized group and by using qualitative measures we believe we had the opportunity to obtain information about the children that will be contextually and culturally rich and provide information about strengths as well as difficulties. We also hoped that the information would complement the data obtained from previous studies and hopefully provide important information in guiding future research, intervention and prevention. The research information can then be used to develop mental health services for children and adolescents (these services are currently rudimentary and inadequate in northern Uganda) and to seek resources to properly address the emotional needs of these children. Our findings from this exploratory research indicate that the children of Gulu district experience significant levels of not only post-traumatic stress disorder but also other mental health problems including anxiety disorders, depression and suicidal behavior in response to a variety of traumatic stressors some of which might be regarded by families as normal manifestations of growing up in this rural district of northern Uganda. Our results suggest that the mental health problems of the children and adolescents resulted from the experiences of loss due to death or separation of parents, domestic violence, lack of food and or school fees, or disputes in the youngsters' relationships at school and in the community. The presentation of depressive disorder with sadness, "many thoughts" and worrying is typical for both adults and children and adolescents. It is also noteworthy that adult caregivers were able to recognise the signs of emotional problems in their children albeit the lack of access to appropriate mental health services in rural areas . Though we did not specifically screen for alcohol and substance use disorders, this appears to be a major potential mental health issue for the children and adolescents in Gulu district. Our findings suggest that children in this resource-poor setting used a variety of psycho-social, traditional and complementary healing approaches to manage their own experiences of distress as reported by Akello et al (2007) and Akello, Richters and Ovuga (2011). As a result of the selection procedure it is possible that the results of this study are skewed toward the mental health needs of participants who were willing to participate in the study. We are aware that our results might not apply to the general population of children and adolescents in northern Uganda generally. However we endeavoured to ensure that participant selection was unbiased and that our findings could form a reasonable basis for further research into the mental health needs of children and adolescents in post-conflict settings.

7. Management of post-traumatic stress disorder

The professional management of post-traumatic disorder in northern Uganda has taken advantage of the special social and cultural situation of the communities in the region. In general, the principles of PTSD management follow the general ones of any psychiatric disorder but with particular emphasis on preventing re-traumatisation and aimed to promote psychosocial functioning within the individual's social milieu.

7.1 Principles of assessment

Most individuals who suffer from PTSD will a) present with symptoms that will not suggest the condition b) come to the health unit late and or c) present to health facilities with physical complications of traumatic experiences. Typically patients will present with multiple somatic and vegetative or psychotic symptoms, behavioural problems (children and adolescents), or symptoms of alcohol or other drug abuse. An adequate assessment of PTSD is made on suspicion of the presence of the condition at all times and progresses through three related stages. Firstly the process and type of assessment is thus influenced by the patient's residential address; circumstances in which the individual lives; and history, timing and type of trauma. There are various types of traumatic events but these can be categorised as either individual (e.g. car-accident, rape, etc) or group (e.g. landslides, floods, war-trauma, volcanic eruptions, plane crashes, rebel attacks, etc). It also depends on the severity of injuries sustained, some of which may be life threatening or at other times minor, e.g. slaps. Secondly assessment aims to determine the nature of traumatic event, and as to when the trauma occurred. PTSD can be acute (including Acute Stress Disorder, PTSD), chronic or delayed. Complex PTSD involves the exposure of the individual to multiple and complex patterns of trauma that are often repeated and or prolonged leading to changes in the victim's personality and general behaviour. Acute PTSD (including acute stress disorder) calls for immediate treatment and sometimes rescue operations e.g. in volcanic eruption, in war or terrorist attacks. Thirdly, assessment aims to determine the need for immediate intervention.

7.2 Immediate treatment

Immediate intervention is contemplated during the acute phase or shortly after exposure to a traumatic event. Immediate intervention is provided based on the principles of crisis intervention.

a. *Assessing risk factors for post-traumatic stress disorder:* Several risk factors for post-traumatic stress disorder have been documented including criminal assault, political detention and torture, rape, childhood physical abuse (Kaimer et al, 2009); acute posttraumatic stress disorder and the presence of premorbid and comorbid psychopathology (Koren et al, 1999); age at first experience of traumatic experience, severity of traumatic experience and availability of social support after traumatic exposure (Engdahl et al, 1997), sex, previous experience of trauma and the subjective appraisal of threat to life (Stallard, Velleman & Baldwin, 1998). Assessment and identification of predictive risk factors for posttraumatic stress disorder and addressing these at the earliest opportunity after exposure to traumatic experience is a vital first step in trauma management.

b. *Assessing the nature and severity of physical injury:* An adequate assessment of the nature, type and severity of injury and associated complications usually involves collaboration with other health care specialists including general surgeons, gynaecologists, neurologists, and general physicians. On the basis of a comprehensive assessment of findings, it will then be possible to plan a comprehensive care program targeting the needs of each trauma victim.
c. *To prevent further re-traumatisation:* This often involves removing the individual from the traumatising situation e.g. from a fire, war-front or from domestically an abusive home (sexually or domestic violence).
d. Prevention of further injury or provision of immediate first aid to care for injuries e.g. to prevent bleeding, to immobilise unstable fractures or treatment for surgical shock: Often, traumatised individuals experience psychological shock and panic. In such emotionally laden situations, victims of trauma cannot make rational decisions. They need support and someone else to make decisions for them; e.g. not to run back in a raving fire in order to rescue someone or property. Removing someone from the scene of trauma to an area of safety, provision of security, ensuring protection from rain, cold, or other harsh environmental condition, providing food and immediate shelter and giving emotional support are all part of crisis intervention. After crisis intervention, one then embarks on planning the longer-term treatments based on an assessment of risk factors for post-traumatic stress disorder.

7.3 Short-term treatment

This involves those treatments necessary to mitigate the effects of the trauma or limit the progression of the psychological sequel into chronic or complicated phase. Intervention follows a thorough psychiatric assessment and then treatment planning involving the individual in which the interventions are individualized depending on the needs of each patient. These interventions include (as deemed necessary):

- Medication
- Psychotherapy
- Counselling
- Management of co-morbid physical and psychiatric disorders.

7.4 Medications

The use of medications in PTSD is for the control of symptoms that include insomnia, agitation, anxiety, panic, depression or those specific to the organs injured e.g. epilepsy, . Anxiety, panic and agitation are especially common and will respond to minor tranquillisers such as alprazolam, diazepam, and clorazepam. Depression, panic disorder and phobias will respond to antidepressants such as fluoxetine, paroxetine, amitryptiline. Specific medications for other health problems may be indicated such as Anticonvulsants for Seizures due to brain injury, such as Phenytoin or carbamazepine or antipsychotics such as haloperidol or chlorpromazine.

7.5 Psychotherapy

Psychotherapy involves talk therapies popularly termed "counselling" in the Ugandan context. Various types psychotherapy have been used by trained counsellors and members of humanitarian agencies in PTSD in Uganda including individual counselling of a supportive nature, group counselling such as Interpersonal Psychotherapy, Cognitive Behaviour Therapy, Narrative Exposure Therapy, Play therapy for children, and Art therapy for both children and adults. Specific issues are dealt with during psychotherapy e.g. helping clients overcome the problems of memory loss and denial related to the traumatic stress experience, exploring social resources available to the client, strategies the client might have used in coping with symptoms of PTSD before seeking professional help, how to come to terms with a shameful trauma such as rape, imprisonment, and how to deal with perpetrators who may be in the victims environment such as police officers, prison guards or rebel abductors in the victim's community, and how the client can reconstruct his/her life so as to continue living positively. For psychotherapy to be successful, the environment for psychotherapy should be neutral so that the client can feel safe to share or receive support in coping with his/her traumatic experiences. The role of the therapist/counsellor is to facilitate the validation of the client's traumatic experience and foster recovery.

7.6 Management of co morbidities

PTSD in northern Uganda tends to be associated with other specific psychiatric illnesses and physical complications (Ovuga, Oyok and Moro, 2008), which need treatment. Co-morbid psychiatric disorders include Depression, Anxiety and Panic disorder, social phobia, sexual disorders and alcohol dependence. Often, these occur in multiple combinations. Specific interventions are directed to these disorders as appropriate e.g. treatment of depression, addictions, and counselling or family interventions for unwanted babies of rape etc. Often there's a need for, or age and gender specific intervention as well as spiritual atonement in line with cultural traditional practice, and the individual needs of specific clients. Specialised surgical interventions include removal of foreign bodies, correction of contractures and deformities and surgery for osteomyelitis to prevent prolonged effects of physical disability.

7.7 Rehabilitation

The aim of rehabilitation in PTSD in northern Uganda is to integrate the victim back into his/her society as a fully functioning individual with dignity. Many of these victims were abducted as young children and missed the opportunity for formal education. Other individuals got institutionalized to camp life in internally displaced persons' camps and require adaptation to life outside camp life. The various types of rehabilitation that are tailored to the individual needs of victims include job acquisition or vocational skills re/training; training for social functioning in the family and community with integrity as a leader; and traditional or social remedies to redress financial losses, material supplies e.g. to repossess one's land upon return from camp life; reconciliation rituals and ceremonies aimed to facilitate the acts of forgiveness for acts committed in the course of the northern Uganda war.

7.8 Prevention

Some forms of PTSD as in landslides, or earthquakes may not be preventable but their long-term impacts on the lives of victims can be mitigated through emergency medical and

psychological interventions. Evidence suggests that immediate intervention prevents the development of long-term psychological effects of trauma of whatever cause. Availability of services for the early detection of landslides and earth tremors with prompt evacuation of civilians from danger spots prevents unnecessary physical and psychological harm and public sensitisation and education about these services is perhaps the most significant step toward preventing the occurrence of posttraumatic stress disorder.
Secondly, road safety based on controlling the use of alcohol and other intoxicants, following appropriate road safety regulations, and taking measures to promote visibility on public roads and access routes reduces or prevents unnecessary motor vehicle accidents. Parenting skills and availability of family services reduces on domestic violence and child abuse. This should also be extended to child guidance and counselling in schools for teachers and children. Strengthening existing social support systems in the face of disasters will help mitigate the long-term harmful effects of traumatic experience.
Measures to prevent crime should form the armamentarium against PTSD. Issues of poverty reduction, the early detection and treatment of severe mental illness in the household and sensitization on security matters might act together to significantly reduce the incidence of violent traumatic events in the lives of the ordinary individual.
Communities in Northern Uganda are keen to prevent the vicious cycles of militarised violence as seen in perpetual wars in Uganda. This can only be by building institutions for respecting observances of Universal Human Rights and as well as participatory democratic governance which is culturally acceptable and understandable by the cultural diversity of peoples in their various groupings and yet with respect and tolerance of others who may be different. The principles and values of Human Rights should be a taught subject in schools from primary school to the highest levels of learning and in all the colleges of the nation as well as in homes as a sign of good education, civility and culture. Furthermore renewed cycles of violence in African countries can only be stopped if governments make peace and reconciliation with respect for the principles of fairness, justice and equal opportunities for every citizen to participate in governance at the top of their policy agendas for national security and stability that support all other government efforts toward good governance.

7.9 Prognosis

The outcome of post-traumatic stress disorder in Uganda is unknown. However clinical experience indicates that most individuals with the disorder recover on two to six sessions of counselling. It is possible that the ubiquitous social support available to people in their communities contributes to the apparent good prognosis for victims of traumatic experiences in rural Uganda. Ovuga et al (2008) have reported that former child soldiers in northern Uganda who returned to their homes without passing through government established reception centres had lower mean scores on the Harvard Trauma questionnaire and the Hopkins Symptom Checklist for depression. Ovuga and colleagues attributed their observation on the possibility that the child soldiers who went directly to their communities had committed fewer atrocities, were more readily received and forgiven by their respective communities, and possibly experienced fewer traumatic experiences than their colleagues who returned home through the government reception facilities.

8. References

Akello G, Reis R, Ovuga E, Rwabukwali EB, Kabonesa C, & Richters R (2007). Primary school children's perspectives of common diseases and medicines used:

implications for school healthcare programmes and priority setting for Uganda, *African Health Sciences*, 7(2): 74-80

Akello G, Richters A, & Ovuga E. (2011) Children's management of complaints symptomatic of psychological distress: A critical analysis of the different approaches in Northern Uganda, African Journal of Traumatic Stress, 1(2): 70-79

American Psychological Association (1992), *Diagnostic and Statistical Manual of Mental and Behavioral Disorders*, American Psychological Association, Washington DC

Anonymous (2007). Invisible wounds from the Congo war, *CMAJ*; 181: (6-7)

Bardin C. Growing up too quickly: Children who lose out on their childhoods. Paediatric Child Health; 2005; 10(5): 264-268.

Bayer, C. P., Klasen, F., Adam, H. (2007). Association of trauma and PTSD symptoms with openness to reconciliation and feelings of revenge among former Ugandan and Congolese child soldiers, *JAMA*; 298 (5): 555-559.

Betancourt T.S. and Khan K.T. (2008). The mental health of children affected by armed conflict: Protective processes and pathways to resilience; Int Rev Psychiatry; 20(3): 317–328. Doi:10.1080/09540260802090363.

Betancourt T.S., William T.P. (2008). Building an evidence base on mental health interventions for children affected by armed conflict, Intervention (Amstelveen). 2008; 6(1): 39–56. doi:10.1097/WTF.0b013e3282f761ff.

Betancourt T.S., Borisova I.I., Williams T.P., Brennan R.T., Whitfield, T.H., Marie de la Soudiere, Williamson J., and Gilman S.E. (2010). Sierra Leone's Former Child Soldiers: A Follow-up Study of Psychosocial Adjustment and Community Reintegration, *Child Dev*; 81(4): 1077–1095. doi:10.1111/j.1467-8624.2010.01455.x.

Derluyn, I., Broekaert, Schuyten, G., De Temmerman, E. (2004). Posttraumatic stress in former Ugandan child soldiers, *The Lancet*; 363: 861-863.

de Jong,J.T., Komproe, I. H., Van Ommeren, M. (2001). Lifetime events and posttraumatic stress disorder in 4 postconflict settings. *JAMA*; 286 (5): 555-562.

Herman, J. L. (1997). *Trauma and Recovery. The aftermath of violence - from domestic abuse to political terror* (Second edi., pp. 115-132). New York: Basic Books.

Jones G.H., and Lovett J.W.T. (1987). *Journal of the Royal College of General Practitioners*; 37, 34-35.

Engdahl B., Dikel T.N., Eberly R., and Arthur A., Jr. (1997). Posttraumatic stress disorder in a community group of former prisoners of war: A normative response to severe trauma, *Am J Psychiatry* 1997; 154:1576–1581

Kaminer D., Grimsrud A., Myer L., Stein D., & Williams D. (2008). Risk for posttraumatic stress disorder associated with different forms of interpersonal violence in South Africa, *Soc Sci Med.*; 67(10): 1589–1595.

Karunakara, U. K., Neuner, F., Schauer, M., Singh, K., Hill, K., Elbert, T., et al. (2004, August). Traumatic events and symptoms of post-traumatic stress disorder amongst Sudanese nationals, refugees and Ugandans in the West Nile, *African health sciences*, 2004; 4(2): 83-93

Koren D., Arnon I., & Klein E., M.D. (1999). Acute stress response and posttraumatic stress disorder in traffic accident victims: A one-year prospective, follow-up study, *Am J Psychiatry*; 156:367–373

Mock N.B., Duale S., Brown L.F., Mathys E., O'Maonaigh H.C., Abul-Husn N.K.L. and Elliott S (2004). Conflict and HIV: A framework for risk assessment to prevent HIV in conflict-affected settings in Africa; *Emerging Themes in Epidemiology*; 1:6 doi:10.1186/1742-7622-1-6

Murray, C. J. L., King, G., Lopez, a D., Tomijima, N., & Krug, E. G. (2002). Armed conflict as a public health problem, *BMJ (Clinical research ed.)*, *324*(7333), 346-9

Murthy R. S., Lakshminarayana R. (2006). Mental health consequences of war: a brief review of research findings. *World Psychiatry*, 5 (1): 25-30

Neuner, F., Schauer, M., Karunakara, U., Klaschik, C., Robert, C., & Elbert, T. (2004). Psychological trauma and evidence for enhanced vulnerability for posttraumatic stress disorder through previous trauma among West Nile refugees. *BMC psychiatry*, *4*, 34. doi: 10.1186/1471-244X-4-34.

Onyut L.P., Neuner F., Schauer E., Ertl V., Odenwald M., Schauer M. and Elbert T. (2005) *BMC Psychiatry*; 5:7 doi:10.1186/1471-244X-5-7

Ovuga, E. (2005). *Depression and Suicidal Behavior in Uganda*. PhD Thesis, Karolinska Institutet and Makerere University, Stockholm and Kampala, 2005

Ovuga E, Boardman J & Wasserman D (2005). The prevalence of depression in two districts of Uganda. *Social Psychiatry Psychiatric Epidemiology*, 40(6): 439-445

Ovuga E, Boardman J & Wasserman D (2005). Prevalence of suicide ideation in two districts of Uganda. *Archives of Suicide Research*, 9(4): 321-332

Ovuga, E., Oyok, T. O., & Moro, E. B. (2008). Post traumatic stress disorder among former child soldiers attending a rehabilitative service and primary school education in northern Uganda. *African health sciences*, *8*(3), 136-41.

Pham P.N., Vinck P. and Stover E. (2009). Returning home: forced conscription, reintegration, and mental health status of former abductees of the Lord's Resistance Army in northern Uganda, *BMC Psychiatry*, 9:23 doi:10.1186/1471-244X-9-23

Roberts, B., Ocaka, K. F., Browne, J., Oyok, T., & Sondorp, E. (2008). Factors associated with post-traumatic stress disorder and depression amongst internally displaced persons in northern Uganda. *BMC psychiatry*, *8*, 38. doi: 10.1186/1471-244X-8-38.

Roberts, B., Felix Ocaka, K., Browne, J., Oyok, T., & Sondorp, E. (2009). Factors associated with the health status of internally displaced persons in northern Uganda. *Journal of epidemiology and community health*, *63*(3), 227-32. doi: 10.1136/jech.2008.076356.

Shanks L. and Schull M.J. (2000). Rape in war: the humanitarian response, *JAMC*; 163 (9)

Stallard P., Velleman R., & Baldwin S. (1998). Prospective study of posttraumatic stress disorder in children involved in road traffic accidents, *BMJ*; 317: 1619-1623

Tonks A., War in Uganda leaves deep psychological scars, *BMJ*; 335: 278-279

Volkan V. (2004). *Blind Trust: Large Groups and Their Leaders in Times of Crisis and Terror*, Pitchstone Publishing, Charlottesville, Virginia

Vinck P., Pham, P.N., Stover, E., Weinstein, H. M. (2007). Exposure to war crimes and implications for peace building in northern Uganda, JAMA; 298 (5):543-553.

Part 4

Post-Traumatic Stress in Special Situations

10

Earthquake and Mental Health

Xueyi Wang and Kezhi Liu
Psychiatric Department of the First Hospital of Hebei Medical University,
The Mental Health Institute of Hebei Medical University,
Brain Ageing and Cognitive Neuroscience,
Key Laboratory of Hebei Province, Shijiazhuang,
China

1. Introduction

Earthquakes, as a nature disaster, not only causes deaths, physical disease, damage to the infrastructure and economic loss, it also keeps long-lasting **Mental Health** effects on individuals involved. There will always be cases of psychological disorders such as Post-Traumatic Stress Disorder (PTSD), depression, cognitive disorder, personality disorders, and so on, especially with individuals dealing pre-existing conditions.

At 3:42 am on July 28 1976, a magnitude 7.8 earthquake struck Tangshan, an industrial city of 1 million people in northern China, which had been built on the unstable soil of the Luanhe River's flood plain. Ninety-three percent of residential buildings and 78 percent of industrial buildings were completely destroyed. This alluvial soil liquefied during the quake, undermining entire neighborhoods. The entire earthquake lasted approximately 14 to 16 seconds and killed at least 242,000 people.

Although earthquakes are among the most common and devastating natural disasters, relatively little attention has been paid to their mental health consequences and associated risk factors long time after earthquake. There have been few studies of post-earthquake psychological problems using randomly selected samples of earthquake survivors. Fortunately, we have done a lot of studies about mental disorders due to Tangshan earthquake in China.

1.1 Earthquake as a "trauma" related to mental health

An earthquake, also known as a quake, tremor or temblor, is the result of a sudden release of energy in the Earth's crust. In its most general sense, the word earthquake is used to describe any seismic event – whether natural or caused by humans – that generates seismic waves. Earthquakes are caused mostly by rupture of geological faults, but also by other events such as volcanic activity, landslides, mine blasts, and nuclear tests. For humans who live on earth, an earthquake is a trauma. "Trauma" has both a medical and a psychiatric definition. Medically, "trauma" refers to a serious or critical bodily injury, wound, or shock. This definition is often associated with trauma medicine practiced in emergency rooms and represents a popular view of the term. In psychiatry, "trauma" has assumed a different meaning and refers to an experience that is emotionally painful, distressful, or shocking, which often results in lasting mental and physical effects.

Although earthquakes are among the most common and devastating natural disasters, relatively little attention has been paid to their mental health consequences and associated risk factors. There have been few studies of post-earthquake psychological problems using randomly selected samples of earthquake survivors. Fortunately, we have done a lot of studies about mental disorders due to Tangshan earthquake.

One of our studies was to explore the long-term effect of Tangshan earthquake on psychosomatic health of paraplegic suffers. Sixty-four paraplegic suffers of Tangshan earthquake and 64 normal controls were interviewed and assessed with self administered questionnaire for psychosomatic health, SCL-90, SAS, SDS, CMI (Cornell Medical Index) and SSRS (Social Support Rating Scale). Six patients (9.38%) were diagnosed as PTSD according to Chinese Classification of Mental Disorders, Second Edition, Revised (CCMD-2-R) in sixty-four paraplegic suffers, however, there was no body who was diagnosed as PTSD in normal controls, the incidence of PTSD in paraplegic suffers was higher than that of normal citizen experienced the earthquake. At present, patients' group had poorer mental health than control reflected by SCL-90.The total score of SCL-90 in paralegic suffers was (143.98±49.22), and the total score of SCL-90 in normal controls was (111.20±23.13), there was significant difference in statistic (t=4.822, P<0.001) The severity of trauma both mentally and physically has great influence on mental health of suffers even after 25 years.

Another study was to investigate the long term effect of earthquake on mental and physical health of sufferers. Eight hundred and fifty eight first rank relatives of those who died in the earthquake 12 years ago formed the study group, as they experienced the earthquake themselves. Eight hundred and thirty-seven inhabitants who experienced the earthquake but did not lose any first rank relatives formed the control group. The research instruments included: SCL-90[study group/controls= (143.98±49.22)/ (111.20±23.13)], SAS [study group/controls= (40.05±9.47)/ (36.61±5.0)], SDS [study group/controls= (49.08±11.36)/ (42.66±11.74)]. The mental health of study group was worse than that of controls. At the same time, hypertension, ischemic brain disease were more common in study group.

Trauma and Disaster

The literature distinguishes between "trauma" and "disaster". Traumas are experiences that threaten individual health and well being, render one helpless in the face of intolerable internal or external danger, overwhelm coping mechanisms, violate basic assumptions about survival, and stress the uncontrollability and unpredictability in the world. Traumas may be caused by an isolated, unanticipated event or long-lasting stressful experience, due to repeated exposure to several extreme external events.

Disasters are relatively sudden, more or less time-limited, and public events that extensively damage properties and lives, engendering a systemic continuously disruptive impact on the social network and basic daily routines of children and families. The community as a whole is compromised in its capacity to negotiate the recovery of its individual members (e.g., massive displacement and relocation). Matters are often made worse when resources are over-stretched and the community's infrastructure is affected. This can result in unemployment, lack of housing and food, poor health and mental health services, school closures, school and job absenteeism, family dysfunction, and displacement of large populations.

Disasters differ in scope and schedule. Some result mainly in loss anddisruption (loss of possessions and housing), whereas others involve also a threat to life. Some last a few seconds (e.g., earthquake), whereas others continue for years (e.g., war). Unlike traumas, disasters are characterized by the immediate, long-lasting and repeated exposure of victims to reminders of the disastrous event. Usually, three types of experience are combined: terror due to a danger to one's life or exposure to grotesque sights; grief following loss (e.g., human lives, basic trust, self-esteem); and the disruption of normal living. On the social level, there are shock, depression and mourning, confusion and social disarray, rage and blaming, crime and delinquent behavior, emergence of mythic ideologies, collapse of formal leadership, emergence of informal popular leadership, and social disintegration into primary affiliations. Children feel the disruption in their family, neighborhood and school. Since the pathological and recovery processes continue long after the disastrous event itself is over, even if it was restricted to a single point in time, theoretical, research and intervention studies should follow both a systemic and a long-term design.

1.2 Earthquake as a "trauma" in China

The most recent large earthquake was a 9.0 magnitude earthquake in Japan, and it was the largest Japanese earthquake since records began. On March 11, 2011, an earthquake struck off the coast of Japan, churning up a devastating tsunami that swept over cities and farmland in the northern part of the country and set off warnings as far away the west coast of the United States and South America. By June 2011, the official count of dead and missing remained above 24,000. Tens of thousands of people remained housed in temporary shelters or evacuated their homes due to the nuclear crisis following this earthquake.

Although the people in China had just experienced the Wenchuan earthquake in 2008, nobody could forget the Tangshan earthquake. At 3:42 am on July 28 1976, a magnitude 7.8 earthquake struck Tangshan, an industrial city of 1 million people in northern China, which had been built on the unstable soil of the Luanhe River's flood plain. Ninety-three percent of residential buildings and 78 percent of industrial buildings were completely destroyed. This alluvial soil liquefied during the quake, undermining entire neighborhoods. The entire earthquake lasted approximately 14 to 16 seconds and killed at least 242,000 people (the official death count). Some observers place the actual toll as high as 700,000, and many more were trapped in the rubble. Coal miners working deep underground in the region perished when the mines collapsed on them.

Survivors were faced with no water, no food, and no electricity. With so much damage, recovery was not easy. Some food was parachuted in, but the distribution was uneven. Water, even just for drinking, was extremely scarce. Many people drank out of pools or other locations that had become contaminated during the earthquake.

Although earthquakes are among the most common and devastating natural disasters, relatively little attention has been paid to their mental health consequences and as social risk factors. Posttraumatic stress disorder is a common outcome of major earthquakes.

2. Earthquake and depression

Depression is a popular topic these days. The New Yorker magazine once estimated that more than fourteen million Americans suffer from major depression every year, with minor depression affecting more than three million. National Public Radio's Depression Out of the Shadows website reports that by 2020 depression will be the second most common health problem in the world. Previous assessments among survivors of earthquake have shown that depression and other mental health problems are common. Depression and posttraumatic stress disorder may arise weeks or months after earthquake. Earthquakes stir up concerns in people not directly affected. They also trigger both a desire to help and a sense of overwhelming hopelessness. This cluster of emotions, helplessness, hopelessness and a sense of being overwhelmed are classic symptoms of depression.

In 1976 a severe earthquake struck Tangshan, China, resulting in 240,000 deaths, thousands of injuries, and widespread destruction of houses and basic services. The United States Geological Survey has termed this event as the worst earthquake in the past four centuries. This catastrophic event might serve as a natural experiment since all Tangshan women who were pregnant at the time were stressed by the quake. We assessed symptoms of depression in young adult offspring exposed to the earthquake prenatally and controls that were not exposed to the earthquake. The pregnant women of Tangshan endured severed stress during the earthquake. Animal and human literature suggests that exposure to prenatal stress can alter the developing hypothalamo-pituitary-adrenal axis and have negative, long-term effects on the offspring. [E.J.H. Mulder,et al. 2002] The animal and human research has demonstrated an association between prenatal stress and adult depression. In our study we found that young adults who were exposed to the earthquake in utero demonstrated a marked increase in severe depression when compared to controls [Lu Lin, Wang Xueyi, Li Jing, et al.1999; Zhang Ben, et al. 2002a]. In addition, the effect was stronger in males than females; males exposed to the earthquake during the second trimester of fetal development exhibited the highest proportion of severe depression [Wang Xueyi, et al.2006, Zhang Ben, et al. 2002b].

In this research, we hypothesized: Firstly, a higher proportion of severe depression will be observed in the earthquake exposed subjects as compared to the non-exposed subjects; Secondly, the effect will be more pronounced in males; Thirdly, subjects exposed to the earthquake during the second trimester of gestation will exhibit higher rates of severe depression than those not exposed or those exposed during the first or third trimester; Fourthly, the second-trimester effect will be stronger in males, and the fifthly, offspring whose mothers reported higher levels of emotional stress due to the earthquake will have higher rates of severe depression.

The purpose of this study was to determine if exposure to a severe maternal stress (major earthquake) in utero increased risk for adult depression. We found that individuals who were exposed to the earthquake in utero demonstrated a marked increase in SD (as measured by the Hamilton depression scale, HAMD) when compared to age and season-of-birth matched controls; in addition, the effect was stronger in males than females. The timing of exposure to the earthquake also proved to be significantly related to the proportion of SD. Males exposed to the earthquake during the second trimester of fetal development exhibited the highest proportion of SD when compared to males exposed during trimesters one and three and females exposed during the first, second, and third trimesters. We also found that the offspring of mothers who were exposed prenatally to a severe earthquake have a lower level of emotional stress. Thus, the mothers who endorsed symptoms such as "After the earthquake, I felt sad, frightened, and/or nervous," had offspring who later reported higher rates of SD. [Zhang Ben, et al.1999]

These findings provide evidence that at least one type of SD may be related to exposure in urero to a stressful event related to maternal experience of earthquake. Furthermore, our results indicate that exposure to maternal stress related to earthquake during the first and second trimesters significantly increases the risk of developing depression in adulthood, which supports previously reported results [Zhang Ben, et al.2000;Wang Xueyi, et al.2005]. Our findings that males exposed during the second trimester exhibited the highest rates of depression and that males overall reported more depression are comparable to previous studies regarding teratogenic and affective disorders following earthquakes [Zhang Ben, et al.2001; Wang Xueyi, et al.2005]. In addition the markedly high rate of depression in the males exposed during second trimester of fetal development provides evidence for a neuro-developmental hypothesis of the etiology of depression [Zhang Ben, et al.1999; Wang Xueyi, et al.2005].

3. Earthquake and schizotypical personality

In Schizotypical personality disorder, people exhibit odd behavior, respond inappropriately to social cues and hold peculiar beliefs. Schizotypical personality disorder occurs in 3% of the general population and occurs slightly more commonly in males than females. People with classic schizotypical personalities are apt to be loners. They feel extremely anxious in social situations, but they're likely to blame their social failings on others. They view themselves as alien or outcast, and this isolation causes pain as they avoid relationships and the outside world. People with schizotypical personalities may ramble oddly and endlessly during a conversation. They may dress in peculiar ways and have very strange ways of viewing the world around them. Often they believe in unusual ideas, such as the powers of Extra Sensory Perception (ESP) or a sixth sense. At times, they believe they can magically influence people's thoughts, actions and emotions. In adolescence, signs of a schizotypical personality may begin as an increased interest in solitary activities or a high level of social anxiety. The child may be an underperformer in school or appear socially out-of-step with peers, and as a result often becomes the subject of bullying or teasing. Schizotypical personality disorder typically begins in early adulthood and is likely to endure, though symptoms may improve with age.

Experiencing trauma is a factor that appears to increase the risk of schizotypical personality disorder. In a sample of 75 women recruited from the community, researchers measured trauma/maltreatment history and symptoms of schizotypical personality disorder, using both questionnaire and interview measures [Howard Berenbaum, et al. 2003]. As hypothesized, individuals with histories of trauma/maltreatment had elevated levels of schizotypical symptoms. Among types of trauma, maltreatment was especially strongly associated with schizotypical symptoms. Although posttraumatic stress disorder symptom severity, depression, dissociation, and difficulty identifying one's emotions were all associated with schizotypical symptoms, they could not account completely for the association between trauma/maltreatment and schizotypical symptoms.

Previous research has demonstrated that prenatal exposure to maternal stress is a possible risk factor for development of schizophrenia-spectrum diagnoses among adult offspring; however, research examining the effects of prenatal stress exposure on sub-threshold psychotic symptoms is lacking. Similarly, there is a paucity of research investigating prenatal stress exposure in relation to anxiety and depression among adult offspring, and how anxiety and depression may contribute to schizophrenia-spectrum symptom outcome among prenatally exposed offspring. The present study examined a large dataset to investigate whether 18-year-old, male and female, Chinese, high school seniors exposed to the 1976 Tangshan earthquake during one of nine months of gestation demonstrated higher levels of schizophrenia-spectrum, anxiety, and depression symptoms than unexposed control participants. This study further examined the relationship between schizophrenia-spectrum, anxiety, and depression symptoms, and investigated the effects of prenatal stress exposure on schizophrenia-spectrum symptoms after controlling for anxiety and depression. Results indicated that prenatal exposure to the Tangshan earthquake did not have an overall effect on schizophrenia-spectrum, anxiety, or depression symptoms [Wang Xueyi, et al.2011; Armstrong, Nikki Panasci, 2009]. However, exposed female participants demonstrated higher negative schizotype scores (SPQ Interpersonal scale) than unexposed females, even when anxiety and depression were controlled statistically [Wang Xueyi, et al.2007]. When anxiety and depression were included in analyses, exposed females also demonstrated higher disorganized schizotype scores (SPQ Disorganized scale) [Wang Xueyi, et al.2007; Armstrong, Nikki Panasci, 2009]. Additionally, females exposed to the earthquake during gestational months one and five produced higher depressions scores than unexposed females of the same gestational months. Finally, anxiety and depression significantly correlated with schizophrenia-spectrum scores; however, their relationships with negative and positive schizophrenia-spectrum symptom scores were relatively similar. An interesting finding in this study was that control group participants demonstrated higher levels of psychopathology symptoms on some measures. A possible explanation for such findings is that although control participants were not exposed to the earthquake, their mothers (who did experience the earthquake a year prior to pregnancy) may have experienced chronic stress that possibly resulted in more disruption to their offspring's stress response system, and ultimately to increased symptoms of psychopathology in their offspring. Future research examining the effects of prenatal exposure to acute vs. chronic stress on schizophrenia-spectrum symptom outcome is suggested.

Our research named “Adult schizotypical personality characteristics of a fetus exposed to Tangshan earthquake in its sixth month of gestation”, aimed to evaluate the fetus exposed to earthquake in their sixth month of gestation with or without high risk for adult schizotypical personality characteristics · The subjects were drawn from the fourteen high schools in the Tangshan area. All 12th grade students who were 18 years were invited to participate in the research project. Discarding any data from subjects whose mothers resided outside the Tangshan area during the 1976 earthquake, 604 subjects who were born from July 28th, 1976 to April 28th, 1977 were selected as the exposure group. The control group consisted of 601 subjects who were born one year after the exposure group from July 28th, 1977 to April 28th, 1978. The recruitment and testing of exposure group took place in December of 1995 and for the control group in December of 1996.This ensured that exposure group and control group were the same age at the time of the assessment. Fully informed consent was obtained in all subjects. Raine's Schizotypical Personality Questionnaire (SPQ-B) was used as a measure of schizotypice personality. The SPQ-B of 22 True-False items made up a total score together with three sub-factors: Cognitive-Perceptual factor, Interpersonal Deficits factor and Disorganization factor. The more mean scores on SPQ-B (0-

20 points), the more possibility of schizotypical personality was. Differences of means were evaluated with t-test in the two groups. Six hundred and three in exposure group and 598 in control group completed the SPQ-B evaluation, and effective data were obtained and all were included in the analysis. Total score and score of Cognitive-Perceptual factor of SPQ-B score in fetus of sixth month of gestation in the exposure group were (9.1± 4.6) and (13.4± 2.1) points, respectively, which were markedly higher than those in the control group [(7.6±3.6),(2.7±1.6) points, t=2.04,2.00, $P<0.05$].There was no significant difference between the Interpersonal Deficits factor and Disorganization facto r($P>0.05$).②Comparison of total score and factor scores of SPQ-B in fetus of sixth month (different weeks) of gestation in the two groups: At week 23 the total score of SPQ-B in the exposure group was remarkably higher than that in the control group (t=2.1, $P<0.05$).Score of Cognitive-Perceptual factor was higher than that in the control group, but there was no significant difference ($P>0.05$). Score of Disorganization factor was distinctly higher than that in the control group (t=2.3, $P<0.05$). There was no significant difference of SPQ-B score at weeks 21, 22 and 24 in the two groups ($P>0.10$). The fetus exposed to the earthquake in their sixth month of gestation may be has high risk for adult Schizotypal personality characteristics. [Wang Xueyi, et al.2007]

4. Earthquake and cognitive function disorders

Trauma has been shown to significantly compromise cognitive development. [Levine, 2007; Perry & Szalavitz, 2006] Cognitive deficits such as poor problem solving, (unable to think things out or make sense of what is happening), low self-esteem (how one thinks of oneself - victim-thinking) and hopelessness (loss of future orientation) have all been clearly linked to traumatic events including earthquake, influenza, and so on [Stein & Kendell, 2004; William Steele, 2007]. There is evidence in two independent studies that the trauma and second-trimester influenza, is associated with deficits in cognitive ability as measured by infant habituation to visual stimuli [Wang Xueyi, et al. 2001; Watson JB, 1999; Van OS, 1998]. In both of these studies, the infants whose mothers suffered an influenza infection during their second trimester of fetal development exhibited impaired habituation to visual stimuli. The test of infant habituation of attention is excellent predictor of later intellectual development school readiness and intelligence quotients. Based on the above two studies, we hypothesized that the trauma is related to cognitive impairment. Our purpose of this investigation was to examine the long-term effects of the severe earthquake (7.8 Richter Scale) that struck Tangshan, China in July 1976 on the offspring of women who were pregnant at the time of the earthquake. The extremely severe stress of the earthquake may have resulted in a physiological response in the pregnant women of Tangshan, which adversely affected their fetuses. To determine if exposure to the earthquake as a fetus results in a negative outcome we administered a test of cognitive functioning to test the following hypotheses: 1). The stress of a severe earthquake during gestation will disrupt neural development producing deficits in cognitive functioning. Thus, the exposed group should have poorer cognitive functioning as compared to the control group. 2) Exposure to a severe stressor during a critical period of fetal brain development (the second trimester of gestation) may result in more pronounced cognitive deficits as compared to those exposed during the first or third trimesters. The exposed group consists of 606 high school seniors who were fetuses at the time of the earthquake. The birth dates of the subjects were used to determine their stage of gestation at the time of the earthquake. The control group was assessed exactly one year after the exposed group and consists of 606 high school seniors, who were born exactly one year after the exposed group. Thus, the control subjects were not exposed to the earthquake as fetuses. The 1212 exposed and control subjects were matched for birth date so there are an equivalent number of subjects representing months one through nine of gestation. Assessment of the control subjects one year after the exposed group, resulted in an exposed and control group that were both 18 years of age at the time of testing. Both the exposed and control subjects were randomly drawn from the seniors students who attended the five high schools in Tangshan, China.

As we all know, that while in the arousal state or, not feeling safe at the sensory level, cognitive functioning and processing is altered. Short-term memory suffers; verbal memory also decreases. From our research, we can draw conclusion that: the subjects exposed to the earthquake during gestation had significantly lower scores on the Raven's Progressive Matrices at age 18 when compared to 18 year-old control subjects who were not exposed to the quake. The prenatal stress of a severe earthquake on a developing fetus is associated with lowered adult cognitive ability. Subjects exposed to a major prenatal stress (the severe earthquake of Tangshan China) during months five through nine had significantly lower score than control subjects who were born in the same months one year later. There were no differences in average Raven's scores for subjects exposed during months one through four when compared to control subjects matched for date of birth. Thus, a prenatal exposure to a severe stressor during months five through nine may adversely affect cognitive functioning at age of 18. It is probable that the stress of the Tangshan earthquake resulted in the elevation of glucocorticoids in the pregnant mothers. [Jin Guixing，Wang Xueyi，Wang Lan, et al. 2011] This elevation of glucocorticoids may have negatively affected the developing fetus. In addition vasoconstriction of the placenta may have occurred in pregnant mothers at the time of the earthquake, which could have had deleterious effects on the developing fetus. [Calvin Hobel, 2003].

Following exposure to trauma，such as earthquake, survivors may become frozen in an activated state of arousal. Research documenting the effects of arousal on cognition has become increasingly available and consistent in its descriptions of the cognitive and behavioral alterations. In the arousal state, changes in the brain are triggered by a variety of stress related functions. One researcher found that victims of trauma had lower memory volume in the left-brain (Hippocampal) area than did the non-abused (http://www.nimh.nih.gov). This left-brain function refers to understanding or processing information. One of these functional alterations takes place in the neocortex. On the contrary, the right brain is involved "in the vital functions that support survival and enable the organism to cope actively and passively with stress" The right hemisphere controls perception analysis of visual patterns and emotions. One study supports these and similar findings that appropriate responses to external changes (stress/crisis) can be altered by activation of the arousal state - the heightened state of fear induced by traumatic exposure.

Disorders of memory constitute one of the diagnostic categories for PTSD due to earthquake in the form of re-experiencing. Trauma-based memory phenomena often involve declarative memory in the form of variably accurate verbal and imaginal recall of the traumatic event. Declarative memory, the form of memory that relates to facts and events, initially involves hippocampal and prefrontal cortical pathways and plays an

important role in conscious recall of trauma-related events. Although declarative memory may account for much of the arousal-based cognitive symptoms of PTSD, procedural memory provides the seemingly unbreakable conditioned link that perpetuates the neural cycle of trauma and dissociation.
To study whether severe stress caused by earthquake had negative effect on fetal cognitive function, Raven's Standard Progress Matrices (RSPM) was used to evaluate cognitive function of 616 young students who experienced earthquake during their fetal stage; 616 controls who did not experience this trauma were assessed with the same instrument. Scores of RSPM of earthquake group were significantly lower than those of controls, especially for those who experienced earthquake in their second or the third trimester (Wang Xuey,et al. 2001). Earthquake has negative effect on cognitive function development of fetus.

5. Earthquake and Post-Traumatic Stress Disorder (PTSD)

Post-Traumatic Stress Disorder has been recognized as a formal diagnosis since 1980. However, as early as the 6th century BC/BCE, reports of battle-associated stress reactions had been reported. One of the first descriptions of PTSD was made by the Greek historian Herodotus. In 490 BC/BCE he described, during the Battle of Marathon, an Athenian soldier who suffered no injury from war but became permanently blind after witnessing the death of a fellow soldier. However, it was called by different names as early as the American Civil War, when combat veterans were referred to as suffering from "soldier's heart." In World War I, symptoms that were generally consistent with this syndrome were referred to as "combat fatigue." Soldiers who developed such symptoms in World War II were said to be suffering from "gross stress reaction," and many troops in Vietnam who had symptoms of what is now called PTSD were assessed as having "post-Vietnam syndrome." PTSD has also been called "battle fatigue" and "shell shock."
PTSD is an emotional illness that is classified as an anxiety disorder and usually develops as a result of a terribly frightening, life-threatening, or otherwise highly unsafe experience, for instance the earthquake. Traumatic events that may trigger PTSD include violent personal assaults, natural or human-caused disasters, accidents, or military combat. The rates of PTSD suffered from earthquake vary from 2 to 87%.

5.1 Symptoms of PTSD due to earthquake

PTSD can cause many symptoms. These symptoms can be grouped into three categories:

1. Re-experiencing symptoms:
    - Flashbacks—reliving the trauma over and over, including physical symptoms like a racing heart or sweating
    - Bad dreams
    - Frightening thoughts.

Re-experiencing symptoms may cause problems in a person's everyday routine. They can start from the person's own thoughts and feelings. Words, objects, or situations that are reminders of the event can also trigger re-experiencing of traumatic events.

2. Avoidance symptoms:
    - Staying away from places, events, or objects that are reminders of the traumatic experience

- Feeling emotionally numb
- Feeling strong feelings of guilt, depression, or worry
- Losing interest in activities that were enjoyable in the past
- Having trouble remembering the dangerous event.

Things that remind a person of the traumatic event can trigger avoidance symptoms. These symptoms may cause a person to change his or her personal routine. For example, after a bad car accident, a person who usually drives may avoid driving or riding in a car.

3. Hyperarousal symptoms:
   - Being easily startled
   - Feeling tense or "on edge"
   - Having difficulty sleeping, and/or having angry outbursts.

Hyperarousal symptoms are usually constant, instead of being triggered by things that remind one of the traumatic event. They can make the person feel stressed and angry. These symptoms may make it hard to do daily tasks, such as sleeping, eating, or concentrating.

The emotional numbing of PTSD may present as a lack of interest in activities that used to be enjoyed (anhedonia), emotional deadness, distancing oneself from people, and/or a sense of a foreshortened future (for example, not being able to think about the future or make future plans, not believing one will live much longer). At least one re-experiencing symptom, three avoidance/numbing symptoms, and two hyperarousal symptoms must be present for at least one month and must cause significant distress or functional impairment in order for the diagnosis of PTSD to be assigned. PTSD is considered of chronic duration if it persists for three months or more.

It's natural to have some of these symptoms after a dangerous earthquake. However, not everyone who lives through an earthquake gets PTSD. In fact, most will not get the disorder. On the contrary, not everyone with PTSD has been through a dangerous earthquake. Sometimes people have very serious symptoms that go away after a few weeks. This is called acute stress disorder, or ASD. When the symptoms last more than a few weeks and become an ongoing problem, they might be PTSD. Some people with PTSD don't show any symptoms for weeks or months.

Most practitioners who examine a child or teenager for PTSD will interview both the parent and the child, usually separately, in order to allow each party to speak freely. Interviewing the child in addition to the adults in his or her life is quite important given that while the child or adolescent's parent or guardian may have a unique perspective, there are naturally things the young person may be feeling that the adult is not aware of. Another challenge for diagnosing PTSD in children, particularly in younger children, is that they may express their symptoms differently from adults.

5.2 PTSD due to earthquake occurred in children and teens

Children and teens can have extreme reactions to the trauma of earthquake, but their symptoms may not be the same as adults as discussion above. In very young children, these symptoms can include:

- Bedwetting, when they'd learned how to use the toilet before
- Forgetting how or being unable to talk

- Acting out the scary event during playtime
- Being unusually clingy with a parent or other adult.

Older children and teens usually show symptoms more like those seen in adults. They may also develop disruptive, disrespectful, or destructive behaviors. Older children and teens may feel guilty for not preventing injury or deaths. They may also have thoughts of revenge. For more information, see the NIMH booklets on helping children cope with violence and disasters (http://www.nimh.nih.gov).

PTSD statistics in children and teens reveal that up to more than 40% have endured at least one traumatic event, resulting in the development of PTSD in up to 15% of girls and 6% of boys. Three to six percent of high school students in the United States, and as many as 30%-60% of children who have survived specific disasters develop PTSD (http://www.nimh.nih.gov). Up to 100% of children who have seen a parent killed or endured sexual assault or abuse tend to develop PTSD, and more than one-third of youths who are exposed to community violence (for example, a shooting, stabbing, or other assault) will suffer from the disorder.

5.3 Causes of PTSD due to earthquake

Virtually any trauma, defined as an event that is life-threatening or that severely compromises the physical or emotional well-being of an individual or causes intense fear, may cause PTSD. Such events often include either experiencing or witnessing a severe accident or physical injury, receiving a life-threatening medical diagnosis, being the victim of kidnapping or torture, exposure to war combat or to a natural disaster, exposure to other disaster (for example, plane crash) or terrorist attack, being the victim of rape, mugging, robbery, or assault, enduring physical, sexual, emotional, or other forms of abuse, as well as involvement in civil conflict. Although the diagnosis of PTSD currently requires that the sufferer has a history of experiencing a traumatic event as defined here, people may develop PTSD in reaction to events that may not qualify as traumatic but can be devastating life events like divorce or unemployment.

5.4 Risk factors and protective factors for PTSD due to earthquake

Issues that tend to put people at higher risk for developing PTSD include increased duration of a traumatic event, higher number of traumatic events endured, higher severity of the trauma experienced, having an emotional condition prior to the event, or having little social support in the form of family or friends. In addition to those risk factors, children and adolescents, females, and people with learning disabilities or violence in the home seem to have a greater risk of developing PTSD after a traumatic event.

While disaster-preparedness training is generally seen as a good idea in terms of improving the immediate physical safety and logistical issues involved with a traumatic event, such training may also provide important preventive factors against developing PTSD. That is as evidenced by the fact that those with more professional-level training and experience (for example, police, firefighters, mental-health professionals, paramedics, and other medical professionals) tend to develop PTSD less often when coping with disaster than those without the benefit of such training or experience.

There are medications that have been found to help prevent the development of PTSD. Some medicines that treat depression, decrease the heart rate, or increase the action of other body chemicals are thought to be effective tools in the prevention of PTSD when given in the days immediately after an individual experiences a traumatic event.

5.5 Treatment for PTSD due to earthquake

Treatments for PTSD usually include psychological and medical interventions. Providing information about the illness, helping the individual manage the trauma by talking about it directly, teaching the person ways to manage symptoms of PTSD, and exploration and modification of inaccurate ways of thinking about the trauma are the usual techniques used in psychotherapy for this illness. Education of PTSD sufferers usually involves teaching individuals about what PTSD is, how many others suffer from the same illness, that it is caused by extraordinary stress rather than weakness, how it is treated, and what to expect in treatment. This education thereby increases the likelihood that inaccurate ideas the person may have about the illness are dispelled, and any shame they may feel about having it is minimized. This may be particularly important in populations like military personnel that may feel particularly stigmatized by the idea of seeing a mental-health professional and therefore avoid doing so.

Teaching people with PTSD practical approaches to coping with what can be very intense and disturbing symptoms has been found to be another useful way to treat the illness. Specifically, helping sufferers learn how to manage their anger and anxiety, improve their communication skills, and use breathing and other relaxation techniques can help individuals with PTSD gain a sense of mastery over their emotional and physical symptoms. The practitioner might also use exposure-based cognitive behavioral therapy by having the person with PTSD recall their traumatic experiences using images or verbal recall while using the coping mechanisms they learned. Individual or group cognitive behavioral psychotherapy can help people with PTSD recognize and adjust trauma-related thoughts and beliefs by educating sufferers about the relationships between thoughts and feelings, exploring common negative thoughts held by traumatized individuals, developing alternative interpretations, and by practicing new ways of looking at things. This treatment also involves practicing learned techniques in real-life situations.

Eye-movement desensitization and reprocessing (EMDR) is a form of cognitive therapy in which the practitioner guides the person with PTSD in talking about the trauma suffered and the negative feelings associated with the events, while focusing on the professional's rapidly moving finger. While some research indicates this treatment may be effective, it is unclear if this is any more effective than cognitive therapy that is done without the use of rapid eye movement.

Families of PTSD individuals, as well as the sufferer, may benefit from family counseling, couple's counseling, parenting classes, and conflict-resolution education. Family members may also be able to provide relevant history about their loved one (for example, about emotions and behaviors, drug abuse, sleeping habits, and socialization) that people with the illness are unable or unwilling to share.

Directly addressing the sleep problems that can be part of PTSD has been found to not only help alleviate those problems but to thereby help decrease the symptoms of PTSD in general. Specifically, rehearsing adaptive ways of coping with nightmares (imagery rehearsal therapy), training in relaxation techniques, positive self-talk, and screening for other sleep problems have been found to be particularly helpful in decreasing the sleep problems associated with PTSD.

Medications that are usually used to help PTSD sufferers include serotonergic antidepressants (SSRIs), like fluoxetine, sertraline, and paroxetine, and medicines that help decrease the physical symptoms associated with illness, like prazosin, clonidine, guanfacine, and propranolol. Individuals with PTSD are much less likely to experience a relapse of their illness if antidepressant treatment is continued for at least a year. SSRIs are the first group of medications that have received approval by the U.S. Food and Drug Administration (FDA) for the treatment of PTSD. Treatment guidelines provided by the American Psychiatric Association (Tori DeAngelis,2008) describe these medicines as being particularly helpful for people whose PTSD is the result of trauma that is not combat-related. SSRIs tend to help PTSD sufferers modify information that is taken in from the environment (stimuli) and to decrease fear. Research also shows that this group of medicines tends to decrease anxiety, depression, and panic (http://www.nimh.nih.gov). SSRIs may also help reduce aggression, impulsivity, and suicidal thoughts that can be associated with this disorder (http://www.nimh.nih.gov). For combat-related PTSD, there is more and more evidence that prazosin can be particularly helpful. Although other medications like duloxetine, bupropion, and venlafaxine are sometimes used to treat PTSD, there is little research that has studied their effectiveness in treating this illness.

Other less directly effective but nevertheless potentially helpful medications for managing PTSD include mood stabilizers like lamotrigine, tiagabine, divalproex sodium, as well as mood stabilizers that are also antipsychotics, like risperidone, olanzapine, and quetiapine. Antipsychotic medicines seem to be most useful in the treatment of PTSD in those who suffer from agitation, dissociation, hypervigilance, intense suspiciousness (paranoia), or brief breaks in being in touch with reality (brief psychotic reactions). The antipsychotic medications are also being increasingly found to be helpful treatment options for managing PTSD when used in combination with an SSRI.

Benzodiazepines (tranquilizers) such as diazepam and alprazolam have unfortunately been associated with a number of problems, including withdrawal symptoms and the risk of overdose, and have not been found to be significantly effective for helping individuals with PTSD [Roxanne Dryden-Edwards,2011].

Our study, "posttraumatic stress disorder in orphans caused by Tangshan earthquake", investigated the morbidity of posttraumatic stress disorder in orphans caused by Tangshan earthquake. Fifty-seven orphans were surveyed using the criteria of Acute Stress Reaction (ASR) and PTSD in Chinese Classification and Diagnostic Criteria for Mental Disorders, the Second Revised Edition. The Self-rating Anxiety Scale, Symptom Checklist 90 (SCL-90), and Minnesota Multiphasic Personality Inventory were used to assess morbidity related to the Tangshan earthquake between the orphans with PTSD and respondents without-PTSD. Twenty seven (47%) cases were diagnosed as ASR and 13 (23%) cases were diagnosed as PTSD among 57 orphans. The orphans caused by Tangshan earthquake may be in the high risk to develop to PTSD.

Another study, "Life Style and Psychosomatic Health in Paraplegic Suffers of Tangshan Earthquake", investigated the relationship between life style and psychosomatic health of paraplegic suffers of Tangshan earthquake. Paraplegic suffers of Tangshan earthquake in a rehabilitation community (RC) and in a paraplegic hospital (PH) were tested with self report psychosomatic health questionnaire, SCL-90, CMI (Cornell medical index) and SSRS (social support rating scale). The two groups were similar in physical injuries and mental trauma caused by the earthquake. But those in RC selected a different life style from 8 years before when RC was founded. After 8 years, those in RC had better psychosomatic health, lower SCL-90 score or CMI score. None of them had PTSD, while 6 of those remained in PH had this diagnosis. This study revealed that election of a more mature way of life is helpful to psychosomatic health of paraplegic patients caused by earthquake.

6. Summary

This chapter has summarized the current status of information on mental disorder caused by experiencing or witnessing a life threatening severe earthquake. The traumatic earthquake was very tragic. Each earthquake phase has different mental health problems. From this chapter, we can conclude that the mental disorders due to the earthquake include depression, cognitive function disorder, PTSD, schizotypical personality.

7. References

[1] Zhang Ben, Wang Xueyi，Sun Hexiang et al. Long Term Effect of Tangshan Earthquake on Mental and Physical Health of It's Suffers. Chinese Mental Health Journal 1998 12 (4): 200-202..

[2] Wang Xueyi, Wang Jingjing, Zhang Ben, et al. Psychological status in adolescents who experienced Tangshan earthquake during fetal period[J].Chinese Journal of Clinical Rehabilitation,2006,10(6):42-45.

[3] Wang XueYi, Shi Shaoxia, Zhang Ben, et al. Adult schizotypal personality characteristics of a fetus exposed to Tangshan earthquake in its sixth month of gestation[J]; Journal of Clinical Rehabilitative Tissue Engineering Research; 2007,11(39):7838-7841.

[4] Zhang Ben, Wang Xueyi, Sun Hexiang, et al. A Study on the Prevalence of Neurosis and Its Cause over 20-years After the Violent Earthquake in Tangshan[J]; Nervous Diseases and Mental Hygjene,2001,1(1):8-11.

[5] Zhang Ben, Zhang Junzeng, Wang Xueyi, et al. A Study Of Mental And Physical Health Of Retired Officer Experience Tangshan Earthquike Of Kailuan Coal Mine[J]; Heath Psychology Journal.1999,7(1):51-54.

[6] Jin Guixing, Wang Xueyi, Wang Lan, et al. Study on memory function impairment and structural MRI in posttraumatic stress disorder. Chin J Nerv Ment Dis 2011,37(5):269-272.

[7] Zhang Ben, Wang Xueyi, Sun Hexiang, et al. Posttraumatic stress disorder in orphans caused by Tangshan earthquake. CHINESE JOURNAL OF PSYCHIATRY, 2000,33:111-114.

[8] Stein, P. & Kendell, J. (2004). Psychological trauma and the developing brain: Neurology based interventions for troubled children. New York, NY: Hawthorne Maltreatment and Trauma Press.

[9] Levine, P. A. (2007). Trauma though a child's eyes: Awakening the ordinary miracle of healing. Berkeley, CA: North Atlantic Books.

[10] Perry, B. & Szalavitz, M. (2006). The boy who was raised as a dog and other stories from a child psychiatrist's notebook: What traumatized children can teach us about loss, love and healing. New York, NY: Basic Books.

[11] Lu Lin, Wang Xueyi, Li Jing, et al. The effects on neurobehavioral development in offspring by maternal stress in gestation. Chinese Journal of Behavioral Medical Science,1999,8(4):241-245.

[12] Wang Xueyi, Zhang Ben,Zhang Baoting,et al. Long-term Effect of Earthquake on Fetal Cognitive Function. Chinese Mental Health Journal,2001,15(1):42-43.

[13] E.J.H. Mulder , P.G. Robles de Medina, A.C. Huizink, et al. Prenatal maternal stress: effects on pregnancy and the (unborn) child. Early Human Development ,70 (2002) 3 -14.

[14] Zhang Ben, Wang Xueyi, Sun Hexiang. Long-term Effects of Tangshan Earthquake on Psychosomatic Health of Paraplegic Suffers. Chinese Mental Health Journal, 16(2002):23-25.

[15] Zhang Ben, Xu Guangming, Wang Xueyi,et al. Life Style and Psychosomatic Health in Paraplegic Suffers of Tangshan Earthquake. Chinese Mental Health Journal, 16(2002):26-29.

[16] Zhang Ben, Wang Xueyi, Sun Hexiang, et al. A study of the prevalence of post-traumatic stress disorder after a violent earthquake in Tanshan. Chin J Psychiatry, 32(1999):106-108.

[17] Wang Xueyi,Zhang Ben,Ma Wenyou,et al. Affective Disorder in Offspring Exposed Prenatally to a Severe Earthquake. Chinese Journal of Health Psychology, 13(2005):31-32.

[18] Howard Berenbaum, Eve M. Valera, and John Q. Kerns. Psychological Trauma and Schizotypal Symptoms. Schizophrenia Bulletin, 29(2003):143-152.

[19] Armstrong, Nikki Panasci. Schizophrenia-spectrum symptoms following prenatal expo sure to an earthquake. UNIVERSITY OF HAWAI'I AT MANOA, 2009, 138.

[20] William Steele, MSW, PsyD. Trauma's Impact on Learning and Behavior: A Case for Interventions in Schools. Reprinted from Trauma and Loss: Research and Interventions,EVISED May 2007

[21] Watson, JB. Prenatal teratogens and development of adult mental illness. Development and Psychopath, 11(1999): 457~466.

[22] Van OS. Prenatal exposure to maternal stress and subsequent schizophrenia. Brit J Psychia, 172(1998): 324~326.

[23] Calvin Hobel, Jennifer Culhane. Role of Psychosocial and Nutritional Stress on Poor Pregnancy Outcome. The American Society for Nutritional Sciences J. Nutr. 133(2003):1709S-1717S.

[24] Post-Traumatic Stress Disorder (PTSD).
http://www.nimh.nih.gov

[25] Tori DeAngelis. PTSD treatments grow in evidence, effectiveness
http://www.apa.org. monitor/jan08/ptsd.aspx(2008)

[26] Roxanne Dryden-Edwards. Posttraumatic Stress Disorder.
http:// www. medicinenet. com/posttraumatic_stress_disorder(2011).

11

PTSD in the Context of Malignant Disease

A.M. Tacón
Texas Tech University,
USA

1. Introduction

Posttraumatic stress disorder or PTSD is an extreme psycho-physiological response disorder that may occur in individuals who are exposed to a potentially traumatic event that can involve the subjectively profound *threat of loss*... of life or limb. To receive a PTSD diagnosis, reactions to the catastrophic or traumatic stressor must involve profound fear, helplessness, or horror. Furthermore, such individuals have to experience symptoms from three separate yet co-occurring symptom clusters. These clusters or domains include: 1) intrusive recollections or the re-experiencing of the event with accompanied intense psychological distress or physiologic reactivity; 2) persistent avoidance of activities, thoughts as well as feelings associated with the traumatic event; and, 3) increased or extreme arousal that may include an exaggerated startle response, hypervigilance, or insomnia (American Psychiatric Association, 1994; Uddin et al., 2010) To justify a PTSD diagnosis, symptoms must be present for at least one month, and impair an individual's interpersonal, occupational, or social functioning (American Psychiatric Association, 1994). PTSD has been described as a specific phenotype that develops as the result of a failure to contain the normal stress response (Yehuda & LeDoux, 2007), resulting in dysregulation of the hypothalamic–pituitary–adrenal or HPA axis, a major stress response system of the body that interacts with the immune system to maintain homeostasis (Wong, 2002). PTSD-affected and unaffected individuals have distinct expression patterns in genes involved in immune activation (Segman et al., 2005; Zieke et al., 2007), and in genes that encode neural and endocrine proteins (Segman et al., 2005; Yehuda et al., 2009). Above all, PTSD is both an external and internal experience; that is, PTSD is an external catastrophic or traumatic event and an internal psycho-physiological experience.

1.1 Trauma goes public

Once upon a time in the west, PTSD was known as a psychiatric disorder associated most frequently with Vietnam War veterans as exemplified by the label of *post-Vietnam syndrome* (Friedman, 1981). This, despite the fact that similar war-related traumatic experiences of American soldiers who served in Vietnam can be found as far back as Homer's epic account of Achilles in *The Iliad* (Shay, 1994). Yet, while long-term psychiatric conditions were witnessed in previous war veterans, PTSD failed to permeate public consciousness until Vietnam. Indeed, Mezey and Robbins point out that PTSD has socio-economic and political implications since veterans are the group most associated with this disorder (2001). With

Vietnam, PTSD exploded into popular culture, becoming a subject for the general public in many films: *Apocalypse Now (1979), Born on the Fourth of July (1989), Casualties of War (1989), The Deer Hunter (1978), First Blood (1982), Full Metal Jacket (1987), Good Morning, Vietnam (1987), the Green Berets (1968), Hamburger Hill (1987), Hanoi Hilton (1987), Jacob's Ladder (1990), the Killing Fields (1984), and Platoon (1986)* (Bealle, 1997). In 1980, posttraumatic stress disorder (PTSD) was recognized as an official classification of a psychiatric disorder in the third edition of the *Diagnostic and statistical manual of mental disorders* (DSM-III) (American Psychiatric Association, 1980). This inclusion of PTSD in DSM-III basically served to legitimize a psychological disorder by re-labeling what had been described in the forgotten past as "soldier's heart," "shell shock," "railway spine, "war neurosis," "traumatic neurosis," "combat trauma," or "combat fatigue" (Bealle, 1997). The growth of psychiatric epidemiology enabled PTSD investigations to include samples of general populations in the United States and in other countries. In the U.S. during the past decade, trauma "goes public" was nowhere more evident than in the aftermath of 9/11 with the attack on the Twin Towers. The public traumatic event of "9/11 changed the picture of PTSD, and transformed it from being simply a mental disorder that psychiatrists deal with to a *public health* issue;" that is, "for the first time, psychiatric leaders pondered how factors such as media coverage, community cohesion, and poverty may affect the public's mental health when mass disaster strikes" (Brandt, 2011; Schuster et al., 2001).

2. PTSD in the context of cancer

The potentially traumatic and protracted nature of cancer's disease course and treatment received acknowledgment in 1994, when the *Diagnostic and statistical manual of mental disorders,* fourth edition (DSM-IV), revised events that may precede a posttraumatic response. Specifically, "life threatening illness" was added to the criteria as a potentially precipitating event for the development of posttraumatic stress disorder or PTSD (American Psychiatric Association, 1994). Previously, criteria has centered upon *acute* events such as war, natural disasters, automobile accidents, and rape. This new inclusion of a *chronic* and possible terminal disease process into the diagnostic criteria, indeed, was a significant change as to precipitating events associated with PTSD. First, diagnostic criteria for PTSD within the context of malignant disease will be summarized. Then, a review of the literature with current PTSD prevalence will follow with prevalence rates for identified cancer types and age groups as adequate data and number of studies permit, risk factors, and lastly, concluding comments.

2.1 Trauma by any other name

Before reviewing diagnostic criteria, the debate about malignant illness as a traumatic event needs to be addressed (e.g., Kwekkeboom & Seng, 2002; Palmer et al., 2004), in order to highlight that variation is not negation of a disorder's existence. Events such as rape, assault, or natural disasters, etc., tend to be singular events restricted to a finite period during which the external agent ceases to be acutely present in real living time. It may be argued that cancer is ambiguous as a stressor event, for malignancy does not fit neatly into an objective timeline with a discrete beginning and ending. One may ask, what event truly was the stressor event? Was it the public *confirmation* of the diagnosis, of which, one may have already decided privately was cancer?... Was it a procedure such as the biopsy?...Was it a treatment regimen? Or was it side effects from treatment or the disease? Regardless, the bottom line is that the threat of death is real and present at all times. Secondly, another variation of the posttraumatic experience related to cancer is the time zone of traumatic-inducing reality. Traditional stressor events, which are thus *re-experienced,* obviously are located in the past, and are primarily retrospective trauma. Cancer-related PTSD, however can be viewed as being bidirectionally traumatic in terms of subjective time: retrospective with memories of the past, yet also prospectively traumatic with a truncated future that is equally as threatening, helpless, and horrifying. The integrity of self is threatened because self is in clear and present---and future danger of not existing. The traumatizing reality that one's long-term plans, hopes or dreams for the future such not seeing a child graduate from high school, get married, etc., or worse---that one will die an agonizing and painful death are forward experiences in the timeline of trauma that personifies a foreshortened future.

Thirdly, another variation of cancer-related PTSD is the fact that the traumazing agent or perpetrator is not external; rather, it is internal in the form of a biological, patho-physiological disease with threat of recurrence, which can lead to a sense of ambivalent betrayal of self. The rude reality is that trauma is experienced in the past, in the present, and in the future tense of experiencing. Moreover, being a cancer *survivor* does not dispell distress of life-threatening trauma or mean that one is safe and sound. In addition to anxiety and depressive symptomatology, pervading concerns about prognosis, treatment options and effects, or ruminating fears about upcoming doctor appointments and disease recurrence, are all common sources of distress that can plague cancer patients in survivorship long after treatment has ended (Andrykowsk, et al., 2008; Montgomery et al., 2003). For example, acute or sub-acute symptoms may erupt with each doctor's visit, routine check-up, or getting a mammogram, etc., from post-traumatic cues of a patient's previous cancer experience. Unfortunately, routinized triggers can become embedded within the healthcare system, to where a cycle of nosocomial re-triggering or institutional re-traumatization is conceivably possible. Indeed, the original traumatic-inducing event may be lost among the chronic cascading triggers---even in the absence of recurrence or a new primary site of malignancy.

The addition of "life threatening" illness to official PTSD criteria is exactly that---life-threatening. The spontaneous potential of traumatic triggers and the pervasive pain from living a disrupted life with PTSD---be it from a traditional event or a life-threatening illness with recurrence---is equivocal when it comes to human physical and mental suffering. In sum, cancer-related PTSD is no less valid or more ambiguous than traditional stressor events, for trauma by any other name---cancer--- *is* trauma despite variation of etiology.

3. DSM in brief: Diagnostic criteria for PTSD

A brief review of general symptom clusters (see Table 1) (Friedman, 2006) and diagnostic criteria A – F (Table 2) (American Psychiatric Association, 1994, 2000), will be presented with tables to make the information more user-friendly. In order to be diagnosed with PTSD due to the potentially [precipitating] traumatic event of being diagnosed with a malignant disease process, all of the six criteria, that is, A – F, must be satisfied.

Cluster	Specific Symptoms
Reexperiencing	• Intrusive recollections • Traumatic nightmares • Flashbacks • Trauma-evoked psychological distress • Trauma-evoked physiological reactions
Avoidant/Numbing	• Avoiding trauma-related thoughts/feelings • Avoiding trauma-related activities/places/people • Amnesia of trauma-related memories • Diminished interest • Detached or estranged feelings • Restricted range of affect • Sense of foreshortened future
Hyperarousal	• Insomnia • Irritability • Difficulty in focusing/concentrating • Hypervigilance • Exaggerated startle reaction

Table 1. Symptom Clusters for PTSD (Friedman, 2006)

Criterion A: Stressor	Exposed to a traumatic event in which *both* of the following were present: **A1.** Person experienced, witnessed or was confronted by an event that involved actual or threatened death or serious injury, or a threat to the physical integrity of self or others. The traumatic event includes diagnosis of a life-threatening illness such as cancer that threatened one's life and/or physical integrity; involved either direct personal experience (such as being the patient) witnessing, confronting, or learning about the illness experience through a family member or close friend—which may pose a threat to the integrity of a significant other **A.2** Response involved intense fear, helplessness or horror
Criterion B: Intrusive Recollection	The traumatic event is re-experienced persistently in at least *one* of the following ways: **B1.** Recurrent and intrusive distressing recollections of the event, including images, thoughts, or perceptions; thus, a distressing cancer event such as a diagnosis, has occurred for one to have distressing recollections about such an event

	B2. Recurrent distressing dreams of the event **B3**. Acting or feeling as if the traumatic event were recurring with sense of re-experiencing it again; dissociative flashbacks **B4**. Intense psychological distress at exposure to internal or external cues that symbolize or resemble an aspect of the traumatic event **B5**. Physiological reactivity at exposure to internal/external cues that symbolize or resemble an aspect of traumatic event
Criterion C: Avoidant/Numbing	Persistent avoidance of stimuli associated with the trauma and numbing of general responsiveness (not present before the trauma, in this case, cancer) as indicated by three or more of the following: **C1**. Efforts to avoid thoughts, feelings, or conversations associated with the trauma **C2**. Efforts to avoid activities, places or people that arouse recollections of the trauma **C3**. Inability to recall an important aspect of the trauma **C4**. Markedly diminished interest or participation in significant activities **C5**. Feeling of detachment or estrangement from others **C6**. Restricted range of affect (for example, unable to have loving feelings) **C7**. Sense of a foreshortened future, (does not expect to have a career, children, marriage, or a normal lifespan
Criterion D: Hyperarousal	Persistent symptoms of increased arousal (not present before the trauma, in this case, cancer) as indicated by two or more of the following: **D1**. Difficulty falling or staying asleep **D2**. Irritability or outbursts of anger **D3**. Difficulty concentrating **D4**. Hypervigilance **D5**. Exaggerated startle response
Criterion E: Duration	Duration of the disturbance (symptoms in B, C, & D) is more than one month.
Criterion F: Functional Significance	The disturbance causes clinically significant distress or impairment in social, occupational or other important areas of functioning.

Table 2. DSM Criteria for PTSD (American Psychiatric Association, 1994, 2000)

3.1 Criteria A - C

The *DSM's* first criterion, A, relates to PTSD's conceptualization as a stress-related response syndrome where the person experienced an event that threatened his/her life or physical integrity (American Psychiatric Association, 1994, 2000). Also, this criterion includes vicarious traumatization, that is, the witnessing of traumatic events as well as hearing traumatic news or unexpected occurrences about loved ones. For example, related PTSD investigations have expanded to include not only the patients themselves, but also, loved ones affected by the experience of cancer in another, for example, parents of children diagnosed with malignancy as well as partners, siblings or significant friends of cancer patients (e.g., Alderfer et al., 2010; Poder, Ljungman, & von Essen, 2008).

Criterion B involves persistent re-experiencing of the cancer experience (including the intrusive thoughts of symptoms, the way diagnosis was communicated, impending death or experience of review visits), for example, recurrent and intrusive memories or images in the form of flashbacks or nightmares. Noticeable physical reactions may present, such as breaking out in a sweat, feeling light-headed or nauseous, having palpitations, or breathing gets fast and shallow. Criterion C is persistent avoidant or emotional numbing strategies that serve the purpose of blocking internal or external stimuli reminiscent of the traumatic event, such as avoiding certain people, places, or perhaps even the music playing in the background when the event was occurring.

3.2 Criteria D - F

Criterion D involves persistent hyper-physiological arousal and-or anxiousness that were not present before the diagnosis or experience of cancer, and might include sleeping problems, irritability and anger outbursts. Another form of arousal unique to cancer-related PTSD is body symptom hypervigilance or being on guard for signs of another tumor. Criterion E indicates that the duration of such symptoms has persisted for a period of at least one month following the cancer-related traumatic event (e.g., being told the diagnosis or given a poor prognosis, etc.), thereby, distinguishing it from acute stress disorder. Finally, Criterion F indicates the extent to which the symptoms impair domains of life functioning, for example, relational or occupational. Please refer to Table 2.

4. PTSD and adult cancer

In general, assessing prevalence rates of PTSD in adults with cancer have consisted primarily of breast cancer patients including a wide range of time between diagnosis and treatment, from several days to over a decade.

4.1 Current search strategy

For the current investigation, a literature search was conducted for PTSD prevalence rates in current adult cancer patients and cancer survivors via the *ISI Web of Knowledge, Medline* and *PsycINFO* databases, and the references of retrieved articles. The search considered only studies published in English. The main search terms were posttraumatic, PTSD, cancer, cancer patients, prevalence, prevalence rates, in various combinations as needed. The search strategy consisted of several levels of filtering out articles not relevant to the purposes here. The articles considered useful for the aim of this paper included studies where current prevalence rates for PTSD were ssessed, and identification of posttraumatic cases were based on official DSM guidelines, that is, DSM criteria had been applied. Also included were studies where participants were current patients at the time of data collection or were cancer survivors of 60 months (the five-year marker) or less since the end of treatment; this was done in an effort to reduce the potential wide range of survivor time variation (i.e., survivors of 13-years, 2-years, etc). No limitations were placed on the number of participants, the study design, or whether a control group was included; studies reporting qualitative data were excluded.

4.2 Results

A total of 11 studies met criteria, which were based on interview data collection (See Table 3). The current PTSD diagnostic gold standard remains a clinical interview that is based on the predefined criteria in the DSM (American Psychiatric Association, 1994).Three of the studies involve participants with a mixture of cancer diagnoses excluding breast malignancy (Akechi et al., 2004; Kangas et al., 2005; Widows et al., 2000). The remaining eight studies investigating adult cancer-related PTSD include breast cancer patients at various points in the disease process (Andrykowski et al., 1998; Gandubert et al., 2009; Green et al., 1998; Luecken et al., 2004; Mehnert & Koch, 2007; Mundy et al., 2000; Okamura et al., 2005; Shelby et al., 2008). Despite advances in breast cancer diagnosis and treatments, it remains a monumental stressor in these women's lives that continues to elicit greater distress than any other medical diagnosis (Shapiro et al., 2001). This distress is now recognized as an integral component of a patient's clinical presentation (Bultz & Carlson, 2006); additionally, depression, anxiety, and posttraumatic stress disorder (PTSD) symptoms occur somewhere between 20 and 66% of women in the first 12 months alone after their diagnosis (Burgess et al., 2005; Vos et al., 2004). Such psychological difficulties in response to the challenge of cancer appear congruent with certain pre-cancer factors that may set the stage for patient vulnerability. Specifically, posttraumatic stress disorder cases have been found to be distinguished by a previous history of violent traumas as well as psychological problems such as anxiety disorders that predate the diagnosis of cancer (e.g., Shelby et al., 2008).

The sample sizes in the eight breast cancer studies range from 37 to 160 participants with a mean of 93 participants and the median of 78 participants. Prevalence rates for current cancer-related PTSD range from 0% to 16.2% with a mean of 4.6% and a median of 2.75%. Sample sizes for the three studies of mixed cancer types range from 82 to 209; the mean of participants in these studies is 131 participants with a median of 102 participants. The prevalence rates for current cancer-related PTSD in these studies range from 0% to 22% with a mean prevalence rate of 9% and a median of 5%.

In sum, these findings are congruent with previous estimates of current cancer-related PTSD prevalence in adult patients, that is, in women with breast cancer or mixed samples of gender with head, neck or lung cancer ranging from 0 to 32% (Hamann et al., 2005). Several points need to be acknowledged. First, the majority of studies---as tends to be the case in the area of psychosocial oncology literature---involved women with breast cancer. Thus, most of the current knowledge regarding PTSD in the context of malignant disease is based upon this cancer population with frequently low sample sizes. Basically, this means that caution is needed, for generalizability of such results do not apply to other disease populations or to those with other types of cancer. Also, the majority of these investigations were cross-sectional studies with time variation as to time of assessment from either diagnosis or treatment. Cross-sectional data prohibits clinical understanding of the development as well as the trajectory patterns of PTSD in those with cancer. The relative lack of prospective or longitudinal data is needed to inform clinicians early in the disease process so that highly distressed patients may be identified and helped as early as possible with this life-threatening illness. Lastly, cross-sectional research does little in the way of determining risk factors relevant to later survival time or predicting the delayed onset of posttraumatic stress. Attention now turns to investigations assessing PTSD in strictly male samples with diagnosed malignancy.

Study	Design	Sample	Time Period	PTSD Prevalence
Akechi et al. (2004)	Cross-sectional	n = 209; males/ females; mixed cancer diagnosis	25 months post-dx	0% (only 100 participants assessed)
Andrykowski et al., (1998)	Cross-sectional	n = 82 females; breast	37 months post-treat	6% current
Gandubert et al., (2009)	Cross-sectional	n = 144; female; breas	1-3 years post-dx	4.9% current
Green et al., (1998)	Cross-sectional	n = 160 females; breast	6.5 months post-treat	2.5% current
Kangas et al., (2005)	Longitudinal	n = 82; males/females head, neck, lung cancer	6 months post-dx	22% current
Luecken et al., (2004)	Cross-sectional	n = 71; females; breast	1-6 months post-dx	3% current
Mehnert & Koch, (2007)	Cross-sectional	n = 127; females; breast	15 days post-dx of initial cancer (77%) ; or recurrent malignancy (23%), 0-67 days	2.4% current
Mundy et al., (2000)	Cross-sectional	n = 37; females; breast	> 100 days post-tx	0% current
Okamura et al., (2005)	Cross-sectional	n = 50; females; breast recurrence	1-6 months post-dx recurrence	2% current
Shelby et al., (2008)	Longitudinal	n = 74; females; breast	dx/surgery - 18 months	1.6% current
Widows et al., (2000)	Cross-sectional	n = 102; males/females; cancer not specified	20.4 month post-bone marrow transplant	5% current

Table 3. Summary of select cancer studies

5. PTSD: Men with malignant disease

While a plethora of studies have assessed prevalence rates for women breast cancer and/or mixed gender studies, a paucity of research exists as to the prevalence rates in strictly male samples with malignant processes. This is surprising since survivors of prostate cancer continue to grow as well as the fact that psychological distress in response to receiving a prostate cancer diagnosis is a recognized phenomenon (Anastasiou et al., 2011; Gwede et al., 2005; Namiki et al., 2007). Furthermore, psychological distress in prostate cancer patients has been found to be related to the following: stage of disease, shorter time since diagnosis, and treatment options or decisions (Gwede et al., 2005). For men, receiving the diagnosis of malignant disease has been found to be associated with responses that include anxiety, denial or distress (Kronenwetter et al., 2005).

Distress related to decision-making is a common experience among men after the diagnosis of prostate cancer; specifically, reports indicate that around 63% report high decision-related distress persisting the first year after treatment for 42% of all men (Steginga et al., 2008). Also, an increased manifestation of traumatic stress symptoms has been found in some cases of newly diagnosed men with localized prostate cancer before the beginning of treatment (Bisson, 2007); on the other hand, low emotional distress has been documented as being present even 2 years after a radical prostatectomy (Perez et al., 2002). Follow-up of males during the disease course of their prostate cancer show that males with prostate cancer may suffer with long-term physical and psychological consequences---to the point of affecting their quality of life (Penson, 2007; Sanda et al., 2008). Indeed, accumulating data within the past 15 years has produced a body of literature investigating health related quality of life (HRQOL) outcomes pertaining to localized prostate cancer (Penson, 2007). This stands in contrast to empirical exploration regarding adjustment in such male cancer survivors. Unfortunately, the literature does not demonstrate ample descriptive investigations as to the course of psychological adjustment for men who have been diagnosed with and treated for prostate cancer (Steginga et al., 2004). No studies could be identified that specifically assessed PTSD in men with testicular or lung cancer.

A recent study by Anastasiou and colleagues (2011), believed to be the only study that has focused on investigating the presence of acute posttraumatic stress disorder (PTSD) in men with malignant disease necessitating a radical prostatectomy, assessed symptoms in 15 men one month after surgery. The men completed the Davidson Trauma Scale rather than being assessed by interview. Analyses determined that 26.7% of the men's scores met scale criteria for acute PTSD, which was found to be independent of the patient's educational level. In sum, despite the fact that prostate cancer is the second most frequently diagnosed cancer in developed countries, and the third most common cause of death in men (Damber & Aus, 2008), little is known about the psychological ramifications for prostate cancer patients after surgery (Burns & Mahalik, 2008; Namiki et al., 2007; Steginga et al., 2004).

6. PTSD and pediatric cancer

The present search revealed few reports of cancer-related PTSD in pediatric patients.

6.1 Results

The search revealed only two studies assessed within the five-year mark with data regarding cancer-related PTSD (Landolt et al.,1998; Pelcovitz et al., 1998). However, due to the lack of studies, a third study that assessed PTSD at 5.3 years, was included (Kazak et al., 2004). The participants were less than 18 years of age at the time of these investigations, which all consisted of mixed cancer diagnoses. The sample sizes were 150, 7, and 23; the mean sample size was 60 with a median of 23 participants. The prevalence rates in these studies for current cancer-related PTSD were documented as ranging from 4.7% to 71%; the mean prevalence rate was 31% with a median of 17%. It must be pointed out that the prevalence rate finding of 71% comes from a study with a small sample size of only seven participants who were identified as "newly diagnosed." Therefore, the possibility that this assessment may have occurred within the first month of diagnosis calls into question the appropriateness and validity of these data. Unfortunately, the small number of studies in this area of current PTSD prevalence as well as the questionable data provided limit the scope of this review; therefore, conclusions about prevalence rates in this population are prohibitive at this time.

7. PTSD: All in the family

Each year in the United States, approximately 14,000 children are diagnosed with cancer (Ries et al., 2008). The diagnosis of cancer is a traumatic experience for both the child and the family, which understandably causes a great of deal of disruption within the family system. The following review will encompass literature regarding posttraumatic stress among 1) parents of children with cancer, 2) the bereaved following the death of a cancer patient, and 3) the siblings of children diagnosed with cancer.

7.1 Parent and child

Having a child diagnosed with cancer is one of the most severe stressors that parents could possibly ever experience; the threat of their child's death and feelings of helplessness converge as the entire family faces the uncontrollable enemy of cancer (Kazak, 1998; Patterson et al., 2004). The change of the A1 event criterion of PTSD in the *DSM-IV* (American Psychiatric Association, 1994) to include "the diagnosis of a life threatening disease" as potentially traumatizing, expanded the range of traumatic effects to include the patient's interpersonal domain. Consequently, an individual's witnessing of or learning about a significant person [to him/her] being diagnosed with a terminal illness is now viewed as potentially traumatizing. Family members and loved ones torturously stand by as their beloved endures periods of devastating illness effects, physical and psycho-spiritual pain, or highly invasive treatments that produce debilitating fatigue---all of which may be followed by uncertain terminal illness and possible death. Family members typically are involved in making hard calls or decisions regarding the patient, which, in itself, is a potential independent risk factor for PTSD (Azoulay et al., 2004, 2005).

The psychological sequelae of childhood cancer in both the children and their parents are well documented, which consists of post-traumatic stress symptoms, anxiety, and depression (Kazak, 2005; Kazak et al., 2005). More specifically, the sequelae as to parental risk for affective and stress reactions range from 9% to 40% for the above disorders for a period of up to 3 years after the child's diagnosis (Kazak et al., 2004; Stoppelbein & Greening, 2007). Barakat and colleagues found that the parents of childhood cancer survivors showed significantly higher levels of post-traumatic stress symptoms (PTSS) than a comparison group of parents (1997). Similarly, Kazak found significant elevation in PTSD scores in parents of survivors in contrast to parents of never-ill children (Kazak et al., 1997). Indeed, the line of research exploring distress in relatives of cancer survivors indicate that the relatives are usually at least as distressed as the cancer survivors (especially mothers)---if not more so (e.g., Couper et al., 2006; Mosher & Danoff-Burg, 2004; Tuinmann et al., 2004). Kazak and colleagues (2004) found current PTSD prevalence for parents of adolescent survivors to be 13.7% for mothers and 9.6% for fathers. Lastly, an interesting finding in the literature demonstrating the parent-child relationship is that the ill child's psychological adjustment to his/her cancer diagnosis and treatment is highly correlated with the parents' adjustment to their child's diagnosis and treatment (Ljungman et al., 2003).

7.2 The bereaved

It has been estimated that for every person that dies five close friends or family members are affected (Zisook et al., 1998), therefore, the bereaved family and friends of cancer patients are at risk of developing psychological traumatic disorders as well as complicated grief. Indeed, the existing PTSD literature for the most part ignores bereaved individuals despite the fact that such individuals meet the A1 stressor criterion of the *DSM* (American Psychiatric Association, 1994). One of the few studies in this area investigated pre- and post-loss bereavement levels of posttraumatic stress symptoms (intrusion and avoidance) in 50 partners of women with metastatic/recurrent breast cancer as well as the relationship of these symptoms to past, current, and anticipatory stressors (Butler et al., 2005). The data indicated that 17 (34%) of the bereaved partners experienced clinically significant PTS symptom levels prior to the patients' deaths; specifically, prior to loss, partners' symptoms were positively associated with their current level of perceived stress and anticipated impact of the loss. However, following the death of the loved one, partners' posttraumatic stress symptoms were predicted by higher pre-loss levels of symptoms, past family deaths, and anticipated impact of the loss.

A more recent study investigated PTSD and PTSD predictors in bereaved individuals who had experienced the loss of a close relative to cancer and were attending counseling (Elklit et al., 2010). A total of 251 bereaved relatives, with ages ranging from 14 to 76 (M = 41.3), were recruited at a counseling service that assisted cancer patients and their relatives. The findings indicated that the prevalence of current PTSD was 40% in this sample. Furthermore, hierarchical logistic regression analysis showed that the following variables moderated the risk for PTSD: full-time employment, perceived control, and a secure attachment style. An extended period of caretaking as well as high levels of somatization and dissociation also were associated with an increased risk of PTSD (Elklit et al., 2010).

7.3 The forgotten children: Siblings

As parents become distressed over the condition of their child diagnosed with cancer, they likely will spend extended periods of time at home or in the hospital attending to the needs of that child. As the stress of the malignant disease process takes its biopsychosocial toll on the child and the parents, the time will come, undoubtedly, when the adults will not be able to physically and emotionally meet the needs of their other children in the family (Alderfer & Kazak, 2006). It is not surprising, then, that the siblings of children with cancer become increasingly at risk for affective, behavioral, and school problems (Aldelfer & Hodges, 2010; Alderfer et al., 2010). Siblings of children with cancer experience feelings of anger, fear, grief, guilt, helplessness, insecurity, jealousy, loneliness, loss, resentment, and shock, (e.g., McGrath, 2001; Nolbris et al., 2007; Woodgate, 2006). Furthermore, investigations examining sibling distress increasingly indicate that a marked subset of siblings with a brother or a sister with cancer display post-traumatic stress (PTS) symptoms.

One study investigated whether 78 adolescent siblings of childhood cancer survivors experience posttraumatic stress; the participants completed self-report measures of anxiety, perceptions of the cancer experience, and posttraumatic stress (Alderfer et al., 2003). The findings showed that close to half of the sample (49%) reported mild posttraumatic stress, and more worrisome, that 32% indicated moderate to severe levels. Additionally, one-fourth of the siblings thought their brother/sister would die just during cancer treatment alone, and more than half of the sample viewed the experience of cancer as scary. These perceptions, not surprisingly, were found to be related to PTS. These siblings reported a greater number of PTS symptoms than did an adolescent comparison group who were not psychologically affected by their sibling's illness, yet did report similar levels of general anxiety. Overall, studies that have investigated PTS reactions in siblings of children with cancer estimate that between 29–38% of the siblings exhibit moderate to severe cancer-related PTS---which can occur years after the treatment for cancer has ended (Alderfer et al., 2003; Packman et al., 2004).

8. Risk factors for cancer – Related PTSD

Many of the risk factors associated with adult cancer-related PTSD mirror the risk factors for PTSD from other traumas such as combat (e.g., Foy et al., 1987). Incidence and severity of cancer-related PTSD are affected by factors that occur before and during the cancer experience (Gurevich et al., 2002).

8.1 Gender, age, and socio-economic status

Pre-cancer factors, for example, gender, age, socio-economic status, and trauma history are associated with cancer-related PTSD. In terms of gender, female patients appear to be at greater risk for developing PTSD than their male counterparts (Kangas et al., 2002). Younger age at diagnosis also is a potential risk factor (Epping-Jordan et al., 1999; Kangas et al., 2002). Specifically, younger women may experience more distress due to the perception that cancer is more threatening to their lives; older women who have lived out the majority of their lifespan, on the other hand, may be less worried about recurrence, and thus display more positive mental adjustment (Glanz & Lerman, 1992). Socio-economic circumstances, specifically, lower income has been found to be associated with developing cancer-related PTSD; that is, it may be the case that fewer material resources combined with the stress of a cancer diagnosis may make the cancer experience more traumatic (Cordova et al., 1995).

8.2 Personality, history, and disease characteristics

Personality characteristics/traits such as neuroticism, pessimism, a pre-cancer history of affective disorders such as depression, or having a history of traumatic experiences may increase a patient's vulnerability to developing PTSD in the context of cancer (Andrykowski & Cordova, 1998; Epping-Jordan et al., 1999; Glanz & Lerman, 1992). Ineffective and maladaptive coping strategies, for example, staunch avoidance regarding the diagnosis of cancer as well as a pre-cancer coping style of avoidance, have been linked with PTSD relating to cancer (e.g., Butler et al., 1999; Jacobsen et al., 2002). It is reasonable to expect that characteristics of the disease process, itself, may contribute to distress and suffering in the form of PTSD. Specifically, research shows that, indeed, risk increases for PTSD in the case of malignant recurrence or the more advanced the stage of the disease (e.g., Andrykowski & Cordova, 1998). Additionally, research suggests a greater likelihood of patients developing cancer-related PTSD when more aggressive or invasive treatments are used (Gurevich et al., 2002).

8.3 Social support

In 1979, Berkman and Syme published their seminal study linking social relationships to mortality. Three decades later, reliable links between social support and better physical health outcomes continue to be demonstrated, with epidemiological studies showing that those with low levels of social support have higher rates of mortality (Berkman *et al.*, 2000; Uchino, 2004). Moreover, research indicates that social support is associated with improved immune function (e.g., Dixon et al., 2001; Lutgendorf et al., 2005). Thus, the possibility that social factors may be related to certain patients manifesting PTSD during the distress of cancer is congruent with this literature. Data have shown social support to be inversely associated with intrusive thoughts (DSM criterion B) and avoidance of cancer reminders of the cancer experience (DSM criterion C) (Andrykowski & Cordova, 1998; Jacobsen et al., 2002). In women with metastatic breast cancer, the size of a woman's emotional support network was found to predict avoidant symptoms in the patient (Butler et al., 1999). In terms of relational quality, a salient aspect of significant others in our lives is that of being supportive during times of personal disclosure, that is, the sharing of intimate information with trusted others. In the context of cancer, negative responses from others after a patient disclosed concerns and fears about her disease have been found to be associated with greater PTSD symptomatology (Cordova et al., 2001).

The issue of quality of life is a major line of research in cancer populations. A study by Lewis and colleagues (2001) investigated the moderating effect of social support on the relationship between cancer-related intrusive thoughts and quality of life. Sixty-four breast cancer survivors (of unspecified time since diagnosis or treatment) completed self-report measures of: social support (disclosure of thoughts/feelings to significant others); cancer-related intrusive thoughts; and, quality of life. After controlling for demographic and treatment variables, the appraised level of social support was found to moderate the negative impact of cancer-related intrusive thoughts on both physical and psychological quality of life measures. Specifically, no significant relationship was found for cancer-related intrusive thoughts and quality of life in women who had endorsed high levels of social support. For women with low levels of perceived support, however, the data revealed a significant and negative relationship between cancer-related intrusive thoughts and participants' quality of life. These data suggest that appraised levels of social support in breast cancer patients may mitigate the impact of traumatic events during a woman's distressing breast cancer experience (Lewis et al., 2001). In sum, research indicates that certain demographic, personal and/or historical characteristics as well as biopsychosocial factors contribute to the multifactorial process of a cancer patient's experience. Data support the notion also that multiple factors appear to influence individual vulnerability and manifestation of posttraumatic symptoms in cancer patients.

9. Conclusion

Overall, findings regarding adult prevalence rates in adults with breast and mixed gender/mixed cancer types are consistent with previous estimates of current cancer-related PTSD prevalence in adult patients, that is, rates ranging from 0 to 32% (Hamann et al., 2005). Several points were highlighted. First, eight of the 11 studies identified in the summary review and listed in Table 3 involved women with breast cancer. Consequently, most of the present state of knowledge about PTSD in cancer patients is based upon this cancer population. Generalizability to other disease populations or to individuals diagnosed with different types of cancer thus is limited. Additionally, all but two of the investigations were cross-sectional studies with variance regarding time from either diagnosis or treatment to the study when the assessments occurred. Cross-sectional data limits clinical usefulness and understanding about the trajectory patterns of PTSD in those with cancer; future studies that are prospective or longitudinal in design with large sample sizes are recommended. These findings stand in contrast to the surprising stark paucity of research as to psychological adjustment such as PTSD and risk or predictive factors in men with prostate cancer who end up needing a prostatectomy. Much more research in the future is needed imperatively in this area to fill this gap in the literature. Similarly, further research investigating PTSD and related factors with large sample sizes in pediatric populations are suggested.

Over the past decade, increasing empirical evidence indicates that having a child diagnosed with a malignant disease process is psychologically traumatic for the parents. Specifically, symptoms of PTSD---the reexperiencing of traumatic events, physiologic arousal, and behaviors to avoid cues related to the cancer experience---have been documented in parents (especially mothers), whose children had completed cancer treatment (e.g., Brown et al., 2004; Manne et al., 2004). One study involving 150 families of adolescent childhood cancer survivors 1 to 10 years after completing cancer treatment found that practically all of the families (99%) had at least one parent who met PTSD symptom criteria for reexperiencing (Kazak et al., 2004). Moreover, 20% of the families were found to have at least one parent who met criteria for being diagnosed with current PTSD.

Research into posttraumatic stress in the bereaved and siblings of children with cancer were reviewed, with limited quantity and quality of studies noted. Large-scale epidemiological studies are needed to comprehensively estimate prevalence rates of posttraumatic stress for these emerging, psychologically affected populations. A severe absence of investigation exists regarding PTSD among the bereaved of a cancer patient, and adequate estimates of prevalence do not exist at this time. It must be emphasized that the absence of sufficient data in certain types of cancer populations---such as males with prostate cancer, the bereaved or sibling groups---limits intervention approaches and clinical efficacy. A plethora of well-designed studies ultimately will enable targeted and successful interventions for those suffering with cancer-related PTSD. Only then, will clinical care truly be capable of providing the needed services for those traumatically affected by cancer. Members of the significant family-and-friends system need not struggle helplessly as they live with and function in the context of a "life-threatening" malignant illness.

In closing, future investigations are needed in order to develop targeted, effective and appropriate psychological interventions to address PTSD in cancer patients as well as to educate them about coping with life in the context of cancer. This will be especially salient in the future, based on current philosophical and medical zeitgeist, as the number of cancer survivors continues to increase.

10. References

Akechi, T., Okuyama, T., Sugawara, Y., Nakano, T., Shima, Y., & Uchitomi, Y. (2004). Major depression, adjustment disorders, and post-traumatic stress disorder in terminally ill cancer patients: associated and predictive factors. *Journal of Clinical Oncology*, 22(10), pp. 1957-1965.

Alderfer, M., & Hodges, J. (2010). Supporting siblings of children with cancer: A need for family-school partnerships. *School Mental Health*, 2, pp. 72-81.

Alderfer, M., & Kazak, A. (2006). Family issues when a child is on treatment for cancer. In: *Comprehensive handbook of childhood cancer and sickle cell disease: A biopsychosocial approach,* R.T. Brown, (Ed.), pp. 53–74, Oxford, New York.

Alderfer, M., Labay, L., & Kazak, A. (2003). Brief report: Does posttraumatic stress apply to siblings of childhood cancer survivors? *Journal of Pediatric Psychology, 28,* 281–286.

Alderfer, M., Long, K., Lown, E., Marsland, A., Osrowski, N., Hock, J., & Ewing, L. (2010). Psychosocial adjustment of siblings of children with cancer: a systematic review. *Psychooncology,* 19, pp. 789-805.

American Psychiatric Association. (1980). *Diagnostic and statistical manual of mental disorders: DSM-III,* 3rd ed. Author, Washington, DC.

American Psychiatric Association. (1994). *Diagnostic and statistical manual of mental disorders* (4th ed.). Author, Washington, DC:

American Psychiatric Association. (2000). *Diagnostic and statistical manual of mental disorders:* (4th ed., text revision). Author, Washington, DC.

Anastasiou, I., Yiannopoulou, K.G., Mihalakis, A., Hatziandonakis, N., Constantinides, C., Papageorgiou, C., &Mitropoulos, D. (2011). Symptoms of acute posttraumatic stress disorder in prostate cancer patients following radical prostatectomy. *American Journal of Men's Health,* 5(1), pp. 84–89.

Andrykowski, M. A., & Cordova, M. J. (1998). Factors associated with PTSD symptoms following treatment for breast cancer: Test of the Andersen model. *Journal of Traumatic Stress,* 11, pp. 189–203.

Andrykowski, M.A., Cordova, M.J., Studts, J.L., & Miller, T.W. (1998). Posttraumatic stress disorder after treatment for breast cancer: prevalence of diagnosis and use of the PTSD Checklist-Civilian Version (PCL-C) as a screening instrument. *Journal of Consulting and Clinical Psychology,* 66(3), pp. 586-590.

Andrykowski, M., Lykins, E., & Floyd, A. (2008). Psychological health in cancer survivors. *Seminar in Oncology Nursing,* 24, pp. 193-201.

Azoulay, E., Pochard, F., Chevret, S., & Adrie, C. (2004). Half the family members of intensive care unit patients do not want to share in the decision-making process: a study in 78 French intensive care units. *Critical Care Medicine, 32,* 1832-1838.

Azoulay, E., Pochard, F., Kentish-Barnes, N., Chevret, S. (2005). Risk of post-traumatic stress symptoms in family members of intensive care unit patients. *American Journal of Respiratory and Critical Care Medicine,* 171(9), pp. 987-94.

Barakat, L. P., Kazak, A. E., Meadows, A. T., Casey, R., Meeske, K., & Stuber, M. L. (1997). Families surviving childhood cancer: A comparison of posttraumatic stress symptoms with families of healthy children. *Journal of Pediatric Psychology,* 22, pp. 843–859.

Bealle, L.S. (1997). Post-Traumatic Stress Disorder: A Bibliographic Essay. *CHOICE,* 34(6), pp. 917-930.

Berkman, L. F., Glass, T., Brissette, I., & Seeman, T. E. (2000). From social integration to health: Durkheim in the new millennium. *Social Science and Medicine,* 51, pp. 843–857.

Berkman, L. F., and Syme, S. L. (1979). Social networks, host resistance, and mortality: A nine-year follow-up study of Alameda county residents. *American Journal of Epidemiology,* 109, pp. 186–204.

Bisson, J. I. (2007). Post-traumatic stress disorder. *Occupational Medicine,* 57, pp. 399-403.

Brandt, M. (2011). "What 9/11 has taught us about PTSD." Retrieved on 9/2/2011 fromhttp://scopeblog.stanford.edu/2011/09/what-911-has-taught-us-about-ptsd/

Brown, R., Madan-Swain, A., & Lambert, R. (2003). Posttraumatic stress symptoms in adolescent survivors of childhood cancer and their mothers. *Journal of Traumatic Stress,* 16, pp. 309-318.

Bultz, B. D., & Carlson, L. E. (2006). Emotional distress: The sixth vital sign - future directions in cancer care. *Psychooncology,* 15, pp. 93–95.

Burgess, C., Cornelius, V., Love, S., Graham, J., Richards, M., & Ramirez, A. (2005). Depression and anxiety in women with early breast cancer: Five year observational cohort study. *British Medical Journal,* 330 (7493), pp. 1–4.

Burns, S. M., & Mahalik, J. R. (2008). Sexual functioning as a moderator of the relationship between masculinity and men's adjustment following treatment for prostate cancer. *American Journal of Men's Health,* 2, pp. 6-16.

Butler, L.D., Field, N.P., Busch, A.L., Seplaki, J.E., Hastings, T.A., & Spiegel, D. (2005). Anticipating loss and other temporal stressors predict traumatic stress symptoms among partners of metastatic/recurrent breast cancer patients. *Psychooncology,* 14(6), pp. 492-502.

Butler, L.D., Koopman, C., Classen, C., & Spieger, D. (1999). Traumatic stress, life events, and emotional support in women with metastatic breast cancer: cancer-related traumatic stress symptoms associated with past and current stressors. *Health Psychology,* 18, pp. 555–560.

Cordova, M. J., Andrykowski, M. A., Kenady, D. E., McGrath, P. C., Sloan, D. A., & Redd, W. H. (1995). Frequency and correlates of posttraumatic-stress-disorder-like symptoms after treatment for breast cancer. *Journal of Consulting and Clinical Psychology,* 63, pp. 981–986.

Cordova, M. J., Cunningham, L. L. C., Carlson, C. R., & Andrykowski, M. A. (2001). Posttraumatic growth following breast cancer: A controlled comparison study. *Health Psychology,* 20, pp. 176–185.

Couper, J., Bloch, S., Love, A., Macvean, M., Duchesne, G. M., & Kissane, D. (2006). Psychosocial adjustment of female partners of men with prostate cancer: A review of the literature. *Psychooncology,* 15, pp. 937–953.

Damber, J. E., & Aus, G. (2008). Prostate cancer. *Lancet,* 371, pp. 1710-1721.

Dixon, D., Kilbourn, K., Cruess, S., Klimas, N., Fletcher, M. A., Ironson, G., Baum, A., Schneiderman, N., & Antoni, M.H. (2001). Social support mediates the relationship between loneliness and Human Herpesvirus-Type 6 (HHV-6) antibody titers in HIV+ gay men following Hurricane Andrew. *Journal of Applied Social Psychology,* 31, pp. 1111–1132.

Elklit, A., Reinholt, N., Nielsen, L., Blum, A., & Lasgaard, M. (2010). Posttraumatic stress disorder among bereaved relatives of cancer patients. *Journal of Psychosocial Oncology,* 28, pp. 399–412.

Epping-Jordan, J.E., Compas, B.E., & Osowiecki, D. (1999). Psychological adjustment in breast cancer: process of emotional distress. *Health Psychology,* 18, pp. 315–326.

Foy, D. W., Resnick, H. S., Sipprelle, R. C., & Carroll, E. M. (1987). Premilitary, military, and postmilitary factors in the development of combat-related posttraumatic stress disorder. *Behavior Therapist,* 10(1), pp. 3 - 9.

Friedman, M.J. (1981). Post-Vietnam syndrome: Recognition and management. *Psychosomatics,* 22, pp. 931-942.

Friedman, M.J. (2006). *Posttraumatic and acute stress disorders*, Compact Clinicals, Kansas City, MO. Gandubert, C., Carriere, I., Escot,C., Soulier, M., Herme`s, A., Boulet, P., Ritchie, K., & Chaudieu, I. (2009). Onset and relapse of psychiatric disorders following early breast cancer: a case–control study. *Psycho-Oncology,* 18, pp. 1029–1037.

Glanz K, Lerman C. 1992. Psychosocial impact of breast cancer: A critical review. *Annals of Behavioral Medicine*, 14, pp. 204–212.

Green, B.L., Rowland, J.H., Krupnick, J.L., Epstein, S.A., Stockton, P., Stern, N.M., Spertus I.L., & Steakley, C. (1998). Prevalence of posttraumatic stress disorder in women with breast cancer. *Psychosomatics,* 39(2), pp. 102-111.

Gurevich, M., Devins, G.M., & Rodin, G.M. (2002). Stress response syndromes and cancer. *Psychosomatics,* 43, pp. 259–281.

Gwede, C. K., Pow-Sang, J., Seigne, J., Heysek, R., Helal, M., Shade K., et al. (2005). Treatment decision-making strategies and influences in patients with localized prostate carcinoma. *Cancer*, 104, pp. 1381-1390.

Hamann, H., Somers, T.J., Smith, A.W., Inslicht, S.S., & Baum, A. (2005). Posttraumatic stress associated with cancer history and *BRCA1/2* genetic testing. *Psychosomatic Medicine,* 67, pp. 766-772.

Jacobsen, P.B., Sadler, I.J., Booth-Jones, M., Soety, E., Weitzner, M.A., & Fields KK. (2002). Predictors ofposttraumatic stress disorder symptomatology following bone marrow transplantation for cancer. *Journal of Consulting and Clinical Psychology, 70,* pp. 235-240.

Jurbergs, N., Long, A., Ticona, L., & Phipps, S. (2009). Symptoms of posttraumatic stress in parents of children with cancer: Are they elevated relative to parents of healthy children? *Journal of Pediatric Psychology,* 34, pp. 4-13.

Kangas, M., Henry, J.L., & Bryant, R.A.(2002). Posttraumatic stress disorder following cancer. *Clinical Psychology Review,* 22, pp. 499–524.

Kangas, M., Henry, J.L., & Bryant, R.A. (2005). Predictors of posttraumatic stress disorder following cancer. *Health Psychology,* 24(6), pp. 579-585.

Kazak, A.E. (1998). Posttraumatic distress in childhood cancer survivors and their parents. *Medical and Pediatric Oncology* (supplement), 1, pp. 60-68.

Kazak, A. (2005). Evidence-based interventions for survivors of childhood cancer and their families. *Journal of Pediatric Psychology,* 30(1), pp. 29–39.

Kazak, A.E., Alderfer, M., Rourke, M.T., Simms, S., Streisand, R., & Grossman, J.R. (2004). Posttraumatic stress disorder (PTSD) and posttraumatic stress symptoms (PTSS) in families of adolescent childhood cancer survivors. *Journal of Pediatric Psychology, 29,* pp. 211-219.

Kazak, A. E., Barakat, L. P., Meeske, K., Christakis, D., Meadows, A. T., & Casey, R. (1997). Posttraumatic stress, family functioning, and social support in survivors of childhood leukemia and their mothers and fathers. *Journal of Consulting and Clinical Psychology,* 65, pp. 120–129.

Kazak, A.E., Boeving, C.A., Alderfer, M.A., Hwang, W.T., & Reilly, A. (2005). Posttraumatic stress symptoms during treatment in parents of children with cancer. *Journal of Clinical Oncology,* 23, pp. 7405-7410.

Kronenwetter, C., Weidner, G., Pettengill, E., Marlin, R., Crutchfield, L., McCormac, P., et al. (2005). A qualitativeanalysis of interviews of men with early stage prostate cancer: The Prostate Cancer Lifestyle Trial. *Cancer Nursing*, 28, pp. 99-107.

Kwekkeboom, K.L., & Seng, J.S. (2002). Recognizing and responding to post-traumatic stress disorder in people with cancer. *Oncology Nursing Forum*, 29(4), pp. 643-650.

Landolt, M.A., Boehler, U., Schwager, C., Schallberger, U., & Nuessli, R. (1998). Post-traumatic stress disorder in paediatric patients and their parents: an exploratory study. *Journal of Paediatrics and Child Health*, 34(6), pp. 539–543.

Lewis, J., Manne, S., DuHamel, K., Johnson, S., Bovbjerg, D., Currie, V., Winkel, G., Redd, W. (2001). Social Support, Intrusive Thoughts, and Quality of Life in Breast Cancer Survivors. *Journal of Behavioral Medicine*, 24, pp. 231-245.

Luecken, L.J., Dausch, B., Gulla, V., Hong, R., & Compas, B. (2004). Alterations in morning cortisol associated with PTSD in women with breast cancer. *Journal of Psychosomatic Research*, 56, pp. 13-15.

Ljungman, G., McGrath, P., Cooper, E., Widger, K., Cecolini, J., Fernandez, C., Wilkins, K. (2003). Psychosocial needs of families with a child with cancer. *Journal of Pediatric Hematology-Oncology*, 25(3), pp. 223–231.

Lutgendorf, S. K., Sood, A. K., Anderson, B., McGinn, S., Maiseri, H., Dao, M., Sorosky, J. I., Geest, K. D., Ritchie, J., & Lubaroff, D. M. (2005). Social support, psychological distress, and natural killer cell activity in ovarian cancer. *Journal of Clinical Oncology*, 23, pp. 7105–7113.

Manne, S., DuHamel, K., & Ostroff, J. (2004). Anxiety, depressive, and posttraumatic stress disorders among mothers of pediatric hematopoietic stem cell transplantation. *Pediatrics* 113, pp. 1700-1708.

McGrath, P. (2001). Findings of the impact of treatment for childhood acute lymphoblastic leukaemia on family relationships. *Child and Family Social Work*, 6, pp. 229–237.

Mehnert, A., & Koch, U. (2007). Prevalence of acute and post-traumatic stress disorder and comorbid mental disorders in breast cancer patients during primary cancer care: a prospective study. *Psychooncology*,16(3), pp. 181-188.

Mezey, G., & Robbins, I. (2001). Usefulness and validity of post-traumatic stress disorder as a psychiatric category. *British Medical Journal*, 323(7312, pp. :561-563.

Montgomery, C., Pocock, M., Titley, K., & Lloyd, K. (2003). Predicting psychological distress in patients with leukaemia and lymphoma. Journal of Psychosomatic Research, 54(4), pp. 289-292.

Mosher, C. E., & Danoff-Burg, S. (2005). Psychosocial impact of parental cancer in adulthood: A conceptual and empirical review. *Clinical Psychology Review*, 25, pp. 365–382.

Mundy, E. A., Blanchard, E. B., Cirenza, E., Gargiulo, J., Maloy, B., & Blanchard, C. G., (2000). Posttraumatic stress disorder in breast cancer patients following autologous bone marrow transplantation or conventional cancer treatments. *Behaviour Research & Therapy*, 38, pp. 1015–1027.

Namiki, S., Saito, S., Tochigi, T., Numata, I., Ioritani, N., & Arai, Y. (2007). Psychological distress in Japanese men with localized prostate cancer. *International Journal of Urology*, 14, pp. 924-929.

Nolbris, M., Enskar, K., & Hellstrom, A. (2007). Experience of siblings of children treated for cancer. *European Journal of Oncology Nursing*, 22, pp. 227–233.

Okamura, M., Yamawaki, S., Akechi, T., Taniguchi, K., & Uchitomi, Y. (2005). Psychiatric disorders following first breast cancer recurrence: prevalence, associated factors and relationship to quality of life. *Japanese Journal of Clinical Oncology* 35(6), pp. 302-309.

Packman, W.L., Fine, J., Chesterman, B., van Zutphen, K., Golan, R., Amylon, M. (2004).Camp Okizu: Preliminary investigation of a psychological intervention for siblings of pediatric cancer patients. *Children's Health Care,* 33, pp. 201-215.

Palmer, S.C., Kagee, A., Coyne, J.C., & DeMichele, A. (2004). Experience of trauma, distress, and posttraumatic stress disorder among breast cancer patients. *Psychosomatic Medicine,* 66(2), pp. 258-264.

Patterson, J.M., Holm, K.E., & Gurney, J.G. (2004). The impact of childhood cancer on the family: A qualitative analysis of strains, resources, and coping behaviors. *Psychooncology,* 13, pp. 390-407.

Pelcovitz, D., Libov, B.G., Mandel, F., Kaplan, S., Weinblatt, M., & Septimus, A. (1998). Posttraumatic stress disorder and family functioning in adolescent cancer. *Journal of Traumatic Stress,* 11(2), pp. 205-221.

Penson, D. F. (2007). Quality of life after therapy for localized prostate cancer. *Cancer Journal,* 13, pp. 318-326.

Perez, M. A., Skinner, E. C., & Meyerowitz, B. E. (2002). Sexuality and intimacy following radical prostatectomy: Patientand partner perspectives. *Health Psychology,* 21, pp. 288-293.

Poder, U., Ljungman, G., & von Essen, L. (2008). Posttraumatic stress disorder among parents of children on cancer treatment: a longitudinal study. *Psychooncology,* 17, pp. 430-437.

Ries, L., Melbert, D., Krapcho, M., Stinchcomb, D.G., Howlader, N., & Horner, M.J. (Eds.) *SEER cancer statistics review, 1975-2005.* Retrieved 8/1/08 from http://seer.cancer.gov/csr/1975_2005/

Sanda, M. G., Dunn, R. L., Michalski, J., Sandler, H. M., Northouse, L., Hembroff, L., et al. (2008). Quality of life andsatisfaction with outcome among prostate-cancer survivors. *New England Journal of Medicine,* 358, pp. 1250-1261.

Schuster, M.A. Stein, B.D., Jaycox, L.H., Collins, R.L., Marshall, G.N., Elliott, M.N., Zhou, A.J., Kanouse, D.E., Morrison, J.L., & Berry, S.H. (2001). A national survey of stress reactions after the September 11, 2001, terrorist attacks. *New England Journal of Medicine,* 345(20), pp. 1507-1512.

Segman, R.H., Shefi, N., Goltser-Dubner, T., Friedman, N., Kaminski, N., & Shalev, A.Y. (2005). Peripheral blood mononuclear cell gene expression profiles identify emergent post-traumatic stress disorder among trauma survivors. *Molecular Psychiatry,*10, pp. 500-513.

Shapiro, S. L., Lopez, A. M., Schwartz, G. E., Bootzin, R., Figueredo, A. J., Braden, C. J., et al. (2001). Quality of life and breast cancer: Relationship to psychosocial variables. *Journal of Clinical Psychology,* 57(4), pp. 501-519.

Shay, J. (1994). *Achilles in Vietnam: Combat Trauma and the Undoing of Character,* Maxwell Macmillan, New York.

Shelby, R., Golden-Kreutz, D., & Andersen, B. (2008). PTSD diagnoses, subsyndromal symptoms, and comorbidities contribute to impairments for breast cancer survivors. *Journal of Traumatic Stress,* 21, pp. 165-172.

Steginga, S. K., Ferguson, M., Clutton, S., Gardiner, R. A., & Nicol, D. (2008). Early decision and psychosocial support intervention for men with localised prostate cancer: An integrated approach. *Supportive Care in Cancer*, 16, pp. 821-829.

Steginga, S. K., Occhipinti, S., Gardiner, R. A., Yaxley, J., & Heathcote, P. (2004). Prospective study of men's psychological and decision-related adjustment after treatment for localized prostate cancer. *Urology*, 63, pp. 751-756.

Stoppelbein, L., & Greening, L. (2007). The risk of posttraumatic stress disorder in mothers of children diagnosed with pediatric cancer and type I diabetes. *Journal of Pediatric Psychology* 32, pp. 223-229.

Tuinmann, M. A., Fleer, J., Hoekstra, H. J., Sleijfer, D. T., & Hoekstra-Weebers, J. (2004). Quality of life and stress response symptoms in long-term and recent spouses of testicular cancer survivors. *European Journal of Cancer*, 40, pp. 1696–1703.

Uchino, B. N. (2004). *Social support and physical health: Understanding the health consequences of relationships,* Yale University Press, New Haven, CT.

Uchino, B.N., Cacioppo, J.T., & Kiecolt-Glaser, K.G. (1996). The relationships between social support and physiological processes: A review with emphasis on underlying mechanisms and implications for health. *Psychological Bulletin,* 119, pp. 488–531.

Uddin, M., Aiello, A.E., Wildman, D.E., Koenen, K.C., Pawelec, G., de Los Santos, R., Goldmann, E., & Galea, S. (2010). Epigenetic and immune function profiles associated with posttraumatic stress disorder. *Proceedings of the National Academy of Sciences*, 107, pp. 9470-9475.

Vos, P. J., Garssen, B., Visser, A. P., Duivenvoorden, H. J., & de Haes, H. C. J. M. (2004). Early stage breast cancer: explaining level of psychosocial adjustment using structural equation modeling. *Journal of Behavioral Medicine,* 27(6), pp. 557–580.

Widows, M.R., Jacobsen, P.B., & Fields, K.K. (2000). Relation of psychological vulnerability factors to posttraumatic stress disorder symptomatology in bone marrow transplant recipients. *Psychosomatic Medicine,* 62, pp. 873-882.

Wong, C.M. (2002) Post-traumatic stress disorder: Advances in psychoneuroimmunology. *Psychiatric Clinics of North America,* 25, pp. 369–383.

Woodgate, R.L. (2006). Siblings' experiences with childhood cancer. *Cancer Nursing* 29, pp. 406–414.

Yehuda, R., Cai, G., Golier, J.A., Sarapas, C., Galea, S., Ising, M., Rein, T., Schmeidler, J., Müller-Myhsok, B., Holsboer, F., & Buxbaum, J.D. (2009). Gene expression patterns associated with posttraumatic stress disorder following exposure to the World Trade Center attacks. *Biological Psychiatry,* 66, pp. 708–711.

Yehuda, R., & LeDoux, J. (2007) Response variation following trauma: A translational neuroscience approach to understanding PTSD. *Neuron,* 56, pp. 19–32.

Zieker, J., Zieker, D., Jatzko, A., Dietzsch, J., Nieselt, K., Schmitt, A., Bertsch, T., Fassbender, K., Spanagel, R., Northoff, H., & Gebicke-Haerter, P.J. (2007) Differential gene expression in peripheral blood of patients suffering from post-traumatic stress disorder. *Molecular Psychiatry,* 12, pp. 116–118.

Zisook, S., Chentsova-Dutton, Y., & Shuchter, S.R. (1998). PTSD following bereavement. *Annals of Clinical Psychiatry,* 10, pp. 157-63.

12

Posttraumatic Stress Disorder after Stroke: A Review of Quantitative Studies

Paul Norman[1], Meaghan L. O'Donnell[2], Mark Creamer[2] and Jane Barton[3]
[1]*Department of Psychology, University of Sheffield,*
[2]*Australian Centre for Posttraumatic Mental Health, University of Melbourne,*
[3]*Sheffield Health and Social Care NHS Trust,*
[1,3]*UK*
[2]*Australia*

1. Introduction

Stroke is a sudden and devastating illness that occurs when the blood supply to the brain is cut off due to a blood clot or when blood vessels supplying the brain burst, thereby damaging or destroying brain cells. Stroke is the third most common cause of death after heart disease and all cancers in both the UK and the US. In the UK approximately 100,000 people have a stroke each year, with stroke causing over 50,000 deaths. Stroke predominantly affects older people, with almost 80% of first-time strokes occurring in people aged 65 years or older. Stroke is a leading cause of severe, long-term disability. Stroke survivors may experience a range of ongoing problems including weakness or paralysis, problems with balance and coordination, speech and language impairments (e.g., aphasia), cognitive and psychological problems, and emotional lability. As well as the direct care costs, which have been estimated to be £2.8 billion per annum, stroke accounts for £1.8 billion per annum in lost productivity and disability and £2.4 billion per annum in informal care costs in the UK. In the US, the direct and indirect costs of stroke for 2010 have been estimated to be $73.3 billion (American Stroke Association, 2011; Stroke Association, 2011).

Elevated levels of psychological distress have been documented after stroke, although research to date has focused almost exclusively on depression and general anxiety symptoms (e.g., Fure et al., 2006; Hackett et al., 2005; Leppavuoir et al., 2003). However, stroke has many of the characteristics of events likely to trigger post-traumatic stress disorder (PTSD) symptoms in that it is unexpected, uncontrollable and potentially life-threatening (Field et al., 2008). Stroke is a "frightening experience" with the symptoms (e.g., weakness or numbness down one side of the body or face, problems with balance and coordination, problems with communication, confusion) appearing suddenly and without warning (Stroke Association, 2011). Thus, in addition to coping with chronic stressors that may arise from ongoing disability, stroke survivors also have to come to terms with the sudden, unexpected and life-threatening nature of the stroke event itself.

The Diagnostic and Statistical Manual of Mental Disorders (DSM-IV) (American Psychiatric Association, 1994, p. 424), defines PTSD as "the development of characteristic symptoms following exposure to an extreme traumatic stressor involving direct personal experience of an event that involves actual or threatened death or serious injury, or other threat to one's physical integrity" (Criterion A1) and is associated with feelings of intense fear, helplessness or horror (Criterion A2). The symptoms includ "persistent reexperiencing of the traumatic event (Criterion B), persistent avoidance of stimuli associated with the trauma and numbing of general responsiveness (Criterion C), and persistent symptoms of increased arousal (Criterion D)". In addition, these symptoms must be present for at least one month (Criterion E) and have a significant negative impact on social, occupational or other areas of functioning (Criterion F). Research on PTSD has traditionally focused on traumas such as war, physical and sexual assaults, and road traffic accidents (Shalev et al., 1993). However, there has been growing recognition that PTSD symptoms may occur after a range of medical events (Tedstone & Tarrier, 2003), including cancer (e.g., Kangas et al., 2005), myocardial infarction (MI) (e.g., Kutz et al., 1994), and subarachnoid haemorrhage (e.g., Berry, 1998). Accordingly, life-threatening illnesses were added as an example of a traumatic event that may lead to the development of PTSD in DSM-IV (APA, 1994).

The experience of PTSD after stroke may have important implications for recovery. For example, PTSD has been related to poorer physical health in the general population (Spitzer et al., 2009) as well as to non-adherence to medication and adverse clinical outcomes in MI patients (Shemesh et al., 2001), worse functional recovery in patients with severe traumatic brain injury (Bryant et al., 2001), and higher levels of disability following hospitalization for physical injuries (e.g., resulting from road traffic accidents) (O'Donnell et al., 2009). PTSD is also likely to have a negative impact on stroke rehabilitation (Williams, 1997). First, the experience of intrusions may place high demands on the already limited cognitive resources of many stroke survivors, which may further restrict their ability to fully process, and come to terms with, the trauma experience. Second, survivors with PTSD are likely to try to avoid reminders of the stroke, which may hinder attempts to integrate them back into the community. Third, survivors with PTSD are likely to engage in catastrophic thinking and to have excessively negative perceptions of possible future harm which may further impede rehabilitation efforts.

Previous reviews have focused on the prevalence and correlates of PTSD following a range of life-threatening physical illnesses (e.g., Pedersen, 2001; Spindler & Pedersen, 2005; Tedstone & Tarrier, 2003). However, these reviews have either been very general in their scope (e.g., Tedstone & Tarrier, 2003) or have focused on specific medical conditions other than stroke (e.g., Spindler & Pedersen, 2005). In relation to the psychological consequences of stroke, reviews to date have only focused on the prevalence (Hackett et al., 2005) and the correlates (Hackett & Anderson, 2005) of depression after stroke. The main aims of the current review were to (i) assess the prevalence of PTSD after stroke, (ii) identify the main correlates of PTSD after stroke, (iii) highlight a range of methodological issues in research on PTSD after stroke, and (iv) make recommendations for future research.

2. Methods

The following electronic databases were searched in order to identify relevant studies to include in the review: Web of Knowledge, PsycINFO, and Medline. The searches were restricted to studies published between 1994 (the year life-threatening illnesses were included as an example of a traumatic event in the DSM) and May 2011. The following search terms were used: (i) stroke and cerebrovascular accident, and (ii) post-traumatic stress disorder, posttraumatic stress disorder, PTSD, Impact of Events Scale, IES, Penn, Post

Traumatic Stress Disorder Checklist, PCL, Posttraumatic Diagnostic Scale, PDS, Clinician-Administered PTSD Scale, CAPS, Structured Clinical Interview for DSM-IV, and SCID. Combinations of these two sets of search terms were searched using the Boolean operator "AND". In addition, the reference lists and citation histories of relevant articles were also examined in order to identify further studies to be included in the review. Studies on adult stroke survivors, with a self-report measure of PTSD symptomatology or a clinical interview to diagnose PTSD, that were published in English in peer-reviewed journals were included in the review. Single case studies, qualitative studies, papers without primary data (e.g., editorials), conference abstracts, dissertations, and studies on childhood stroke or subarachnoid haemorrhage were excluded.

The searches identified 411 articles. After applying the inclusion and exclusion criteria detailed above, 10 articles reporting 9 studies were included in the review; one study was reported in two articles (Sagen et al., 2009, 2010). The following data were extracted from each study (see Table 1): date of publication, country of origin, study design, recruitment site, number of patients screened and excluded, main exclusion criteria, response rate, first or recurrent stroke, sample size, age, gender, stroke location, time since stroke, assessment of PTSD, prevalence of PTSD, and significant correlates of PTSD symptom severity.

3. Results and discussion

3.1 Prevalence of PTSD after stroke

Seven studies reported the prevalence of PTSD after stroke. There was considerable variability in the estimated prevalence of PTSD after stroke, which ranged from 3% (Sagen et al., 2010) to 31% (Bruggimann et al., 2006). This variation is likely to be due to differences in assessment methods, time since stroke, and recruitment procedures (e.g., exclusion criteria, response rates). Nonetheless, despite the heterogeneity in study designs and reported prevalence rates, research to date indicates that stroke survivors are at risk of developing PTSD in line with work on other life-threatening illnesses (Tedstone & Tarrier, 2003). For example, Spindler and Pedersen (2005) reported that the estimated prevalence rate for PTSD after heart disease ranged from 0% to 38%. The estimated prevalence of PTSD after stroke is higher than that found in large-scale community studies which have reported PTSD prevalence rates of less than 1% among older adults (Creamer & Parslow, 2008; Maercker et al., 2008; van Zelst et al., 2003).

3.2 Correlates of PTSD after stroke

Eight studies examined associations between the severity of PTSD symptoms and potential risk factors. In addition, one study also compared stroke survivors with and without PTSD (Sembi et al., 1998), reporting that those with PTSD had higher levels of neuroticism, anxiety and depression and lower levels of psychological well-being. However, the PTSD group was very small (n = 6). A range of significant correlates of PTSD symptom severity has been reported, including demographic variables such as age (Sampson et al., 2003; Sharkey, 2007) and gender (Bruggimann et al., 2006), stroke details including the number of previous strokes (Merriman et al., 2007), time since stroke (Merriman et al., 2007) and post-stroke disability (Wang et al., 2011), personality variables such as neuroticism (Sembi et al., 1998), negative affect (Merriman et al., 2007), emotionalism (Eccles et al., 1999) and alexthymia

(Wang et al., 2011), psychological distress including psychiatric morbidity (Wang et al., 2011) as well as anxiety and depression (Bruggimann et al., 2006; Field et al., 2008; Merriman et al., 2007; Sembi et al., 1998), and cognitive appraisals about the stroke (Bruggimann et al., 2006; Field et al., 2008; Merriman et al., 2007; Sharkey, 2007). A number of additional risk factors have been found to have non-significant associations with PTSD symptom severity, including neurological impairment (Bruggimann et al., 2006), lesion site/hemisphere (Bruggimann et al., 2006, Merriman et al., 2007), memory of stroke (Bruggimann et al., 2006), dissociation (Merriman et al., 2007), and consciousness (Field et al., 2008). The strongest and most consistent correlates of PTSD symptom severity have been anxiety, depression and negative cognitive appraisals about the stroke. Considering each correlate in turn, the significant correlations between generalized anxiety and PTSD symptom severity are not unexpected given that PTSD is an anxiety disorder and that several symptoms overlap in the diagnostic criteria for the two disorders. The significant correlations with depression suggest that there might be high levels of psychological co-morbidity following stroke, with many stroke survivors experiencing both mood and anxiety disorders. For example, co-morbidity between anxiety and depression after stroke has been reported to be in the range of 11-18% (Astrom, 1996; Barker-Collo, 2007; Leppavuori et al., 2003; Sagen et al., 2009). In relation to co-morbidity with PTSD, Wang et al. (2011) reported that 93% of their sample of stroke survivors with PTSD scored above the GHQ-28 cut-off for psychiatric morbidity at one month post-stroke, although this figure fell to 50% at three months post-stroke. More generally, large community surveys have revealed that 80-85% of people diagnosed with PTSD also meet the diagnostic criteria for at least one other psychiatric condition (Brady, Killeen, Brewerton, & Lucerini, 2000; Creamer, Burgess, & McFarlane, 2001). While these figures highlight the breadth of psychopathology that may develop following trauma exposure, they may also reflect the lack of specificity of the current PTSD diagnostic criteria (Spitzer, First, & Wakefield, 2007). The significant correlations between negative cognitive appraisals about the stroke and PTSD symptom severity are consistent with psychological models of PTSD (e.g., Ehlers & Clark, 2000; Foa & Rothbaum, 1998) that emphasize that the way in which the trauma is interpreted and processed is important in the development and persistence of PTSD. However, closer inspection of the items used in some of the studies reveals that they may be confounded with Criterion A2 of the DSM-IV (APA, 1994) which states that an event must evoke feelings of intense fear, helplessness or horror to qualify as a traumatic event. Thus, items assessing feelings of hopelessness and helplessness (Bruggimann et al., 2006), fear (Merrimann et al., 2007) and horror (Sharkey, 2007) have been related to the severity of PTSD symptoms after stroke. In contrast, Field et al. (2008) focused on negative cognitions about the self and the world, that do not exhibit this overlap with Criterion A2.

3.3 Methodological issues

It is difficult to draw strong conclusions regarding the prevalence and correlates of PTSD after stroke because the majority of studies suffer from a number of important methodological limitations. These include a reliance on self-report measures of PTSD, small sample sizes, a preponderance of cross-sectional designs, a lack of representative samples, the assessment of a limited set of potential risk factors and a failure to fully consider the impact of specific features of stroke, medical events and older adults on PTSD symptomatology.

Authors (Date), Country	Study Design, Recruitment and Main Exclusion Criteria	Sample	Assessment of PTSD	Prevalence of PTSD	Significant Correlates of PTSD Symptom Severity
Sembi et al. (1998) UK	Cross-sectional Recruitment: outpatient stroke prevention clinic, outpatient elderly day care hospital, and inpatient stroke rehabilitation ward *N* patients screened and excluded not reported Response rate after exclusions = 77% Exclusions: dysphasia	First-ever stroke or transient ischemic attack *N* = 61 Age *M* = 66 years % male not reported Stroke details not reported	IES, cut-off score = 30 Penn, cut-off score = 35 CAPS Time since stroke not reported	21% (IES) 7% (Penn) 10% (CAPS) - only those who scored above cut-offs on the IES or Penn were interviewed using the CAPS	IES: physical disability, neuroticism, GHQ-28, anxiety, depression. Penn: neuroticism, GHQ-28, anxiety, depression.
Eccles et al. (1999) UK	Cross-sectional Recruitment: inpatient hospital wards *N* patients screened = 177 *N* patients excluded = 112 (63%) Response rate after exclusions not reported Exclusions: poor physical health, cognitive impairment, communication problems	Stroke *N* = 65 Age *M* = 72 years % male = 31% Stroke details not reported	IES Administered within 1 month of stroke	Not reported	Emotionalism.
Sampson et al. (2003) UK	Cross-sectional Recruitment: inpatient stroke units, older adult medical wards, standard medical wards *N* patients screened = 150 *N* patients excluded = 73 (49%) Response rate after exclusions = 76% Exclusions: cognitive impairment, dysphasia, physically unwell, hearing impairment	Stroke *N* = 54 Age *Median* = 72.5 years % male not reported Stroke details: right hemisphere stroke *n* = 30, left hemisphere stroke *n* = 22	PCL-S, cut-off score = 44 Median time in hospital = 43.5 days	6%	Age (negative *r*).

Table 1. Summary of Studies Examining the Prevalence and Correlates of PTSD after Stroke

Authors (Date), Country	Study Design, Recruitment and Main Exclusion Criteria	Sample	Assessment of PTSD	Prevalence of PTSD	Significant Correlates of PTSD Symptom Severity
Bruggimann et al. (2006) Switzerland	Cross-sectional Recruitment: postal questionnaire N patients screened = 142 N patients excluded = 37 (26%) Response rate after exclusions = 52% Exclusions: persistent moderate or severe neurologic deficit, major psychiatric illness prior to stroke, neurologic comorbidity	First-ever non-severe stroke $N = 49$ Age $M = 51$ years % male = 67% Stroke details: frontal $n = 11$, temporal $n = 6$, parietal $n = 13$, occipital $n = 6$, basal ganglia $n = 13$, cerebellum $n = 16$	IES, cut-off score = 30 1 year post-stroke	31%	Gender (female), education (negative r), subjective trauma appraisals, depression, anxiety.
Merriman et al. (2007) UK	Cross-sectional Recruitment: inpatient stroke wards N patients screened = 108 N patients excluded = 50 (46%) Response rate after exclusions = 95% Recruitment: postal questionnaire to discharged patients N patients screened and excluded not reported Response rate after exclusions = 52% Exclusions: dysphasia, acute medical problems	Stroke $N = 102$ Age $M = 74$ years % male = 56% Stroke details not reported	PDS Time since stroke $M = 123.01$ days	31% (fulfilling Criteria B, C and D)	Time since stroke (negative r), number of previous strokes, anxiety, depression, negative affect (trait measure), trauma appraisals.
Sharkey (2007) UK	Cross-sectional Recruitment: not reported in sufficient detail Exclusions: severe communication difficulties and cognitive impairments	First-ever stroke $N = 34$ Age $M = 73$ years % male = 59% Stroke location not reported	IES Penn CAPS Time since stroke $M = 62$ weeks	3% received a PTSD diagnosis using the CAPS and cut-off scores on the IES and Penn	IES: feeling horrified immediately after stroke, stroke-specific quality of life (negative r). Penn: age (negative r), fear of another stroke, stroke-specific quality of life (negative r).

Table 1. (Continued)

Table 1. (Continued)

Authors (Date), Country	Study Design, Recruitment and Main Exclusion Criteria	Sample	Assessment of PTSD	Prevalence of PTSD	Significant Correlates of PTSD Symptom Severity
Field et al. (2008) UK	Prospective Recruitment: inpatient stroke wards, followed-up by postal questionnaire *N* patients screened not reported Response rate after exclusions = 90% Response rate to time 2 questionnaire = 86% Exclusions: cognitive impairment (e.g., aphasia), acute medical problems	Stroke $N = 81$ at time 1 $N = 70$ at time 2 Age $M = 71$ years % male = 53% Stroke details not reported	PDS Time since stroke at time 1 $M = 19.94$ days Time 2 follow-up 3 months later	Not reported	Cross-sectional correlations at time 1: age (negative *r*), anxiety, depression, negative appraisals about the self, negative appraisals about the world. Prospective correlations (time 1 to time 2): anxiety, depression, negative appraisals about the self, negative appraisals about the world, PDS.
Sagen et al. (2009, 2010) Norway	Cross-sectional Recruitment: inpatient stroke unit *N* patients screened not reported *N* patients excluded ≥ 81 Response rate after exclusions at recruitment = 84% Response rate after exclusions at 4 months = 69% Exclusions: aphasia, cognitive impairment , transient ischemic attack	Stroke $N = 104$ Age $M = 65$ years % male = 59% Stroke details: cerebral infarction $n = 99$, cerebral haemorrhage $n = 5$	SCID 4 months post-stroke	3%	Not assessed

Authors (Date), Country	Study Design, Recruitment and Main Exclusion Criteria	Sample	Assessment of PTSD	Prevalence of PTSD	Significant Correlates of PTSD Symptom Severity
Wang et al. (2011) UK	Longitudinal Recruitment: stroke rehabilitation unit *N* patients screened = 191 *N* patients excluded = 91 (48%) Response rate after exclusions at recruitment = 90% Response rate to time 2 questionnaire = 87% Exclusions: dysphasia, language problems, history of mental health problems	Stroke *N* = 90 at time 1 *N* = 78 at time 2 Age *M* = 75 years % male = 48% Stroke details: cerebral infarction *n* = 72, cerebral haemorrhage *n* = 11	PDS Time since stroke at time 1 *M* = 47 days Time 2 follow-up 2 months later	30.0% at time 1 23.1% at time 2	Cross-sectional correlations at time 1: GHQ, physical disability, alexithymia (identifying and describing feelings). Prospective correlations (time 1 to time 2): GHQ, time since stroke, physical disability, alexithymia (identifying and describing feelings), PDS.

Table 1. (Continued)

3.3.1 Assessment of PTSD

Most studies have used self-report measures to assess the prevalence of PTSD, including the Impact of Events Scale (IES; Horowitz et al., 1979), the Penn Inventory of PTSD (Penn; Hammarberg, 1992), the Post Traumatic Stress Disorder Checklist (PCL-S; Weathers et al., 1993) and the Posttraumatic Diagnostic Scale (PDS; Foa et al., 1997). A major limitation of most of these measures is that they only assess the severity of PTSD symptoms (Criteria B, C and D) and fail to consider the length of time symptoms have been present (Criterion E), the impact of symptoms on daily functioning (Criterion F), or powerful emotional reactions (Criterion A2). Indeed, one of the most frequently used measures, the IES (Horowitz et al., 1979), only assesses two symptom clusters: intrusive thoughts and avoidance. Such measures are therefore likely to over-estimate the prevalence of PTSD as they do not assess all the DSM-IV criteria (A-F) for a PTSD diagnosis (APA, 1994).

Clinical diagnostic interviews were used in three studies to provide a diagnosis of PTSD. The Clinician-Administered PTSD Scale (CAPS; Blake et al., 1992) was used in two studies in conjunction with self-report measures. Thus, Sembi et al. (1998) first screened stroke survivors using the IES and Penn self-report scales; those scoring above cut-off points on both measures were then interviewed using the CAPS. Similarly, Sharkey (2007) used a "multi-modal" assessment of PTSD using the CAPS in conjunction with the IES and Penn. The Structured Clinical Interview for DSM-IV (SCID; First et al., 1995) was used in one study (Sagen et al., 2009, 2010) to provide a PTSD diagnosis. There was some evidence that studies employing diagnostic interviews to assess PTSD reported lower prevalence estimates than those using self-report PTSD measures. The prevalence rates reported in studies that included a clinical diagnostic interview ranged from 3% (Sagen et al., 2010) to 10% (Sembi et al., 1998), whereas the frequencies reported in studies only employing self-report measures ranged from 6% (Sampson et al., 2003) to 31% (Bruggiman et al., 2006).

3.3.2 Sample sizes

The samples sizes for studies included in the review were small, ranging from 34 (Sharkley, 2007) to 104 (Sagen et al., 2009, 2010). This has important consequences for research on the prevalence of PTSD after stroke. Small sample sizes are likely to lead to large confidence intervals and unreliable prevalence estimates. In addition, they increase the probability that outliers may have a disproportionate impact on prevalence rates (O'Donnell et al., 2003). For example, with a sample size of 50, each additional PTSD diagnosis increases the estimated prevalence rate by 2%. This issue is likely to be exacerbated when cut-off scores on self-report measures, rather than clinical interviews, are used to provide a PTSD diagnosis. Small sample sizes also impact on research on the correlates of PTSD after stroke. In particular, they are likely to lead to many analyses being under-powered, thereby increasing the probability of Type II errors (Cohen, 1992). In addition, outliers may have a disproportionate influence on the strength of correlations, leading to potentially spurious findings.

3.3.3 Study design

One of the most limiting aspects of research on the prevalence of PTSD after stroke is the lack of longitudinal studies. To date, all bar one study (Wang et al., 2011) have employed cross-sectional designs. Moreover, the time since stroke at which PTSD was assessed varied considerably between studies. For example, the average time since stroke ranged from 43.5 days (Sampson et al., 2003) to 62 weeks (Sharkey, 2007). Moreover, there was also considerable variability within studies. For example, the standard deviation for time since stroke reported by Sharkey (2007) was 26.3 weeks, and the time since stroke reported by Merriman et al. (2007) ranged from 27 to 365 days. Only three studies assessed PTSD at fixed time points post-stroke. Sagen et al. (2009, 2010) assessed PTSD at four months post-stroke, Bruggimann et al. (2006) assessed PTSD at one year post-stroke, and Wang et al. (2011) assessed PTSD at approximately one and three months post-stroke, although the mean time since stroke at time 1 was 47.1 days with a standard deviation of 26.0 days. It is therefore difficult to compare reported prevalence rates or to provide an accurate point prevalence of PTSD. As a result, there are no accurate data on the natural course of post-stroke PTSD over time. The preponderance of cross-sectional designs also limits conclusions regarding the direction of relationships between risk factors and PTSD after stroke. For example, the experience of PTSD may result in excessively negative appraisals about the stroke, rather than negative appraisals determining PTSD as proposed in cognitive models of PTSD (Brewin & Holmes, 2003). As result, prospective designs in which potential risk factors are assessed shortly after stroke and related to the development of PTSD at a later time point are essential. To date, only two studies have employed such a design (Field et al., 2008; Wang et al., 2011). Moreover, only three studies (Field et al., 2008; Merrimann et al., 2007; Wang et al., 2011) have employed multivariate analyses to examine associations between potential risk factors and PTSD symptom severity; all other studies only examined bivariate associations.

3.3.4 Sample representativeness

The number of stroke survivors excluded from participating in the studies was either not reported (Field et al., 2008; Sembi et al., 1998; Sharkey, 2007) or was considerable, ranging from 26% (Bruggimann et al., 2006) to 63% (Eccles et al., 1999). The main exclusion criteria included cognitive impairment, communication difficulties (e.g., aphasia), and poor physical health. Stroke survivors were usually recruited from inpatient stroke wards (Field et al., 2008; Merriman et al., 2007; Sagen et al., 2009, 2010; Sembi et al., 1998), stroke rehabilitation units (Wang et al., 2011) or general hospital wards (Eccles et al., 1999; Sampson et al., 2003). A couple studies also recruited participants using postal questionnaires (Bruggimann et al., 2006; Merriman et al., 2007). Response rates (after exclusions) among studies recruiting from inpatient wards were generally high, ranging from 76% (Sampson et al., 2003) to 95% (Merriman et al., 2007). Studies employing postal questionnaires obtained lower response rates (52%) (Bruggimann et al., 2006; Merriman et al., 2007).

High exclusion rates raise questions regarding the representativeness of the samples recruited as stroke survivors who experienced more severe strokes are likely to have been excluded. None of the studies reported whether the samples were representative of the populations from which they were drawn. As a result, generalisability is limited. Moreover, it is possible that these exclusion criteria may themselves be risk factors for the development of PTSD. If so, this would imply that most studies have underestimated the prevalence of PTSD after stroke.

3.3.5 Assessment of risk factors

Studies on the correlates of PTSD after stroke have focused on a limited set of variables and have not assessed variables that have been identified as having strong associations with PTSD symptomatology in response to other traumas (Brewin et al., 2000; Ozer et al., 2003). For example, Ozer et al. (2003) conducted a meta-analysis of seven potential predictors of PTSD diagnosis or symptoms and found that more distal characteristics related to the individual or their life history (e.g., prior trauma, family history of psychopathology) had smaller correlations with PTSD symptomatology than more proximal psychological factors (e.g., perceived life threat, dissociation). Such findings are in line with current psychological models of PTSD that emphasize the importance of appraisal and memory processes in the development of PTSD (Brewin & Holmes, 2003). Few studies on PTSD after stroke have drawn on such models to guide the selection of independent variables. Ehlers and Clark's (2000) cognitive model, which according to Brewin and Holmes (2003) provides the most detailed account of PTSD, proposes that PTSD is likely to develop and persist when the trauma and/or its sequelae is processed in such a way that leads to a sense of serious current threat, as a result of (i) making excessively negative appraisals and (ii) disturbances in autobiographical memory.

Considering negative appraisals, only one study has tested the Ehlers and Clark (2000) model in relation to stroke. Field et al. (2008) reported that negative cognitions about the self (e.g., "I am inadequate") and about the world (e.g., "The world is a dangerous place"), assessed shortly after the stroke (M = 20 days), were significantly correlated with the severity of PTSD symptoms both cross-sectionally and prospectively three months later. However, the prospective correlations became non-significant after controlling for the effect of initial PTSD symptoms. Considering disturbances in autobiographical memory, Ehlers and Clark (2000) propose that the overwhelming experience of a traumatic event may disrupt peritraumatic cognitive processing resulting in trauma memories that are disorganised and poorly elaborated. This, in turn, may make trauma memories more vulnerable to triggering by matching cues, thereby increasing the frequency of reexperiencing symptoms. Three aspects of cognitive processing during the trauma have been related to poorly elaborated/organised trauma memories and subsequent PTSD (Halligan et al., 2003); namely, (i) engaging in surface level, or data-driven processing (e.g., "It was just like a dream of unconnected impressions following each other"), (ii) a lack self-referential processing (e.g., "I felt as if it was happening to someone else"), and (iii) dissociation (e.g., reduced awareness of the self, time and/or environment at the time of the trauma). Halligan et al. (2003) reported that measures of these memory processes, assessed within three months after assault, were predictive of the severity of PTSD symptoms at three and six months follow-up. Only two studies have examined memory variables after stroke. Bruggiman et al. (2006) reported no differences in the symptom severity scores of survivors with fragmented versus complete memories of their stroke, whereas Merriman et al. (2007) reported that peritraumatic dissociation was related to the number, but not the severity, of PTSD symptoms after stroke.

A related strand of work has noted that people with PTSD have difficulty recalling specific autobiographical memories (e.g., "When I watched the football on the television last Sunday") in response to cue words (e.g., "happy"). Instead, they recall abstract or more general memories that cover several different events or time points (e.g., "When I watch football on the television"). This tendency has been termed overgeneral memory bias (Williams & Broadbent, 1986). An inability to retrieve specific autobiographical memories may prevent the trauma memory from being integrated with other autobiographical memories and with the person's schemas about the self and the world (Kleim & Ehlers, 2008), thereby contributing to the development of PTSD. A number of studies have reported that PTSD is associated with overgeneral memory bias (Bryant et al., 2008; Schönfeld & Ehlers, 2006; Schönfeld et al., 2007). Dalgleish et al. (2008) have presented evidence to suggest that people with PTSD may avoid retrieving specific personal information as a means of affect regulation, in support of a functional avoidance account of overgeneral memory bias (Williams et al., 2007). Thus, intentional memory searches may be stopped prematurely at an abstract level in order to avoid retrieving potentially distressing material related to the trauma. To date, no studies have examined the relationship between overgeneral memory bias and PTSD after stroke.

3.3.6 Stroke details

Few studies have reported details of the type (e.g., ischemic vs. haemorrhagic), hemisphere (left vs. right), site (e.g., frontal, temporal, etc.) or severity of the stroke. Sampson et al. (2003) reported details of the hemisphere of stroke, Sagen et al. (2009, 2010) reported whether the stroke was a cerebral infarction or a cerebral haemorrhage, whereas Bruggiman et al. (2006) and Wang et al. (2011) provided detailed information regarding the location of the stroke and neurological deficit. Moreover, these aspects of the stroke have rarely been related to PTSD symptom severity. One exception is Bruggiman et al. (2006) who reported that lesion site and neurological deficit were unrelated to PTSD symptomatology, although it should be noted that the sample included only non-severe strokes.

Psychobiological models of PTSD (Charney et al., 1993) and neuroimaging evidence (Lanius et al., 2006) suggest that the development of PTSD is related to impaired functioning of the medial prefrontal cortex which limits regulation of the amygdala. Neural networks involving these areas have been implicated in fear processing. As a result, damage to such networks may be uniquely related to anxiety disorders (Rauch, 2003). Bryant et al. (2010) reported that traumatic injury survivors who also sustained a mild traumatic brain injury (which tends to be associated with damage to frontal regions of the brain) were more likely to develop anxiety disorders, including PTSD, but not depressive disorders. Other research on stroke has noted that left hemispheric, subcortical and large lesions are associated with verbal memory deficits (Godefroy et al., 2009; Schoten et al., 2009) which, more generally, have been related to PTSD (Brewin et al., 2007; Johnsen & Asbjørnsen, 2008). Such deficits may lead to poorer processing of trauma memories thereby contributing to the development of PTSD.

Many stroke survivors are asleep or unconscious during their stroke. Whether or not PTSD symptoms can develop under such circumstances has been the subject of much debate (e.g., Harvey et al., 2003; Klein et al., 2003). For example, it has been argued that individuals who are amnesic of the trauma event, by definition, cannot meet Criterion A2 of DSM-IV for PTSD (i.e., experience feelings of intense fear, helplessness or horror in response to the trauma event) (O'Donnell et al., 2003). Studies have produced conflicting results on this issue (e.g., Bryant et al., 2009; Caspi et al., 2005; Creamer, O'Donnell, & Pattison, 2005). For example, Bryant et al. (2009) found that longer periods of post-traumatic amnesia were related to less severe intrusive memories one week post-injury, suggesting a protective effect. Thus, individuals who experience post-traumatic amnesia may have fewer or incomplete mental representations of the trauma which are less likely to be triggered by matching cues. However, post-traumatic amnesia was unrelated to the severity of PTSD symptoms at three months, suggesting that it does not protect against the development of PTSD over longer time periods. In relation to stroke, Field et al. (2008) reported that PTSD symptom severity at three months was unrelated to whether the survivor was conscious or not at the time of their stroke. There are a number of ways in which PTSD could develop following post-traumatic amnesia or impaired consciousness. First, it is possible that impaired consciousness does not last throughout the traumatic event, and that PTSD symptoms may develop in relation to those aspects of the trauma experience that individuals are able to encode (Creamer et al., 2005). Second, individuals may retrospectively reconstruct memories of the trauma experience, for example from witnesses' reports, which subsequently develop into intrusive memories or flashbacks (Bryant et al., 2009). Third, processing of the trauma experience may occur at an implicit level during periods of impaired consciousness (Bryant, 2001).

3.3.7 Chronic stressors

Stroke is a leading cause of severe disability. Survivors typically experience a range of ongoing problems (e.g., weakness or paralysis, cognitive impairment, communication difficulties, problems with balance and coordination). One important question is the extent to which PTSD symptom severity reflects the impact of these ongoing stressors, rather than reactions to the stroke event itself. For example, in relation to MI, Shemesh et al. (2001) found that patients who experienced ongoing physical symptoms (e.g., angina) reported more intrusion and avoidance PTSD symptoms than those who were asymptomatic. In addition, in relation to stroke, Wang et al. (2011) reported that the level of physical disability was related to the severity of PTSD symptoms at three months post-stroke. There are a number of ways in which chronic stressors may contribute to the severity of PTSD symptoms. First, chronic stressors may erode individuals' resources, or their ability, to deal with their psychological reactions to the acute stressor (Adams & Boscarino, 2006). Second, chronic stressors may evoke reminders of the stroke which may, more directly, act as triggering cues for the reexperiencing symptoms of PTSD. Third, the experience of ongoing disability may be perceived by the individual to signify permanent negative change. Fourth, some disabilities experienced after stroke, including cognitive and language impairments, may impede the person's ability to fully process and integrate trauma memories with other autobiographical material. For example, cognitive impairment has been related to difficulties in recalling specific autobiographical memories in the elderly (Phillips & Williams, 1997) and in stroke survivors (Sampson et al., 2003). Fure et al. (2006) reported that cognitive impairment was related to elevated levels of anxiety in stroke patients; however, to date, no studies have examined the relationship between cognitive impairment and PTSD. In addition, Thomas and Lincoln (2008) reported that stroke survivors with aphasia had higher levels of emotional distress at one and six months post-stroke, with more detailed analyses revealing that this was the result of expressive, but not receptive, communication impairments. It is possible that stroke survivors with aphasia may also experience more PTSD symptoms.

3.3.8 PTSD and older adults

The majority of stroke survivors are older adults, with almost 80% of first-ever strokes occurring in people aged 65 years or older (Stroke Association, 2011). Knowledge regarding the prevalence and determinants of PTSD as well as its phenomenology in older adults, more generally, is limited (Averill & Beck, 2000; Cook & O'Donnell, 2005). Moreover, the majority of previous research on PSTD in older adults has focused on holocaust survivors, combat veterans and survivors of natural disasters, rather than on survivors of life-threatening illnesses such as stroke (Cook & O'Donnell, 2005). Some studies have highlighted differences between younger and older adults in the experience and/or reporting of PTSD symptoms (Acierno et al., 2002; Davidson et al., 1990; Fontana & Rosenheck, 1994), although other work has suggested that their PTSD reactions are quite similar (Bleich et al., 2005; Chung et al., 2005; Kohn et al., 2005). In relation to stroke, a number of studies have reported negative correlations between age and the severity of PTSD symptoms (Field et al., 2008; Sampson et al., 2003; Sharkey, 2007), although other studies have reported non-significant correlations (Bruggimann et al., 2006; Merriman et al., 2007). Nonetheless, there are a number of specific factors that may need to be considered when assessing PTSD in older people (Cook & O'Donnell, 2005). First, older adults are likely to have experienced multiple lifetime traumas (Creamer & Parslow, 2008) which may compound the impact of the current trauma, and vice-versa (Bechtle-Higgins & Follette, 2002). Second, older adults are more likely to suffer from cognitive impairments and dementia which may impact on their ability to fully process trauma memories. Third, older adults may have a more accepting attitude to illness and its psychological consequences.

3.3.9 PTSD and medical events

The assessment of PTSD symptoms following medical events, such as stroke, is complicated by the possibility that such symptoms may be confounded with the effects of physical illness and/or its treatment. For example, some of the hyperarousal (e.g., disturbed sleep, irritability, difficulty concentrating) and avoidance (e.g., diminished interest, detachment) symptoms that are used in the diagnosis of PTSD are also common problems experienced by survivors as a consequence of their stroke. Similarly, psychogenic amnesia is also included in the diagnostic criteria for PTSD. However, stroke is often associated with periods of amnesia that may have an organic (i.e., physical), rather than psychogenic, origin. The difficulty in differentiating between organic versus psychogenic causes of specific symptoms has implications for the assessment of PTSD. If such symptoms are simply taken to be part of PTSD rather having an organic origin, this is likely to lead to inflated estimates of the prevalence of PTSD. This effect is likely to be amplified when self-report measures are used to assess PTSD as alternative explanations for such symptoms cannot be explored.

Further, Mundy and Baum (2004) have argued that the nature of PTSD intrusion symptoms for medical events might be qualitatively different to those for other traumatic events. Whereas the focus of intrusions for more traditional traumas, such combat injuries and assaults, is on the past events, intrusions for medical events may also be future-oriented focusing on concerns about treatment, disease recurrence and ongoing functional impairment. Thus, in addition to having flashbacks to the traumatic event (in the past), individuals who have survived a life-threatening illness such as stroke may also have intrusive negative thoughts about the future (e.g., "Will I live to see my grandchildren grow up?", "Will I be able to work again?"). To date, there has been no phenomenological studies on the experience of PTSD after stroke. If the intrusions experienced by stroke survivors are found to be predominantly future-oriented, this would raise serious questions as to whether such clinical presentations are best thought of as PTSD - which, by definition, is *post*-trauma and characterized by being haunted by past horror - rather than anxiety about the (future) consequences of the stroke.

3.4 Recommendations for future research

On the basis of the review of studies on the prevalence and correlates of PTSD after stroke, five main recommendations for future research are made focusing on (i) the measurement of PTSD, (ii) study design, (iii) sample sizes, (iv) sample representativeness, and (v) the assessment of risk factors.

3.4.1 Measurement of PTSD

In order to provide accurate prevalence rates of PTSD after stroke, it is essential that structured clinical interviews are routinely employed. The Clinician-Administered PTSD Scale (Weathers et al., 2001) is widely regarded as the measure of choice for PTSD assessment as it is standardized, and can be used to provide both a PTSD diagnosis and a continuous measure of symptom severity. The PTSD module of the SCID (First et al., 1995) is also recommended for PTSD diagnosis, although it fails to provide a measure of PTSD symptom severity. When administering a structured clinical interview the researcher should consider whether the symptoms of PTSD, such as disturbed sleep, difficulty concentrating and amnesia, are better accounted for by alternative explanations (e.g., physical effect of the stroke, medication use, hospital environment) before they are categorized as having a psychogenic origin (O'Donnell et al., 2003). Moreover, this type of more in-depth questioning will also help the researcher to distinguish PTSD from generalized anxiety disorder and/or major depression.

3.4.2 Study design

Future research on the prevalence of post-stroke PTSD should employ longitudinal designs with multiple assessments of PTSD at fixed time points after stroke in order to provide accurate point prevalence rates and to chart the natural course of PTSD after stroke. Similarly, research on the correlates of PTSD after stroke should utilize prospective designs in which potential risk factors are assessed shortly after stroke (e.g., within one month) and related to PTSD caseness and symptom severity at subsequent time points while controlling for the effects of initial PTSD symptoms.

3.4.3 Sample sizes

Large samples are essential to establish the prevalence of psychiatric disorders in new populations (O'Donnell et al., 2003). Future research on the prevalence of PTSD after stroke should therefore aim to recruit larger sample sizes than have been recruited to date. The actual sample size required to accurately estimate different prevalence rates in a population can be calculated (Daniel, 1999). Naing et al. (2006) recommend that the precision of the estimate should be ±5% when the expected prevalence rate is greater than 10%, and half the expected prevalence rate when less than 10%. Current prevalence estimates for PTSD after stroke range from 3% to 31%. The sample size required to estimate a 3% prevalence rate with 95% confidence intervals at 1.5% precision (i.e., between 1.5% to 4.5%) is 497, whereas the sample size required to estimate a 31% prevalence rate with 95% confidence intervals at 5% precision (i.e., between 26% and 36%) is 329. The adoption of multi-site studies are likely to aid the recruitment of such sample sizes and also address possible population differences across sites. In addition, studies on the predictors of PTSD after stroke need to be sufficiently powered to assess the impact of a full range of independent variables. Tabachnick and Fidell (2007) recommend that, in order to have adequate statistical power, the sample size for a regression analysis should be at least $50 + 8k$ (where k = number of independent variables). For example, for a regression analysis with 25 independent variables the sample size should be at least 250.

3.4.4 Sample representativeness

Future studies should provide more information on the representativeness of their samples. Ideally, where there are common care pathways for stroke victims as in the UK, consecutive admissions to stroke units/wards should be recruited. The resultant sample should then be compared with the patient population from which it was drawn in order to assess its representativeness. Given the severity of stroke and the ensuing levels of disability, it is likely that many stroke survivors will be unable to give informed consent and/or complete self-report measures or clinical diagnostic interviews. This is likely to affect the representativeness of the sample and restrict the extent to which the findings can be generalized to all stroke survivors. Studies should therefore provide detailed information on their recruitment procedures and exclusion criteria. In addition, future work should attempt to amend recruitment and assessment procedures in order, as far as possible, to recruit stroke survivors with communication and cognitive impairments into studies on post-stroke PTSD. For example, research on depression and aphasia (Thomas & Lincoln, 2008) has used visual analogue scales (Brumfitt & Sheeran, 1999) to assess emotional distress in stroke survivors with communication difficulties. Future work should therefore also focus on developing measures of PTSD symptom severity that can be completed by stroke survivors with communication difficulties and/or cognitive impairments.

3.4.5 Assessment of risk factors

Future research should assess a comprehensive range of potential risk factors, including stroke details, when assessing the predictors of post-stroke PTSD. In particular, future research should draw upon current models of PTSD (Brewin & Holmes, 2003) to assess the impact of more proximal psychological variables that have been found to be the strongest correlates of PTSD symptomatology across a range of traumas in meta-analytic reviews (e.g., Brewin et al., 2000; Ozer et al., 2003). Future research should routinely employ prospective designs and conduct multivariate analyses in which proximal psychological variables (e.g., appraisals, memory processes) are assessed shortly after stroke (e.g., within one month) and are related to the subsequent development of PTSD and/or symptom severity at later time points, while controlling for the influence of more distal factors (e.g., stroke details, demographics) and initial PTSD symptoms. In this way, future studies may assess the extent to which the effects of distal variables are mediated by these more proximal variables, thereby increasing our understanding of the mechanisms, or processes, underlying the development of PTSD after stroke.

3.5 Ethical considerations

When studying psychological reactions to life-threatening illnesses, such as stroke, that are associated with severe levels of disability and instability in the patient's medical condition, researchers need to be cognisant of ethical as well as scientific considerations (Tedstone & Tarrier, 2003). Particular attention needs to be paid to issues of informed consent given the cognitive and communication impairments experienced by many stroke survivors. PTSD has become a popular diagnosis over recent years (Summerfield, 2001), as evidenced by the increasing range of events, including life-threatening illnesses, that have been the focus of PTSD research (Tedstone & Tarrier, 2003). While such research has increased our understanding of psychological reactions to life-threatening illnesses, researchers should be aware of the risk of pathologising normal reactions to a traumatic event that may naturally remit over time (Middleton & Shaw, 2000).

4. Conclusions

Post-stroke traumatic stress is an important but relatively neglected psychological consequence of stroke. It would be valuable to have reliable and accurate prevalence data from clinical diagnostic interviews with large, representative samples of stroke survivors collected over several time points. In addition, further work is required on the assessment of potential risk factors for the development of PTSD. This should include assessment of a full range of risk factors, including variables from current models of PTSD, shortly after stroke that can be related to subsequent PTSD caseness and symptom severity at later time points. A better understanding of the risk factors for PTSD after stroke has important clinical implications for the management of stroke survivors. It may assist in better differentiating the organic effects of stroke from the behavioural and psychological symptoms of the psychiatric disorder. More importantly, there is now a large body of evidence to guide the effective treatment of PTSD (Ponniah & Hollon, 2009). Appropriate use of such interventions has the potential to improve the quality of life, and reduce the care costs, of this population.

5. References

Acierno, R., Brady, K.L., Gray, M., Kilpatrick, D.G., Resnick, H., & Best, C.L. (2002). Psychopathology following interpersonal violence: A comparison of risk factors in older and younger adults. *Journal of Clinical Geropsychology*, 8, pp. 13-23.

Adams, R.E., & Boscarino, J.A. (2006). Predictors of PTSD and delayed PTSD after disaster: The impact of disclosure and psychosocial resources. *Journal of Mental Disease*, 194, pp. 485-493.

American Psychiatric Association. (1994). *Diagnostic and Statistical Manual of Mental Disorders* (4th ed.). American Psychiatric Association, Washington D.C..

American Stroke Association (2011). About stroke. Retrieved 10 June 2011. Available from <http://www.strokeassociation.org/STROKEORG/AboutStroke/About-Stroke_UCM_308529_SubHomePage.jsp>

Astrom, M. (1996). Generalized anxiety disorder in stroke patients: A 3-year longitudinal study. *Stroke*, 27, pp. 270-275.

Averill, P. M., & Beck, J. (2000). Post-traumatic stress disorder in older adults: A conceptual review. *Journal of Anxiety Disorders*, 14, pp. 133-156.

Barker-Collo, S.L. (2007). Depression and anxiety 3 months post stroke: Prevalence and Correlates. *Archives of Clinical Neuropsychology*, 22, pp. 519-531.

Bechtle-Higgins, A., & Follette, V.M. (2002). Frequency and impact of interpersonal trauma in older women. *Journal of Clinical Geropsychology*, 8, pp. 215-226.

Berry, E. (1998). Post-traumatic stress disorder after subarachnoid haemorrhage. *British Journal of Clinical Psychology*, 37, pp. 365-367.

Blake, D., Weathers, F., Nagy, L., Kaloupek, D., Klauminzer, G., Charney, D., & Keane, T. (1992). *Clinician Administered PTSD Scale*. National Centre for Post Traumatic Stress Disorder, Boston.

Bleich, A., Gelkopf, M., Melamed, Y., & Solomon, Z. (2005). Emotional impact of exposure to terrorism among young-old and old-old Israeli citizens. *American Journal of Geriatric Psychiatry*, 13, pp. 705-712.

Brady, K.T., Killeen, T.K., Brewerton, T., & Lucerini, S. (2000). Comorbidity of psychiatric disorders and posttraumatic stress disorder. *Journal of Clinical Psychiatry,* 61, pp. 22-32.

Brewin C.R., Kleiner, J.S., Vasterling, J.J., & Field, A.P. (2007). Memory for emotionally neutral information in posttraumatic stress disorder: A meta-analytic investigation. *Journal of Abnormal Psychology,* 116, pp. 448-463.

Brewin, C.R., & Holmes, E.A. (2003). Psychological theories of posttraumatic stress disorder. *Clinical Psychology Review,* 23, pp. 339-376.

Brewin, C.R., Andrews, B., & Valentine, J.D. (2000). Meta-analysis of risk factors for posttraumatic stress disorder in trauma-exposed adults. *Journal of Consulting and Clinical Psychology,* 68, pp. 748-766.

Bruggimann, L., Annoni, J.M., Staub, F., von Steinbuchel, N., van der Linden, M., & Bogousslavsky, J. (2006). Chronic posttraumatic stress symptoms after nonsevere stroke. *Neurology,* 66, pp. 513-516.

Brumfitt, S., & Sheeran, P. (1999). *Visual Analogue Self-Esteem Scale.* Winslow Press, Bichester.

Bryant, R.A. (2001). Posttraumatic stress disorder and mild brain injury: Controversies, causes and consequences. *Journal of Clinical and Experimental Neuropsychology,* 23, pp. 718-728.

Bryant, R.A., Creamer, M., O'Donnell, M.L., Silove, D., Clark, R.C., & McFarlane, A.C. (2009). Post-traumatic amnesia and the nature of post-traumatic stress disorder after mild traumatic brain injury. *Journal of the International Neuropsychological Society,* 15, pp. 862-867.

Bryant, R.A., Marosszeky, J.E., Crooks, J., Baguley, I.J., & Gurka, J.A. (2001). Posttraumatic stress disorder and psychosocial functioning after severe traumatic brain injury. *Journal of Nervous and Mental Disease,* 189, pp. 109-113.

Bryant, R.A., O'Donnell, M.L., Creamer, M., McFarlane, A.C., Clark, R.C., & Silove, D. (2010). The psychiatric sequelae of traumatic injury. *American Journal of Psychiatry,* 167, pp. 312-320.

Bryant, R.A., Sutherland, K., & Guthrie, R.M. (2007). Impaired specific autobiographical memory as a risk factor for posttraumatic stress after trauma. *Journal of Abnormal Psychology,* 116, pp. 837-841.

Caspi, Y., Gil, S,., Ben-Ari, I.Z., Koren, D., Aaron-Peretz, J., & Klein, E. (2005). Memory for the traumatic event is associated with increased risk for PTSD: A retrospective study of patients with traumatic brain injury. *Journal of Loss and Trauma,* 10, pp. 319-335.

Charney, D.S., Deutch, A.Y., Krystal, J.H., Southwick, S.M., & Davis, M. (1993). Psychobiologic mechanisms of posttraumatic stress disorder. *Archives of General Psychiatry,* 50, pp. 294-305.

Chung, M.C., Dennis, J., Easthope, Y., Farmer, S., & Werrett, J. (2005). Differentiating posttraumatic stress between elderly and younger residents. *Interpersonal and Biological Processes,* 68, pp. 164-173.

Cohen, J. (1992). A power primer. *Psychological Bulletin,* 112, pp. 155-159.

Cook, J.M., & O'Donnell, C. (2005). Assessment and psychological treatment of posttraumatic stress disorder in older adults. *Journal of Geriatric Psychiatry and Neurology,* 18, pp. 61-71.

Creamer, M., & Parslow, R. (2008). Trauma exposure and posttraumatic stress disorder in the elderly: A community prevalence study. *American Journal of Geriatric Psychiatry,* 16, pp. 853-856.

Creamer, M., Burgess, P., & McFarlane, A.C. (2001). Post-traumatic stress disorder: Findings from the Australian National Survey of Mental Health and Well-Being. *Psychological Medicine,* 31, pp. 1237-1247

Creamer, M., O'Donnell, M.L., & Pattison, P. (2005). Amnesia, traumatic brain injury, and posttraumatic stress disorder: A methodological inquiry. *Behaviour Research and Therapy,* 43, pp. 1383-1389.

Dalgleish, T., Rolfe, J., Golden, A.M., Dunn, B.D., & Bamard, P.J. (2008). Reduced autobiographical memory specificity and posttraumatic stress: Exploring the contributions of impaired executive control and affect regulation. *Journal of Abnormal Psychology,* 117, pp. 236-241.

Daniel, W.W. (1999). *Biostatistics: A Foundation for Analysis in the Health Sciences.* Wiley, New York.

Davidson, J.R.T., Kudler, H.S., Sunders, W.B., & Smith, R.D. (1990). Symptoms and comorbidity patterns in World War II and Vietnam veterans with posttraumatic stress disorder. *Comprehensive Psychiatry,* 31, pp. 62-170.

Eccles, S., House, A., & Knapp, P. (1999). Psychological adjustment and self reported coping in stroke survivors with and without emotionalism. *Journal of Neurology, Neurosurgery, and Psychiatry,* 67, pp. 125-126.

Ehlers, A., & Clark, D.M. (2000). A cognitive model of posttraumatic stress disorder. *Behaviour Research and Therapy,* 38, pp. 319-345.

Field, E.L., Norman, P., & Barton, J. (2008). Cross-sectional and prospective associations between cognitive appraisals and posttraumatic stress disorder symptoms following stroke. *Behaviour Research and Therapy,* 46, pp. 62-70.

First, M., Spitzer, R.L., Gibbon, M., & Williams, J.B.W. (1995). *Structured Clinical Interview for DSM-IV Axis I disorders.* New York State Institute, New York.

Foa, E.B., & Rothbaum, B. (1998). *Treating the Trauma of Rape: Cognitive Behavioral Therapy for PTSD.* Guilford Press, New York.

Foa, E.B., Cashman, L., Jaycox, L., & Perry, K. (1997). The validation of a self-report measure of post-traumatic stress disorder. The post-traumatic diagnostic scale. *Psychological Assessment,* 9, pp. 445-451.

Fontana, A., & Rosenheck, R. (1994). Traumatic war stressors and psychiatric symptoms among World War II, Korean and Vietnam veterans with posttraumatic stress disorder. *Psychology and Aging,* 9, pp. 27-33.

Fure, B., Wyller, T.B., Engedal, K., & Thommessen, B. (2006). Emotional symptoms in acute ischemic stroke. *International Journal of Geriatric Psychiatry,* 21, pp. 382-387.

Godefroy, O., Roussel, M., Leclerc, X., & Leys, D. (2006). Deficit of episodic memory: Anatomy and related patterns in stroke patients. *European Neurology,* 61, pp. 223-229.

Hackett, M.L., & Anderson, C.S. (2005). Predictors of depression after stroke: A systematic review of observational studies. *Stroke,* 36, pp. 2296-2301.

Hackett, M.L., Chaturangi, Y., Parag, V., & Anderson, C.S. (2005). Frequency of depression after stroke: A systematic review of observational studies. *Stroke,* 36, pp. 1330-1340.

Halligan, S.L., Michael, T., Clark, D.M., & Ehlers, A. (2003). Posstraumatic stress disorder following assault: The role of cognitive processing, trauma memory, and appraisals. *Journal and Consulting and Clinical Psychology*, 71, pp. 419-431.

Hammarberg, M. (1992). Penn inventory for PTSD: Psychometric properties. *Psychological Assessment*, 4, pp. 67-76.

Harvey, A.G., Brewin, C.R., Jones, C., & Kopelman, M.D. (2003). Coexistence of posttraumatic stress disorder and traumatic brain injury: Towards a resolution of the paradox. *Journal of the International Neuropsychological Society*, 9, pp. 663-676.

Horowitz, M., Wilner, N.R., & Alvarez, W. (1979). Impact of event scale: A measure of subjective stress. *Psychosomatic Medicine*, 4, pp. 209-218.

Johnsen, G.E., & Asbjørnsen, A.E. (2008). Consistent impaired verbal memory in PTSD:A meta-analysis. *Journal of Affective Disorders*, 111, pp. 74-82.

Kangas, M., Henry, J.L., & Bryant, R.A. (2005). Predictors of posttraumatic disorder following cancer. *Health Psychology*, 24, pp. 579-585.

Kleim, B., & Ehlers, A. (2008). Reduced autobiographical memory predicts depression and posttraumatic stress disorder after recent trauma. *Journal of Consulting and Clinical Psychology, 76*, pp. 231-242.

Klein, E., Caspi, Y., & Gil, S. (2003). The relation between memory of the traumatic event and PTSD: Evidence from studies of traumatic brain injury. *Canadian Journal of Psychiatry*, 48, pp. 28-33.

Kohn, R., Levav, I., Garcia, I.D., Machua, E., & Tamashiro, R. (2005). Prevalence, risk factors and aging vulnerability for psychopathology following a natural disaster in a developing country. *International Journal of Geriatric Psychiatry*, 20, pp. 835-841.

Kutz, I., Shabtai, H., Solomon, Z., Neumann, M., & David, D. (1994). Posttraumatic stress disorder in myocardial infarction patients - prevalence study. *Israel Journal of Psychiatry and Related Sciences*, 31, pp. 48-56.

Lanius R.A., Bluhm, R., Lanius, U., & Pain, C. (2006). A review of neuroimaging studies in PTSD: Heterogeneity of response to symptom provocation. *Journal of Psychiatric Research*, 40, pp. 709-729.

Leppavuori, A., Pohjasvaara, T., Vataja, R., Kaste, M., & Erkinjuntti, T. (2003). Generalized anxiety disorders three to four months after ischemic stroke. *Cerebrovascular Diseases*, 16, pp. 257-264.

Maercker, A., Forstmeier, S., Enzler, A., Krüsi, G., Hörler, E., Maier, C., & Ehlert, U. (2008). Adjustment disorders, posttraumatic stress disorder and depressive disorders in old age: Findings from a community survey. *Comprehensive Psychiatry*, 49, pp. 113-120.

Merriman, C., Norman, P., & Barton, J. (2007). Psychological correlates of PTSD symptoms following stroke. *Psychology, Health and Medicine*, 12, pp. 592-602.

Middleton, H., & Shaw, I. (2000). Distinguishing mental illness in primary care. *British Medical Journal*, 320, pp. 1420-2421.

Mundy, E., & Baum, A. (2004). Medical disorders as a cause of psychological trauma and posttraumatic stress disorder. *Current Opinion in Psychiatry*, 17, pp. 123-127.

Naing, L., Winn, T., & Rusli, B.N. (2006). Practical issues in calculating sample size for prevalence studies. *Archives of Orofacial Sciences*, 1, pp. 9-14.

O'Donnell, M.L., Creamer, M., Bryant, R.A., Schnyder, U., & Shalev, A. (2003). Posttraumatic disorders following injury: An empirical and methodological review. *Clinical Psychology Review,* 23, pp. 587-603.

O'Donnell, M.L., Holmes, A.C., Creamer, M.C., Ellen, S., Judson, R., McFarlane, A.C., et al. (2009). The role of post-traumatic stress disorder and depression in predicting disability after injury. *Medical Journal of Australia,* 190, pp. S71-S74.

Ozer, E.J., Best, S., Lipsey, T.L., & Weiss, D.S. (2003). Predictors of posttraumatic stress disorder and symptoms in adults: A meta-analysis. *Psychological Bulletin,* 129, pp. 52-73.

Pedersen, S.S. (2001). Post-traumatic stress disorder in patients with coronary artery disease: A review and evaluation of the risk. *Scandinavian Journal of Psychology,* 42, pp. 445-451.

Phillips, S., & Williams, J.M.G. (1997). Cognitive impairment, depression and the specificity of autobiographical memory in the elderly. *British Journal of Clinical Psychology,* 36, pp. 341-347.

Ponniah, K., & Hollon, S.D. (2009). Empirically supported psychological treatments for adult acute stress disorder and posttraumatic stress disorder. *Depression and Anxiety,* 26, pp. 1086-1109.

Rauch, S.L. (2003). Neuroimaging and neurocircuitry models pertaining to the neurosurgical treatment of psychiatric disorders. *Neurosurgery Clinics of North America,* 14, pp. 213-223.

Sagen, U., Finset, A., Moum, T., Mørland, T., Vik, T.G., Nagy, T., & Dammen, T. (2010). Early detection of patents at risk for anxiety, depression and apathy after stroke. *General Hospital Psychiatry,* 32, pp. 80-85.

Sagen, U., Vik, T.G., Moum, T., Mørland, T., Finset, A., & Dammen, T. (2009). Screening for anxiety and depression after stroke: Comparison of the Hospital Anxiety and Depression Scale and the Montgomery and Åsberg Depression Rating Scale. *Journal of Psychosomatic Research,* 67, pp. 325-332.

Sampson, M.J., Kinderman, P., Watts, S., & Sembi, S. (2003). Psychopathology and autobiographical memory in stroke and non-stroke hospitalized patients. *International Journal of Geriatric Psychiatry,* 18, pp. 23-32.

Schönfeld, S., & Ehlers, A. (2006). Overgeneral memory extends to pictorial retrieval cues and correlates with cognitive features in posttraumatic stress disorder. *Emotion,* 6, pp. 611-621.

Schönfeld, S., Ehlers, A., Bollinghaus, I., & Rief, W. (2007). Overgeneral memory and suppression of trauma memories in post-traumatic stress disorder. *Memory,* 15, pp. 339-352.

Schouten, E.A., Schiemanck, S.K., & Brand, M.W.M. (2009). Long-term deficits in episodic memory and ischemic stroke: Evaluation and prediction of verbal and visual memory performance based on lesion characteristics. *Journal of Stroke and Cerebrovascular Diseases,* 18, pp. 128-138.

Sembi, S., Tarrier, N., O'Neill, P., Burns, A., & Farragher, B. (1998). Does post-traumatic stress disorder occur after stroke: A preliminary study. *International Journal of Geriatric Psychiatry,* 13, pp. 315-322.

Shalev, A., Schreiber, S., Galai, T., & Melmed, R. (1993). Post-traumatic stress disorder following medical events. *British Journal of Clinical Psychology,* 32, pp. 247-253.

Sharkey, M. (2007). Post-traumatic stress symptomatology following stroke. *PSIGE Newsletter,* 99, pp. 14-17.

Shemesh, E., Rudnick, A., Kaluski, E., Milovanov, O., Salah, A., Alon, D., Dinur, I., Blatt, A., Metzkor, M., Golik, A., Verd, Z., & Cotter, G.A. (2001). Prospective study of posttraumatic stress symptoms and nonadherence in survivors of a myocardial infarction (MI). *General Hospital Psychiatry,* 23, pp 215-222.

Spindler, H., & Pedersen, S.S. (2005). Posttraumatic stress disorder in the wake of heart disease: Prevalence, risk factors, and future research directions. *Psychosomatic Medicine,* 67, pp. 715-723.

Spitzer, C., Barnow, S., Völzke, H., John, U., Freyberger, H.J., & Grabe, H.J. (2009). Trauma, posttraumatic stress disorder, and physical illness: Findings from the general population. *Psychosomatic Medicine,* 71, pp. 1012-1017.

Spitzer, R.L., First, M.B., & Wakefield, J.C. (2007). Saving PTSD from itself in DSM-V. *Journal of Anxiety Disorders,* 21, pp. 233-241.

Stroke Association. (2011). Facts and figures about stroke. Retrieved 10 June 2011. Available from <www.stroke.org.uk/media_centre/facts_and_figures/index.html>

Summerfield, D. (2001). The invention of post-traumatic stress disorder and the social usefulness of a psychiatric category. *British Medical Journal,* 322, pp. 95-98.

Tabachnick, B.G., & Fidell, L.S. (2007). *Using Multivariate Statistics.* Allyn and Bacon, Boston.

Tedstone, J.E., & Tarrier, N. (2003). Post-traumatic stress disorder following medical illness and treatment. *Clinical Psychology Review,* 23, pp. 409-448.

Thomas, S.A., & Lincoln, N.B. (2008). Predictors of emotional distress after stroke. *Stroke,* 39, pp. 1240-1245.

van Zelst, W.H., de Beurs, E., Beekman, A.T.F., Deeg, D.J.H., & van Dyck, R. (2003). Prevalence and risk factors of posttraumatic stress disorder in older adults. *Psychotherapy and Psychosomatics,* 72, pp. 333-342.

Wang, X., Chung, M.C., Hyland, M.E., & Bahkeit, M. (2011). Posttraumatic stress disorder and psychiatric co-morbidity following stroke: the role of alexithymia. *Psychiatric Research,* 188, pp. 51-57.

Weathers, F. W., Keane, T. M., & Davidson, J. R. (2001). Clinician-Administered PTSD Scale: A review of the first ten years of research. *Depression and Anxiety,* 13, pp. 132-156.

Weathers, F., Litz, B., Herman, D., Huska, J., & Keane, T. (1993). The PTSD Checklist (PCL): Reliability, validity, and diagnostic utility. *Annual Convention of the International Society for Traumatic Stress Studies,* San Antonio, Texas, October 1993.

Williams, G.N. (1997). Post-traumatic stress disorder: Implications for practice in spinal cord injury rehabilitation. *SCI Psychosocial Process,* 10, pp. 40-42.

Williams, J.M.G., & Broadbent, K. (1986). Autobiographical memory in suicide attempters. *Journal of Abnormal Psychology,* 95, pp. 144-149.

Williams, J.M.G., Barnhofer, T., Crane, C., Hermans, D., Raes, F., Watkins, E., & Dalgleish, T. (2007). Autobiographical memory specificity and emotional disorder. *Psychological Bulletin,* 133, pp. 122-148.

Part 5

Stress Management Training

13

The Potential of Stress Management Training as a Coping Strategy for Stressors Experienced in Theater of Operation: A Systematic Review

Stéphane Bouchard[1], Tanya Guitard[2], Mylène Laforest[3], Stéphanie Dumoulin[2], Julie Boulanger[1] and François Bernier[4]
[1]Université du Québec en Outaouais, Gatineau, Québec,
[2]Université du Québec à Montréal, Montréal, Québec,
[3]Defence Research and Development Canada - Valcartier, Valcartier, Québec,
[4]Ottawa University,
Canada

1. Introduction

This chapter provides a literature review on Stress Management Training (SMT) as a potential tool to help military personnel cope with stressors experienced in the theater of operations. It is hoped that SMT techniques can be used to prepare soldiers for potential highly stressful situations in an effort to diminish their negative reactions to stress. The ultimate long-term prospective benefits would be that training military personnel with SMT would increase resilience and lower the incidence of post-traumatic stress disorder (PTSD).

There are several definitions of stress, but essentially it can be considered an affective state that occurs in response to perceived demands and challenges in the environment with which one feels unable to cope [1]. A variety of stress management techniques have been developed over the years in order to help individuals prevent, eliminate or cope with stress. All these techniques have the objective to modify factors associated with stress (behavioral, cognitive, physiological, emotional and environmental).

Early references to SMT date back to the work of Gottlieb, Strite and Koller et al. [2] who applied stress reduction strategies in behavioral medicine. SMT now represents an extremely diverse set of strategies and our literature review confirmed that notion several times. Authors include almost any available techniques, from Yoga [3] to prayer [4], along with exposure to feared situations [5], cognitive restructuring [6], problem solving [7], etc.

In general, SMT can be defined as the application of any set of techniques aiming to improve the way people cope with stress. Coping represents efforts to manage demands, conflicts and pressures that drain, or exceed, a person's resources [1]. Murphy and Sauter [8] offered to better integrate the applications of SMT strategies to contemporary notions of prevention by dividing SMT into primary, secondary and tertiary interventions. Primary interventions focus on changing the sources of the stress response (e.g., by modifying the environment) before stress becomes a problem, while secondary interventions aim at reducing the severity of symptoms associated with stress (much like secondary prevention, before non-clinical symptoms crystallize into disorders). Finally, tertiary interventions represent the application of SMT to treat mental and physical disorders. According to Murphy and Sauter [8], the most common stress management interventions are secondary programs aimed at the individual level and involve instruction in techniques to manage and cope with the stress associated with current problems.

Since the breath of SMT encompasses such techniques as relaxation, cognitive restructuring, problem solving, social skills training, planning behavioral changes and exposure to stressful situations, other stress management programs relying on these techniques also fall under the broad definition of SMT, such as Stress Inoculation Training [9] and Anxiety Management Training [10]. As opposed to SMT, where there is no coherent set of techniques and official definition, SIT and AMT are far from umbrella categories of various psychological techniques. SIT and AMT represent consistent intervention programs with a number of defining strategies that are carefully selected among those usually found in SMT. We therefore decided to include AMT and SIT in our literature review since they represent subtypes of SMT.

Stress inoculation training is a set of cognitive-behavioral techniques developed as a treatment by Donald Meichenbaum around the same years as SMT was gaining popularity [9]. The aim of SIT is to help individuals cope with the consequences of being exposed to stressful events and on a preventative basis to "inoculate" individuals to current and future stressors. Although it is made to be tailored to the client's need, the application of the SIT program follows a semi-structured and clearly outlined format [11] that unfolds in three distinct phases: (a) conceptualization, (b) skills acquisition and rehearsal, and (c) application and follow through. The term inoculation is used to simulate the concept of immunization through progressive exposure. The individual uses techniques such as imagery and behavior rehearsal, role play, modeling and progressive exposure to stressful situations. Techniques for relapse prevention and attribution of success to one's own efforts are also used.

Anxiety management training was developed by Richard Suinn [12] and research on its use was blooming at the same time as SMT and SIT [13]. It is therefore not surprising that, as cognitive-behavior techniques, they share common roots and principles. However, its focus on learning relaxation and generalizing it to daily stressors is much stronger. AMT was first developed for the treatment of what was defined at the time as "free floating anxiety". It was geared more toward clinical anxiety than was SMT and SIT. Suinn's basic philosophy was that patients could be taught to: (a) detect emotional, cognitive and physical signs associated with the onset of anxiety, and (b) react to these signs in manners that would make them disappear. One specific aspect of AMT is that patients are not required to find the causes or stimuli that precipitate their anxiety; they are essentially taught to focus on recognizing the presence of anxiety and its symptoms. Once anxiety-related cues are felt, the patient learns to use relaxation skills in order to alleviate the anxiety. Later on in therapy, the patient learns to identify the cognitive and physiological signs of anxiety arousal sooner. Even if AMT has been created to treat patients suffering from an anxiety disorder, it has been used in other contexts, such as enhancing performance and reducing general stress, and therefore deserves to be included in the current literature review.

This chapter sets out to report an extensive search in peer-reviewed scientific journals, analyze the published empirical data, and organize the results in such a way that studies could be examined based on their relevance to confirm with empirical evidences that SMT, or some of its strategies, is an effective approach to cope with acute stressors such as those encountered in theater of operations.

2. Method

A search for scientific papers was conducted using the Scopus database (which includes the following databases and more: PsychLit, PsychInfo, Medline, PubMed) with the following search terms (written without quotes): stress management training, stress inoculation training and anxiety management training. To reduce the risk of missing relevant papers the search was not limited to keywords but open to keywords, title and abstract. The search was performed with publication date ranging from 1950 (the oldest paper found in our search dates from 1958) up to 2009. Information available only from websites, dissertations and conferences were not considered. Taking into account the fact that some papers included two or more of the search terms SMT, SIT and AMT, the literature search resulted in 3 611 papers published in peer-review journals.

As intended, a manual examination of each of these results showed that our search strategy was extremely broad. The majority of the 3 611 papers (89.5%) were rejected because they either: (a) were irrelevant to SMT, SIT or AMT (usually because the search terms were not written within quotes), (b) did not include any[1] quantitative or qualitative data (e.g., theoretical paper, description of projects yet to be realized, clinical descriptions, policy position papers), (c) were in languages other than English or French, and (d) were limited to the development of psychometric tests. Meta-analyses and literature reviews based on systematic search of published papers were not rejected. However, their reference lists were crossed-checked to confirm we had not missed any relevant articles. The tedious process of systematic paper selection led to 350 articles, 200 falling under the general umbrella of SMT, 55 on the variation or subtype of SMT called SIT and 95 on the variation or subtype of SMT called AMT.

After reading the 350 papers addressing the broad definition of SMT techniques, they were divided into five categories, presented in increasing order of relevance to the purpose of helping military personnel to cope with acute stressors such as those experienced in theater of operations (see Table 1): (a) improving physical and medical conditions, (b) treatment of anxiety disorders and other mental disorders, (c) control of already existing stress-related issues (i.e., not clinical diagnoses), (d) preventing the consequences of traumatic events, and (e) development of strategies to cope more efficiently with future stressful situation (i.e., primary prevention). In reviewing the articles, special attention was devoted to studies on military personnel and similar populations (e.g., police SWAT teams, firefighters).

The following pages will present the results of this extensive search on SMT. Given the extremely wide variety of SMT techniques, SIT and AMT are considered to fit in the broad inclusive description of SMT. Results for all these techniques will thus be presented together. For the sake of brevity and clarity, in the first four categories only the most relevant studies will be discussed or cited as examples (for an exhaustive list of the studies compiled in this chapter, contact the first author). Category five includes studies that are clearly relevant to the purpose of our work on mental readiness training to cope with acute stress. Due to their relevance to the aim of our endeavor on preventing psychological injuries, studies in category five will be described in more details.

---

[1] This criterion was relaxed for papers applying SMT with populations similar to military personnel (see Category 5).

Categories	Total: broad SMT definition	SIT	AMT	Other SMT techniques
1. Improving physical and medical conditions	124	13	20	91
2. Treatment of anxiety and other mental disorders	61	15	39	7
3. Control of already existing stress-related issues	140	21	36	83
4. Preventing the consequences of traumatic events	13	0	0	13
5. Development of strategies to cope more efficiently with future stressful situations related to sports, military personnel and other stressors.	12	6	0	6
Total	350	55	95	200

Table 1. Number of peer-reviewed papers found in the literature search on SMT.

3. Results

3.1 Category 1: Improvement of physical and medical conditions

Findings on the impact of SMT on physical indices and medical conditions provide objective manifestations of the efficacy of training people to use skills to cope with stress. Most SMT programs have been developed to deal with medical illness and were found in publications dealing with behavioral medicine. Among the 91 scientific papers, 39 examined the efficacy of using SMT to impact on cardiovascular and coronary heart diseases, and most report statistically significant results [14]. Other papers also revealed positive results with medical problems such as cancer [15], HIV [16], diabetes [17], asthma [18], arthritis [19] and acute pain [20].

A more structured form of SMT, AMT, has been studied in 20 peer-reviewed papers. Studies using AMT to help cope with the psychological consequences of having a serious medical condition showed more potent and lasting results, notably for coping with having HIV [21] and cancer. Other studies found a statistically significant impact of AMT on physiological parameters such as glucose level in diabetic patients [22] and systolic / diastolic blood pressure [23-25]. In most applications of AMT to medical conditions, the basic treatment program was slightly adapted to include strategies tailored specifically to the medical condition under study (e.g., pain management [26]).

SIT was also demonstrated to be effective in coping with pain, such as third-degree burns [27], performance of athletes after a surgery [28], dental treatment [29], preparing for surgery [30], and experimental pain [31, 32]. The efficacy of SIT on physiological parameters has also been reported in hypertensive patients [33, 34]. Like AMT, SIT has been tested with success to help patients cope with stress and anxiety related to a medical condition, such as open-heart surgery [35], leukemia [36] and multiple sclerosis [37].

3.2 Category 2: Treatment of anxiety and other mental disorders

The purpose of the current literature review is to document how stress management strategies can be used to help military personnel cope with stressful situations in theatre of operation, not to treat existing anxiety disorders. Nevertheless, one cannot ignore that we found 61 scientific papers on that topic. Among all SMT techniques, AMT has clearly been the tool most often studied in regard to the treatment of anxiety disorders and other mental disorders found in the DSM-IV [38], with 39 papers. Most studies (n = 29) were conducted with people suffering from an anxiety disorder: a third of them targeted generalized anxiety disorder [39-41], while others were conducted with patients suffering from all types of anxiety disorders, ranging from posttraumatic stress disorder [6, 42] to specific phobias [43, 44]. For most of these disorders, at least one randomized controlled trial was conducted with reliably diagnosed patients and long-term follow-up. There is strong evidence to claim that AMT can have a favourable impact on anxiety disorders, including PTSD. AMT has also been used with patients suffering from other mental disorders, such as schizophrenia [45, 46] and alcoholism [47], with statistically significant impact on associated anxiety symptoms.

SIT has been used in 15 published studies to treat anxiety disorder or symptoms of anxiety in people suffering from mental disorders such as schizophrenia (e.g., in comparison with drug treatment [48]) or addictions [49]. Ten studies were conducted on the treatment of PTSD [50-52] and five on specific phobia [53, 54]. For example, in a randomized controlled trial Foa, Rothbaum, Riggs and Murdock [42] compared SIT to prolonged exposure, minimal support (active control condition) and waiting list (passive control condition) for rape victims suffering from PTSD. Results were statistically superior to the other two control conditions at post-treatment and gains were maintained at follow-up. There are only a limited number of outcome studies using SIT with clinical populations, but their results clearly support the efficacy of this approach to psychological injuries that are severe enough to warrant the clinical diagnosis of PTSD.

Much less research has been conducted on the use of more vaguely defined sets of SMT strategies. Our literature search found seven studies conducted on learning stress management skills in different populations suffering from schizophrenia [55-57], substance abuse [58, 59], attention deficit disorder [60], and ambulatory psychosomatic patients [61]. Four of these studies are randomized controlled trials with rigorous designs, acceptable sample and long-term follow-up. For example, it can be safely stated that for people with chronic schizophrenia, training in stress management clearly provides skills for coping with acute work and daily-life stressors and reduces the likelihood of subsequent acute exacerbation of symptoms with needs for hospitalization. It is also useful for substance abuse and ADHD as tools to better regulate stress.

3.3 Category 3: Control of already existing stress-related issues (i.e. non clinical diagnoses)

Intervening on general, non pathological, anxiety symptoms is the most frequent application of the broad set of SMT techniques. Researchers have published 136 studies on controlling already existing stress-related problems and non-clinical anxiety. Some of these studies did not focus on efficacy but even if it was not the aim of their study, they collected meaningful pre/post data and thus were not excluded from our literature search.

A total of 33 studies have focused on using the broad range of SMT strategies with student populations, with 11 studies using essentially the AMT protocol for school related or exam stressors [62, 63]. In a classic experiment, Suinn and Richardson [12] successfully treated 24 students suffering from math anxiety. Additional studies were conducted with university students, seven studies used the SIT protocol [64, 65] and 15 studies used various other SMT strategies, mostly relaxation.

The most frequent use of broad SMT strategies is for coping with work-related stress. Applications to the workplace of various SMT strategies, like relaxation, breathing retraining and repeating coping self-statements, SIT and AMT have been used with numerous types of professionals. Among those, six studies were conducted with high-risk jobs such as policemen or maintenance worker [66, 67]. Richardson and Rothstein [68] also published a meta-analysis of 36 carefully designed studies using SMT in the workplace and demonstrated that it is clearly effective. The most interesting aspect of their study is the dismantling and assessment of the effectiveness of specific strategies. They regrouped broadly defined SMT strategies into five types: cognitive-behavioral (such as SIT and AMT), relaxation training, organizational changes interventions, holistic / multimodal approaches, and alternative strategies (such as biofeedback and meditation). Structured cognitive-behavioral intervention, namely SIT and AMT, were the most effective strategies, with an average effect size of 1.17, followed by alternative strategies ($d = .91$). Other strategies were significantly less effective. These results echoed a previous less rigorous review conducted by Murphy [69] on 64 studies collected based on broader selection criteria.

As mentioned previously, six studies were conducted with people whose work involved high-risk situations. The randomized controlled trial by Peters and Carlson [67] demonstrated convincingly that SMT can be effective but the study by Le Scanff and Taugis [66] deserves to be mentioned in more detail given the similarity between their sample and the military context of this chapter. Le Scanff and Taugis [66] developed and applied a SMT program for the French police Special Forces units. Their seven-day pilot program was built to include corrective solutions for important organizational problems and therefore includes many strategies that may not apply to the training of military personnel. Apart from organizational one, the following SMT strategies were used: identifying stress factors and cues, learning coping skills (progressive muscle relaxation, deep breathing, concentration/centering, releasing tension in specific muscle groups, imagery), follow-up on problems experienced while applying the SMT strategies, reinforce the use of efficient coping skills, and develop better communication and assertiveness skills. Sadly, the authors [66] adopted a limited and unsystematic qualitative approach to document the impact of their program. Empirical data were not systematically collected pre or post implementation with their sample of 150 male police officers. Only global interests towards the training sessions were assessed. It revealed that trainees appreciated the program, felt they had learned something and reported that the program broadened their perspective and understanding of stress. One important factor stands out of their analysis and is pertinent to our work: virility. They defined virility as being able to reestablish order and domination, or to inflict pain and suffering on another person, without expressing doubt or feeling. They noted that, for their participants, admitting to feelings of anxiety was considered akin to being afraid and not being a real man, and could interfere with professional efficiency. This observation is interesting for our own work with military personnel. It is in line with subtle factors that must be built in SMT programs delivered to military personnel working in theaters of operations [70, 71].

3.4 Category 4: Preventing the consequences of traumatic events

Several papers on SMT actually address what is frequently referred to as debriefing, which is an attempt to mitigate the psychological impact of recent traumatic events. There are several different kinds of debriefing[2]. They vary in number of phases, focus of discussion and degree of structure provided in the intervention. Group psychological debriefing is one of the most common early interventions with military units [72]. There has been much debate about the usefulness of debriefing and several studies suggested that it may even be detrimental to participants [73, 74].

Despite the large number of position papers advocating the use of debriefing, every controlled study using adequate measures that we found in our literature search concluded that debriefing was no more effective than the control conditions. For example, Marchand, Guay, Boyer, Iucci, Martin and St-Hilaire [75] looked at the impact of debriefing intervention for victims of armed robbery by randomly assigning 75 victims to either critical incident stress debriefing or a control group. They found no evidence of the usefulness of debriefing to prevent PTSD or attenuate posttraumatic symptoms. The results remain the same after controlling for the severity of depressive mood.

3.5 Category 5: Development of strategies to cope more efficiently with future stressful situations

Our extensive search of the Scopus database journals did not revealed any published study on the use of SMT to cope with stressors such as those experienced in theatre of operations. Nevertheless, it led to the identification of 12 papers reporting empirical results on applications of SMT that should be meaningful to assess whether SMT can be used to help military personnel develop effective coping skills while dealing with acute stressors. We will begin with papers dealing with military or other life-threatening stressors, followed by papers on sports psychology.

One study by Rice and Gerardi [76] was conducted with military personnel. They did not train participants to use SMT techniques for themselves and unfortunately they did not report any results, so at first glance their article may appear less relevant. But they trained occupational therapists to deal with stress related issues in their work with soldiers in a theatre of operation. The philosophy of their program is based on SMT and illustrates well several differences that will be found between papers in this category and those presented in the preceding fourth one.

In this program several SMT strategies are used, such as detecting signs of stress, skills training, exercises, role play, progressive exposure to stressful situations, and fostering a feeling of control. The training focuses on detecting, and intervening with, soldiers manifesting symptoms of combat fatigue. The program is described in detail in Rice and Gerardi's [76] paper, and includes training schedules, casualty role-play scenarios in increasingly stressful situations, practicing critical incident stress debriefing and other clinical tasks performed in theater of operations, as well as learning how to function under stressful conditions. Great emphasis is put on concepts such as progressively practicing newly acquired skills, over learning basic skills so they become automatic, and relying on experience for complex situations. The program brings trainees to perform their work in situations that are increasingly stressful, moving from knowledge acquisition in a safe, non-threatening context, through knowledge integration and finally into high fidelity application in a realistic environment.

---

[2] Note. Because of its methodology, our literature search should not be considered a comprehensive review on debriefing.

Unfortunately, no results are provided on the effectiveness of the program. The authors stated they expect that providing coping skills and practicing them in progressively stressful situations should prevent occupational therapists from feeling overwhelmed or helpless and increase performance in their duties.
Another SMT program has been described by Sheehan [77] for training new FBI agents in coping with stress. The program consists essentially of psychoeducation by teaching future agents about the impact of stress and that they cannot avoid this emotion. They receive information on coping strategies and how other experienced agents deal with stress. They are also lectured on the difference between chronic and traumatic stress. Unfortunately, the author did not report any empirical results on the impact of the program. The interest of this program is the use of simple SMT strategies that the author hopes can be used during acute stress caused by objective threats, as opposed to more complex SIT and AMT strategies. It is also part of a global approach focusing not only on the individual but also involving actions at the organizational level. The program highlights clearly three important steps of most SMT approaches: detecting signs of stress, psychoeducation and applying specific coping strategies. Unfortunately, it remains unclear the extent to which the trainees actually practiced the coping strategies and whether it was effective.
Kamiyama, Yamami, Sato, Aoyagi, Kyoya, Mizuno et al. [78] published a brief report on a SMT program for marine hazard rescuers. They recruited 28 professionals performing rescue operations for marine disasters and accidents. Participants were randomly assigned to a group receiving: (a) a SMT program based on psychoeducation about stress, relaxation and autogenic training, or (b) only psychoeducation about stress. Both interventions were delivered in five weekly 90-minute sessions. Outcome was assessed with self-report questionnaires on anxiety and depression, and with physiological parameters assessed in blood samples. After the fifth session participants were sent in a (real, not simulated) rescue mission following a devastating earthquake. Statistical analyses confirmed that participants who received the enhanced SMT program scored better on the anxiety, depression and physiological measures compared to the control group that received only basic psychoeducation. This study possesses several strengths, such as the use of both self-report and biological markers of stress, a credible control group and random assignment. Even if the lack of a follow-up precludes concluding that the program had a long lasting effect, it is clear that some SMT strategies can help people working in high-risk situations cope more efficiently with stress.
In another paper with professionals working in stressful situations, Hytten, Jensen and Skauli [79] report studies with smoke divers and with free fall lifeboat passengers. In both cases, the SIT program was designed to prepare future oil workers for catastrophes and increase their chances of survival. Participants were recruited for smoke diving (i.e., a task some trained firefighters perform using an oxygen mask and full body gear) among oil industry "regular" employees receiving basic safety course. They were randomly assigned to a control group (n = 43) and an experimental group (n = 44). The experimental group received a one-hour training session based on the SIT protocol and the control group did not receive any SMT training. On the day following training all 87 smoke divers went to a bunker and participated in a fire simulation where they had to crawl in a narrow labyrinth filled with fire smoke, in total darkness. Participants were constantly watched by instructors and could call for help during the simulation. Those who received SIT training required significantly less help from instructors but, contrary to expectations, they reported significantly more anxiety than the control group. No difference was found on salivary cortisol response, a well known biological marker of the stress response.

A second study is reported in the same article [79], this time on the training of oil industry personnel to use a freefall lifeboat. On offshore oil and gas platforms, rapid evacuation in cases of emergencies rely on the use of boats that slide out from a ramp and hit the water away from the platform. This is a stressful experience, especially when falling from the height of an oversea oil platform. After random assignment, 21 participants received one hour of SIT training and the remaining 41 control participants received no additional training at all. On the following day, four consecutive free dives were performed. Results revealed no statistical significant difference between the two conditions on self-report and salivary cortisol measures. However, participants who received SIT training reported higher acceptance of using freefall lifeboats than the control group.

Dealing with the pressure of sport competition is far different from being in a theatre of operations and stressors are not life threatening. However, it is worth examining the SMT strategies used by athletes because SMT was used while athletes were required to perform specific tasks while under stress. Mace and Carroll [80] studied gymnasts to see if SIT could increase athlete's performance by reducing negative beliefs during competitions. In 1989, after encouraging results in pilot case studies, they reported an experimental study with 18 female gymnasts performing a bench sequence [81]. Participants were randomly assigned to two conditions: (a) seven SIT sessions of training in relaxation, imagery and using coping self-statements, or (b) seven training sessions during which they practiced a series of coordination exercises but no psychological stress management training was given to them. Outcome was assessed with several measures, including self-report, heart rate frequency (the most common biological marker of stress and anxiety) prior to the performance, independent observer's ratings of distress and scores provided by qualified gymnastics judges who rated video recording of the participant's performance. Pre/post comparisons revealed that athletes who received SIT training were significantly less anxious during their performance ($F(1, 16) = 12.55$, $p < 0.01$) and obtained significantly better scores by the expert judges than those in the control condition. No difference was found in the heart-rate measure.

The same team tested how SIT could be used to control the stress experienced by rock climbers during rappelling (also known as abseiling, [82]). Half of the twenty volunteers were randomly assigned to a SIT group and the other half to a no training control group. Following SIT training, participants were invited to complete their descent down a rope in rappelling from the roof of a 21.2 m building. Self-reported stress, overt signs of distress assessed by an independent observer and heart rate frequency were measured prior to the descent. The SIT group showed significantly less self-reported stress ($F(1, 18) = 9.49$, $p < 0.01$), distress (F 1, 18 = 14.67, $p < 0.01$) and fewer behavioural signs of distress as judged by the observer (F 1, 18 = 27.77, $p < 0.01$). However, as with the previous study, there were no significant differences between the groups in terms of heart rate.

Finally, another study in sports psychology reported positive results of using SIT on the performance of golf players [83], and two studies had been found on the reduction of injuries among athletes. Kolt, Hume, Smith and Williams [84] could not find any significant impact on the frequency of injuries of among their 22 gymnasts assigned to a SMT or a control condition, but Perna, Antoni, Baum, Gordon and Schneidermann [85] found that 34 athletes randomly assigned to a SMT program experienced significant reductions in the number of illness and injury days as compared athletes in the control group.

4. Discussion

Overall, studies on the application of the broad strategies used in SMT show that it is most of the time effective and for an extensive category of difficulties. For instance, it sometimes has a physical impact on medical conditions and is clearly effective to help patients cope with associated psychological reactions. Most studies, but not all, support the effectiveness of SMT. When it comes to dealing with chronic physical illness, strategies having an enduring impact on patient's life and coping style might also be more effective than brief interventions.

The same can be said when looking at mental disorders. Our literature review showed that SIT and AMT are effective tools for PTSD and, to some extent, for other disorders such as generalized anxiety disorder. The broad set of SMT intervention are powerful enough to treat mental disorders, as long as they are structured an include ingredients at the core of SIT and AMT like exposure, cognitive restructuring and homework assignments. Softer techniques often included in SMT, like relaxation and basic coping skills, can be learned and mastered effectively by people who suffer from significant life impairment like schizophrenia or alcoholism in order to deal with stress-related issues instead of treating the disorder itself.

As for the control of non-clinical anxiety symptoms, a large number of studies successfully used SMT to train people to control their stress. These studies represent a secondary prevention approach, where people are already dealing with stress that has not yet reached a clinical level of significance. Broadly defined SMT interventions are effective to deal with stress in the workplace, with academic stressors and for healthcare professionals. Structured approaches like SIT and AMT, as well as biofeedback, appear to be superior. Adapting the coping strategies and how they are presented to the trainees may be inevitable when it comes to applying them to specific contexts such as school and police work. Along those lines, some authors have noted the need to also adapt SMT programs to attitudes of trainees towards stress and emotion regulation.

On another application of SMT, the studies analyzed almost invariably mention that people attending debriefing programs appreciated the experience and were under the subjective impression that it had been beneficial to them. However, empirical data, especially when collected in rigorously designed studies, do not support its efficacy. The selection criteria for the articles were not designed to target debriefing, therefore our analysis may be incomplete. Nevertheless, the limited efficacy of debriefing to reduce the incident of PTSD has been confirmed several times and in other more comprehensive reviews [73, 86] and the consensus is that debriefing trauma victims is not an effective approach, at least when the goal is to reduce the incidence of mental disorders.

Finally, a few interesting studies could be used to appraise the usefulness of SMT in the training of military personnel to cope with acute stressors. Their results suggest that broadly defined SMT strategies could be effective in preparing individuals to cope with a highly specific upcoming stressor. Studies with military personnel and other people facing life-threatening stressors are scarce, and the breadth of stressors they are likely to experience might be too great to provide effective training options for some of the training programs (e.g., SIT). However, the existing research does suggest that some SMT strategies could be effective, even for life-threatening situations. Their efficacy might be increased if the SMT strategies are structured, sufficiently long to be well learned, and practiced until they are well mastered in stressful situations. Attitudes towards using such coping skills may be a factor to take into account when designing the training protocols. In the case of Hytten et al.'s [79] work, it is possible that the use of SIT may not have been optimal. SIT involves strategies that should be learned over many sessions, accompanied with extensive practices and includes several techniques that may be more appropriate for dysfunctional primary and secondary appraisal than dealing with the adequate appraisal of an objective life-threatening stressors. In any case, Hytten et al.'s [79] paper suggests that a brief, one hour, SMT training is probably not sufficient to learn how to cope effectively with objective life-threatening stressors. Longer programs, with extensive practice, may be required.

Overall, our goal was to assess whether or not SMT could be used to help military personnel develop effective coping skills while in the theatre of operation. Studies reported in the present chapter point towards a positive answer. Many specific strategies have been shown to be useful, from tactical breathing [87] to cognitive restructuring [10] and exposure [11]. However, a challenge may reside in the low motivation of soldiers in using and practicing psychological tools that are viewed as making a person weak or unmanly. It would therefore be strategic to find ways to help military personnel apply SMT without the negative perceptions. One way of accomplishing that could be to combine SMT with virtual reality [88,89,90]. Although virtual reality is sometime viewed as requiring considerable technological equipment, studies are being conducted to assess the capability of video games and a television screen in helping soldiers control their anxiety. With this technology, it is believed that the negative perception would be reduced and soldiers could then benefit from stress management training in a way that would prove beneficial for their health and mission without appearing weak.

5. Authors' notes

This project was supported by a grant from the Canada Research Chairs program awarded to the first author and stems from a contract from the Canadian Forces. Portions of this paper were included in an internal research report presented by the first author to the Canadian Forces [Bouchard, S. (2009). *Foundations for Stress Management Training of Traumatic Stressors Using Virtual Reality*. Defence R & D Canada - Valcartier. Contract report CR 2009-170]. Corresponding address: Stéphane Bouchard, Dept de Psychoéducation et de psychologie, Université du Québec en Outaouais, C.P. 1250 Succ "Hull", Gatineau, Québec, J8X 3X7. E-mail: stephane.bouchard@uqo.ca. The opinions expressed in this publication reflect those of the authors and do not necessarily represent the opinion of the Canadian Forces or the Department of National Defence.

6. References

[1] Lazarus RS, Folkman S: Stress, appraisal, and coping. New York, Springer, 1984.

[2] Gottlieb H, Strite LC, Koller R, Madorsky A, Hockersmith V, Kleeman M, Wagner J: Comprehensive rehabilitation of patients having chronic low back pain, Arch Phys Med Rehabil 1977; 58: 101-08.

[3] Parshad O: Role of yoga in stress management, West Indian Med J 2004; 53: 191-94.

[4] Oman D, Flinders T, Thoresen CE: Integrating spiritual modeling into education: A college course for stress management and spiritual growth, The International Journal for the Psychology of Religion 2008; 18: 79-107.

[5] Milliken T, Clements P, Tillman H: The impact of a stress management on nursing productivity and retention, Nursing Economics 2007; 25: 203-10.
[6] Amstadter AB, McCart MR, Ruggiero KJ: Psychosocial interventions for adults with crime-related PTSD, Professional Psychology: Research and Practice 2007; 38: 640-51.
[7] Timmerman IGH, Emmelkamp PMG, Sanderman R: The effects of a stress-management training program in individuals at risk in the community at large, Behav Res Ther 1998; 36: 863-75.
[8] Murphy LR, Sauter SL: The U.S.A perspective: Current issues and trends in the management of work stress, Australian Psychologist 2003; 38: 151–57.
[9] Meichenbaum D: Cognitive-Behavior Modification: An Integrative Approach. New York, Plenum Press Publishing Corporation, 1977.
[10] Suinn R: Seven steps to peak performance. Toronto, Hans Huber, 1986.
[11] Meichenbaum D: Stress inoculation training. New York, Pergamon Press, 1985.
[12] Suinn RM, Richardson F: Anxiety management training: A nonspecific behavior therapy program for anxiety control, Behavior Therapy 1971; 2: 498-510.
[13] Suinn RM, Bloom LJ: Anxiety management training for pattern A behaviour, J Behav Med 1978; 1: 25-35.
[14] Bundy C, Carroll D, Wallace L, Nagle R: Stress management and exercise training in chronic stable angina pectoris, Psychology and Health 1998; 13: 147-155.
[15] Krischer MM, Xu P, Meade CD, Jacobsen PB: Self-administered stress management training in patients undergoing radiotherapy, J Clin Oncol 2007; 25: 4657-62.
[16] Berger S, Schad T, Von Wyl V, Ehlert U, Zellweger C, Furrer H, Regli D, Vernazza P, Ledergerber B, Battegay M, Weber R, Gaab J: Effects of cognitive behavioral stress management on HIV-1 RNA, CD4 cell counts and psychosocial parameters of HIV-infected persons, AIDS 2008; 22: 767-75.
[17] Hains AA, Davies WH, Parton E, Totka J, Amoroso-Camarata J : A stress management intervention for adolescents with type 1 diabetes, Diabetes Educator 2000; 26: 417-24.
[18] Hockemeyer J, Smyth J: Evaluating the feasibility and efficacy of a self-administered manual-based stress management intervention for individuals with asthma: Results from a controlled study, Behav Med 2002; 27: 161-72.
[19] Multon KD, Parker JC, Smarr KL, Stucky RC, Petroski G, Hewett JE, Wright GE, Rhee SH, Walker SE: Effects of stress management on pain behavior in rheumatoid arthritis, Arthritis Care and Research 2001; 45: 122-28.
[20] Swann P: Stress management for pain control, Physiotherapy 1989; 75: 295-98.
[21] Kemppainen J, Eller LS, Bunch E, Hamilton MJ, Dole P, Holzemer W, Kirksey K, Nicholas PK, Corless IB, Coleman C, Nokes KM, Reynolds N, Sefcik L, Wantland D, Tsai YF: Strategies for self-management of HIV-related anxiety. AIDS Care 2006; 18: 597-607.
[22] Rose MI, Firestone P, Heick HMC, Faught AK: The effects of anxiety management training on the control of juvenile diabetes mellitus, J Behav Med 1983; 6: 381-95.
[23] Bloom LJ, Cantrell D: Anxiety management training for essential hypertension in pregnancy, Behavior Therapy 1978; 9: 377-82.

[24] Canino E, Cardona R, Monsalve P, Pérez Acuña F, López B, Fragachan F: A behavioral treatment program as a therapy in the control of primary hypertension, Acta Cient Venez 1994; 45: 23-30.

[25] Jorgensen RS, Houston BK, Zurawski RM: Anxiety management training in the treatment of essential hypertension, Behav Res Ther 1981; 19: 467-74.

[26] Quillen MA, Denney DR: Self-control of dysmenorrheic symptoms through pain management training, J Behav Ther Exp Psychiatry 1982; 13: 123-30.

[27] Wernick RL, Jaremko ME, Taylor PW: Pain management in severely burned adults: A test of stress inoculation, Journal of Behavioral Medicine 1981; 4: 103-09.

[28] Ross MJ, Berger RS: Effects of stress inoculation training on athletes' postsurgical pain and rehabilitation after orthopedic injury, J Consult Clin Psychol 1996; 64: 406-10.

[29] Law A, Logan H, Baron RS: Desire for control, felt control, and stress inoculation training during dental treatment, J Pers Soc Psychol 1994; 67: 926-36.

[30] Wells JK, Howard GS, Nowlin WF, Vargas MJ: Presurgical anxiety and postsurgical pain and adjustment: Effects of a stress inoculation procedure, J Consult Clin Psychol 1986; 54: 831-35.

[31] Milling LS, Breen A: Mediation and moderation of hypnotic and cognitive-behavioural pain reduction, Contemporary Hypnosis 2003; 20: 81-97.

[32] Milling LS, Levine MR, Meunier SA: Hypnotic enhancement of cognitive-behavioral interventions for pain: An analogue treatment study, Health Psychol 2003; 22: 406-13.

[33] Amigo I, Buceta JM, Becona E, Bueno AM: Cognitive behavioural treatment for essential hypertension: A controlled study, Stress Medicine 1991; 7: 103-08.

[34] Durán Bouza M, Simón MA, Seoane JM : An evaluation of pharmacological treatment combined with stress inoculation training in the management of oral lichen planus, Psychology and Health 2002; 17: 793-99.

[35] Blythe BJ, Erdahl JC: Using stress inoculation to prepare a patient for open-heart surgery, Health Soc Work 1986; 11: 265-74.

[36] Jay SM, Elliott CH: A stress inoculation program for parents whose children are undergoing painful medical procedures, J Consult Clin Psychol 1990; 58: 799-804.

[37] Foley FW, Bedell JR, LaRocca NG, Scheinberg LC, Reznikoff M: Efficacy of stress-inoculation training in coping with multiple sclerosis, J Consult Clin Psychol 1987; 55: 919-22.

[38] American Psychiatric Association: Diagnostic and statistical manual of mental disorders (4th ed. - Text revision). Washington, DC, Author, 2000.

[39] Bond AJ, Wingrove J, Valerie-Curran H, Lader MH: Treatment of generalised anxiety disorder with a short course of psychological therapy, combined with buspirone or placebo, J Affect Disord 2002; 72: 267-71.

[40] Blowers C, Cobb J, Mathews A: Generalised anxiety: A controlled treatment study, Behav Res Ther 1987; 25: 493-502.

[41] Jannoun L, Oppenheimer C, Gelder M: A self-help treatment program for anxiety state patients, Behavior Therapy 1982; 13: 103-11.

[42] Foa EB, Rothbaum BO, Riggs DS, Murdock TB: Treatment of posttraumatic stress disorder in rape victims: A comparison between cognitive-behavioral procedures and counselling, J Consult Clin Psychol 1991; 59: 715-23.

[43] Rothbaum BO, Anderson P, Zimand E, Hodges L, Lang D, Wilson J: Virtual reality exposure therapy and standard (in vivo) exposure therapy in the treatment of fear of flying, Behavior Therapy 2006; 37: 80-90.
[44] Anderson PL, Zimand E, Hodges LF, Rothbaum BO: Cognitive behavioural therapy for public-speaking anxiety using virtual reality for exposure, Depress Anxiety2005; 22: 156-58.
[45] Dodd H, Wellman N: Staff development, anxiety and relaxation techniques: a pilot study in an acute psychiatric inpatient setting, Journal of Psychiatric and Mental Health Nursing 2000; 7: 443-48.
[46] Brown S: (1983). Coping skills training: Attitude toward mental illness, depression, and quality of life 1 year later, Journal of Counseling Psychology 1983; 30: 117-20.
[47] Ormrod J, Budd R: A comparison of two treatment interventions aimed at lowering anxiety levels and alcohol consumption amongst alcohol abusers, Drug Alcohol Depend 1991; 27: 233-43.
[48] Holcomb WR: Stress inoculation therapy with anxiety and stress disorders of acute psychiatric inpatients, J Clin Psychol 1986; 42: 864-872.
[49] Awalt RM, Reilly PM, Shopshire MS: The angry patient: An intervention for managing anger in substance abuse treatment, J Psychoactive Drugs 1997; 29: 353-58.
[50] Cahill SP, Rauch SA, Hembree EA, Foa EB: Effect of cognitive-behavioral treatments for PTSD on anger, Journal of Cognitive Psychotherapy: An International Quarterly 2003; 17: 113-31.
[51] Karam EG, Fayyad J, Karam AN, Tabet CC, Melhem N, Mneimneh Z, Dimassi H: Effectiveness and specificity of a classroom-based group intervention in children and adolescents exposed to war in Lebanon, World Psychiatry 2008; 7: 103-09.
[52] Resick PA, Jordan CG, Girelli SA, Kotsis-Hutter C, Marhoefer-Dvorak S: A comparative outcome study of behavioural group therapy for sexual assault victims, Behavior Therapy 1988; 19: 385-401.
[53] Moses III AN, Hollandsworth Jr. JG: Relative effectiveness of education alone versus stress inoculation training in the treatment of dental phobia, Behavior Therapy 1985; 16: 531-37.
[54] Jaremko ME: The use of stress inoculation training in the reduction of public speaking anxiety, J Clin Psychol 1980; 36: 735-42.
[55] Lee HL, Tan HKL, Ma HI, Tsai K: Effectiveness of a work-related stress management program in patients with chronic schizophrenia, Am J Occup Ther 2006; 60: 435-41.
[56] Norman RMG, Malla AK, McLean TS, McIntosh EM, Neufeld RWJ, Voruganti LP, Cortese L: An evaluation of a stress management program for individuals with schizophrenia, Schizophr Res 2002; 58: 293-303.
[57] Stein F, Nikolic S: Teaching stress management techniques to a schizophrenic patient, Am J Occup Ther 1989; 43: 162-69.
[58] Rohsenow DJ, Smith RE, Johnson S: Stress management training as a prevention program for heavy social drinkers: Cognitions, affect, drinking, and individual differences, Addict Behav 1985; 10: 45-54.
[59] Charlesworth EA, Dempsey G: Trait anxiety reductions in a substance abuse population trained in stress management, J Clin Psychol 1982; 38: 764-69.
[60] Gonzalez LO, Sellers EW: The effects of a stress-management program on self-concept, locus of control, and the acquisition of coping skills in school-age children

diagnosed with attention deficit hyperactivity disorder, Journal of Child and Adolescent Psychiatry Nursing 2002; 15: 5-15.
[61] Stormer-Labonte M, Machemer P, Hardinghaus W: A meditative stress-management-program for psychosomatic patients, Psychother Psychosom Med Psychol 1992; 42: 436-444.
[62] Crockford D, Holt-Seitz A, Adams B: Preparing psychiatry residents for the certification exam: A survey of residency and exam experiences, Can J Psychiatry 2004; 49: 690-95.
[63] Heyne D, King NJ, Tonge BJ, Cooper H: School refusal epidemiology and management, Paediatric Drugs 2001; 3: 719-32.
[64] Sheehy R, Horan JJ: Effects of stress inoculation training for 1st-year law students, International Journal of Stress Management 2004; 11: 41-55.
[65] Schiraldi GR, Brown SL: Primary prevention for mental health: Results of an exploratory cognitive-behavioral college course, The Journal of Primary Prevention 2001; 22: 55-67.
[66] Le Scanff C, Taugis J : Stress management for police special forces, Journal of Applied Sport Psychology 2002; 14: 330-43.
[67] Peters KK, Carlson JG: Worksite stress management with high-risk maintenance workers: A controlled study, International Journal of Stress Management 1999; 6: 21-44.
[68] Richardson KM, Rothstein HR: Effects of occupational stress management intervention programs: A meta-analysis, Journal of Occupational Health Psychology 2008; 13: 69-93.
[69] Murphy LR: Stress management in work settings: A critical review of the health effects, American Journal of Health Promotion 1996; 11: 112-35.
[70] Routhier C: Military Resilience Training Program (MRTP). Valcartier, SQFT, 2007.
[71] Thompson MM, McCreary DR: Enhancing mental readiness in military personnel. In: Military Life: The Psychology of Serving in Peace and Combat, pp 54-79. Edited by Britt TW. Westport, Praeger Security International, 2006.
[72] Adler AB, Bartone PT: International survey of military mental health professionals, Milit Med 1999; 164: 788-92.
[73] Bisson JI: Single-session early psychological interventions following traumatic events, Clin Psychol Rev 2003; 23: 481-99.
[74] Litz BT, Gray M, Bryant RA, Adler AB: Early intervention for trauma: current status and future directions, Clinical Psychology: Science and Practice 2002; 9: 112-34.
[75] Marchand A, Guay S, Boyer R, Iucci S, Martin A, St-Hilaire MH: A randomized controlled trial of an adapted form of individual critical incident stress debriefing for victims of an armed robbery, Brief Treatment and Crisis Intervention 2006; 6: 122-29.
[76] Rice VJ, Gerardi SM: Part II. Work hardening for warriors: Training military occupational therapy professionals in the management of combat stress casualties, Work 1999; 13: 197-209.
[77] Sheehan SS: Stress management in the federal bureau of investigation: Principles for program development, International Journal of Emergency Mental Health 1999; 1: 39-42.

[78] Kamiyama K, Yamami N, Sato K, Aoyagi M, Kyoya M, Mizuno E, Uemura M, Kawamoto Y, Okuda M, Togawa S, Shibayama M, Hosaka T, Mano Y: Effects of a structured stress management program on psychological and physiological indicators among marine hazard rescues, Journal of Occupational Health 2004; 46: 497-99.

[79] Hytten K, Jensen A, Skauli G: Stress inoculation training for smoke divers and free fall lifeboat passengers, Aviat Space Environ Med 1990; 61: 983-88.

[80] Mace R, Carroll D: Stress inoculation training to control anxiety in sport: Two case studies in squash, Br J Sports Med 1986; 20: 115-17.

[81] Mace R, Carroll D: The effects of stress inoculation training on self-reported stress, observer's rating of stress heart rate and gymnastics performance, J Sports Sci 1989; 7: 257-66.

[82] Mace R, Carroll D, Eastman C: Effects of stress inoculation training on self-report, behavioural and psychophysiological reactions to abseiling, J Sports Sci 1986; 4: 229-36.

[83] Larsson G, Cook C, Starrin B: A time and cost efficient stress inoculation training program for athletes: A study of junior golfers, Scandinavian Journal of Sports Sciences 1988; 10: 23-8.

[84] Kolt GS, Hume PA, Smith P, Williams MM: Effects of a stress-management program on injury and stress of competitive gymnasts, Percept Mot Skills 2004; 99: 195-207.

[85] Perna FM, Antoni MH, Baum A, Gordon P, Schneiderman N: Cognitive behavioral stress management effects on injury and illness among competitive athletes: A randomized clinical trial, Annals of Behavioral Medicine 2003; 25: 66-73.

[86] Van Emerik AAP, Kamphuis JH, Hulsbosch AM, Emmelkamp PMG: Single session debriefing after psychological trauma: A meta-analysis, Lancet 2002; 360: 766-71.

[87] Grossman, D., & Christensen, L. On Combat: The Psychology and Physiology of Deadly Conflict in War and in Peace. PPCT Research Publications.

[88] Stetz MC, Long CP, Schober WV, Cardillo CG, Wildzunas RM : Stress assessment and management while medics take care of the VR wounded, Annual Review of Cybertherapy and Telemedicine 2007; 5: 165-72.

[89] Thompson MM, McCreary DR: Enhancing mental readiness in military personnel, Human Dimensions in Military Operations - Military leaders' strategies for addressing stress and psychological support. Meeting Proceedings, pp 4.1-4.12. Neuilly-sur-Seine, RTO, 2006.

[90] Bouchard, S., Guitard, T., Bernier, F., & Robillard, G. (2011). Virtual Reality and the Training of Military Personnel to Cope with Acute Stressors. In S. Brahnam & L. C. Jain (Eds.) *Advanced computational intelligence paradigms in healthcare 6: Virtual reality in psychotherapy, rehabilitation, and assessment, Ch. 6.* (pp.109-124). Berlin: Éditions Springer-Verlag Berlin Heidelberg.